Technology in Mediterranean and European Lands, 600–1600

T0356453

TECHNOLOGY IN MOTION
Pamela O. Long and Asif Siddiqi, *Series Editors*

Published in cooperation with the Society for the History of Technology (SHOT), the Technology in Motion series highlights the latest scholarship on all aspects of the mutually constitutive relationship between technology and society. Books focus on discrete thematic or geographic areas, covering all periods of history from antiquity to the present around the globe. These books synthesize recent scholarship on urgent topics in the history of technology with a sensitivity to challenging perspectives and cutting-edge analytical approaches. In combining historical and historiographical approaches, the books serve both as scholarly works and as ideal entry points for teaching at multiple levels.

Technology in Mediterranean and European Lands, 600–1600

Pamela O. Long

Johns Hopkins University Press
Baltimore

© 2025 Johns Hopkins University Press
All rights reserved. Published 2025
Printed in the United States of America on acid-free paper
9 8 7 6 5 4 3 2 1

Johns Hopkins University Press
2715 North Charles Street
Baltimore, Maryland 21218
www.press.jhu.edu

Library of Congress Cataloging-in-Publication Data

Names: Long, Pamela O., author.
Title: Technology in Mediterranean and European lands, 600-1600 /
   Pamela O. Long.
Description: Baltimore : Johns Hopkins University Press, 2025. | Series:
   Technology in motion | Includes bibliographical references and index.
Identifiers: LCCN 2024033460 | ISBN 9781421451220 (paperback) | ISBN
   9781421451237 (ebook)
Subjects: LCSH: Technology—Mediterranean Region—History—To 1500. |
   Technology—Europe—History—To 1500. | Technology—Mediterranean
   Region—History—16th century. | Technology—Europe—History—
   16th century. | Technology—Social aspects—Mediterranean Region—
   History—To 1500. | Technology—Social aspects—Europe—History—To
   1500. | Technology—Social aspects—Mediterranean Region—History—
   16th century. | Technology—Social aspects—Europe—History—16th
   century. | Technological innovations—Mediterranean Region—
   History—To 1500. | Technological innovations—Europe—History—To
   1500. | Technological innovations—Mediterranean Region—History—
   16th century. | Technological innovations—Europe—History—16th
   century. | Technological innovations—Social aspects—Mediterranean
   Region—History—To 1500. | Technological innovations—Social
   aspects—Europe—History—To 1500. | Technological innovations—
   Social aspects—Mediterranean Region—History—16th century. |
   Technological innovations—Social aspects—Europe—History—16th
   century.
Classification: LCC T17 .L66 2025 | DDC 303.48/30940902—dc23/eng
   /20241219
LC record available at https://lccn.loc.gov/2024033460

A catalog record for this book is available from the British Library.

Special discounts are available for bulk purchases of this book. For more
information, please contact Special Sales at specialsales@jh.edu.

For
Bob Korn
Allison Rachel Korn
Marco Yunga Tacuri
Lucas Samay Yunga Korn
and
Tiago Asha Yunga Korn

Contents

Preface and Acknowledgments

In a sense this book began with two booklets that I wrote (published in 2000 and 2003) in the series Historical Perspectives on Technology, Society, and Culture—a joint venture of the American Historical Association and the Society for the History of Technology. The series moved to Johns Hopkins University Press (and remained cosponsored by the Society for the History of Technology) and is now called Technology in Motion. This institutional relocation gave me the opportunity to rethink both form and substance. This book is in no sense a revision but has become a completely new work with only traces of those earlier efforts. I undertook it because I believe in the importance of synthesis as a significant part of the historian's craft—synthesis for the benefit of ourselves as specialists in one particular area and for a wider readership.

As a coeditor of the series in both its iterations, I have been fortunate to work with two great editors. The first was Bob Post, with whom I first served as coeditor in the earlier series and whose evenhanded and finely honed editing skills serve as an enduring model. In the years following, Asif Siddiqi has been a wonderful editorial colleague, an astute critic, and a pleasure to work with. His reading of the manuscript has improved it immeasurably, as have the readings by my husband, Bob Korn; my sister, poet and author Priscilla Long; and the anonymous readers. This book has also benefited greatly from Bob Korn's expertise in Photoshop.

One of the pleasures—but also one of the difficulties—of writing this book was sifting through and reading literally hundreds of articles and books, some of which I put in the bibliography, some of which, often regretfully, I had to leave out. A full bibliography for a book such as this would be at least twice as long as the book itself. Still, the bibliography that is here represents the essential scaffolding on which this book rests. The references themselves contain hundreds of further references that point to the great depth and excitement of the topics taken up in this book.

I first created the proposal for this book in the dark days of COVID-19 when I was a resident member of the Institute for Advanced Study in Princeton, New Jersey—often under quarantine. Marcia Tucker, librarian of the Historical Studies and Social Science Library, made my work possible by helping me retrieve numerous articles and books with unflagging energy and good cheer, whether delivering in person, by email, or on a little cart outside the library door during those times that I could not enter. I thank her warmly as well as others at the Institute, including Yve-Alain Bois, Myles W. Jackson, and Francesca Trivellato.

For a semester while finishing this book, I was the Robert Janson-LaPalme Scholar-in-Residence in the Department of Art and Archaeology at Princeton University. I thank Robert Janson-LaPalme, Rachel DeLue, Caroline Mangone, and Carolyn Yerkes, as well as graduate students Anna Speyart and John White, for help and support. I also thank the participants of seminars, workshops, and the conference "The Lure of Machines" for stimulating discussions concerning material culture and machines. Later, I was a virtual fellow at the Linda Hall Library in Kansas City, Missouri, for three months. I thank the Library for providing important resources, including images, and especially Benjamin Gross for encouragement and provocative conversations about machines. I thank the staff at the Library of Congress in Washington, DC. Without this great library at my doorstep, this book could not even have been started, let alone finished.

Numerous friends and colleagues have helped me along the way including Chiara Bariviera, Mateusz Falkowski, Tony Grafton, Morten Hansen, Ann Huppert, Bill North, Dennis Romano, Pamela Smith, Erhan Tamur, and Elly Truitt.

I greatly appreciate the Leonard Hastings Schoff and Suzanne Levick Schoff Memorial Fund at the University Seminars, Columbia University, for their help with this publication. Material in this work was presented to the University Seminar on the Renaissance, and I thank the conveners of the seminar, Cynthia Pyle and Alan Stewart.

At Johns Hopkins University Press, I am first of all indebted to Matt McAdam, whose idea it was to adopt the series and whose encouragement and patience I much appreciate. I warmly thank Matt as well as Adriahna Conway, Charles Dibble, and Susan Matheson, whose astute copyediting greatly improved the manuscript.

Finally, Bob Korn has helped with computer difficulties, images, travel, and myriad other issues with good humor and forbearance. This book is dedicated to him and to our daughter, Allison; our son-in-law, Marco; and to Lucas and Asha who make the world go around—still!

On Dates and Calendars

For the centuries that are the focus of this book, people who lived in those times used various calendars. Two of the most important were the Julian calendar and the Hijri calendar. The Julian calendar was a solar calendar created by Julius Caesar in 45 BCE and used by most of the Roman and post-Roman world until 1582 when Pope Gregory XIII instituted the reformed Gregorian calendar, which most of the world uses today. The Hijri calendar was a lunar calendar adopted by the Islamic world in the Gregorian year 622 CE. It is the year that the founder of Islam, Muhammad, migrated from Mecca to Medina, in a journey called the Hijrah, to establish the first Muslim community. To simplify matters, the dates used in this book are those of the present-day (Gregorian) calendar.

Technology in Mediterranean and European Lands, 600–1600

Introduction

Technology, a Human Practice

> As for those [crafts] that constitute a primary objective, they are three,
> namely land-ploughing, tailoring, and building. . . . And given that the
> human being has been created also in need of sustenance and food, and
> that such sustenance and food cannot be obtained but through the grains
> of plants and the fruits of trees, necessity has therefore required the
> labours of tillage and planting. And given that the industry of plough-
> ing and implanting required the raising of the land and the digging of
> watercourses, this would not have been realized without the ploughing
> shovels and the pulling bullocks. . . . While the use of ploughing shovels
> and pulling bullocks would not be realizable without carpentry and black-
> smithing, so necessity called for setting these up. And since the craft
> of blacksmithing requires a metallurgy industry and other crafts, all of
> these then become subordinate and of service to the labours of tillage
> and planting.
>
> IKHWĀN AL-ṢAFĀ, "EPISTLE 8 ON THE PRACTICAL CRAFTS,"
> TRANS. NADER EL-BIZRI

The above quote is excerpted from a tenth-century compendium written by an esoteric group called the Ikhwān al-Ṣafā (the Brethren of Purity) based in Basra and Baghdad in the eastern Mediterranean—a famous, philosophically oriented text called the *Epistles of the Brethren of Purity*. In "Epistle 8," the Ikhwān ranked the crafts hierarchically, depending on their degree of necessity (for instance, po-sitioning plowing, essential for food production, higher than metallurgy).[1] More than a thousand years later, in a very different historical and cultural context, and without accepting their hierarchical rankings, the Ikhwān's recognition that the technical arts and crafts are interconnected still resonates.

This book treats the history of technology from 600 to 1600 in Mediterranean and European lands. Successive chapters discuss food production, hydraulic tech-nologies, building construction and urbanism (including urban sanitation and water supplies), transportation and communication, crafts (such as pottery) and industries (such as mining), and finally, instruments (for example clocks) and machines, including weapons. These divisions are necessary for explanatory coherence but, as the Ikhwān emphasized, in living practice, crafts and technological process are not separated in this way. Rather they are connected to and often dependent on each other.

The premise of this book is that technology is embedded in human societies, in human culture, and in the physical environment. Technology's history concerns fabrication and the human manipulation of the physical world to produce and make things, both perishable such as wheat and olive oil, and nonperishable such as ceramic pots. This is a short book that covers a thousand years and a large geographic territory. It is a synthesis in which I explicate broad historical issues while also shining spotlights on local practices. For these centuries and geographical regions, I attempt to provide a guide to the specific topics and to the historical issues and debates that have surrounded them.

Over the past several decades, the discipline of the history of technology has changed. Many historians, including myself, have turned away from a fascination with inventions and their "origins"; from a focus on technological "progress" leading to industrialization, modernization, and Anglo-European culture (often associated with an implicit or explicit European superiority); and from the idea of genius.[2] In this book, I portray a complex view of process rather than "progress." The transmission of technological practice and knowledge is not part of a simplistic linear history. It involves a complex mix of local practices, assimilation to new or changing circumstances, the development of new meanings, and new forms of use.

The term "technology" is anachronistic for the time period being considered here and, as Eric Schatzberg especially has shown, it has a complex history both linguistically and conceptually. Further, as Schatzberg's critics have noted, the idea of technology is by no means an exclusively Western concept. Rather, diverse conceptualizations grew up in other regions such as East Asia, for example.[3] Indeed, various technologies, whatever they were called, including craftwork, the use of tools, systems of agriculture, and of the production of physical objects, were and are essential for human survival and are coextensive with human evolution and with all of human history, including the premodern centuries.

Technologies are not developed by individuals in isolation. Rather, they are collective in their origins, development, and use. They exist within societies that possess specific familial and social forms, gender relationships, cultural characteristics, structures of power, and (often conflicting) economic interests. Historians have used the notion of "distributed cognition" to indicate the ways in which complicated tasks, including technological projects, are not completed by one person but involve the communication among and work of people with differing competencies, knowledge bases, ideas, and practical skills.[4]

Individual artisans acquire skill, usually through apprenticeship, formal and informal, and through practice. Skill, as Richard Sennett has put it, is "trained practice." The skill required to make crafted things is usually evident in the finished product, whether it be a beautifully carved, ivory devotional object or any of thousands of other kinds of hand-crafted objects, from paintings to silk cloth to iron tools. Skill requires practiced handwork, but it also requires reasoning. The common notion that thinking and mental work are separate from (and superior to) handwork must be rejected. Technical practices and craftwork require both skilled handwork and thought, a fact obscured by the centuries-old and still prevalent view that assigns writing and scholarship higher status than making and handwork.[5] For one thing, neuroscientists are now discovering the ways in which thinking and knowledge are embodied—the brain does not think by itself but in conjunction with the hands and other body parts. In addition, thinking includes sensations and emotions.[6]

Societies and cultures have sometimes deemed craftwork as low status, *banausikos* (banausic) as the ancient Greeks would call it, but there were countervailing views. The ancient Greeks, exemplified by Aristotle, thought artisans far below the "free" citizen. In contrast, the Hellenistic tradition of engineering, represented by Hero of Alexandria (fl. 60s CE) valued technical work (and, I think, workers). The ancient architect Vitruvius (c. 40s BCE—c. 15 CE), whose *De architectura* included discussions of buildings, but also machines and hydrology, believed that the architect should be knowledgeable in handwork (*fabrica*) and reasoning (*ratiocinatio*). In the medieval Islamic world, as Margaret Graves has brilliantly shown, although there was a tradition of social disdain for artisans, the significance of such disparagement has been exaggerated. There was also a tradition of deep respect and the integration of craft production with theological and philosophical values in what she calls "the intellect of the hand."[7]

Elspeth Whitney has investigated the twelfth- and thirteenth-century inclusion of the "mechanical arts" (a category that included skilled craftwork) alongside the seven liberal arts—grammar, rhetoric, and dialectic (or logic), called the *trivium*; and geometry, arithmetic, music, and astronomy, called the *quadrivium*. Particularly important was the *Didascalicon* of Hugh of St. Victor (c. 1096–1141), who named seven mechanical arts—fabric-making, armaments, commerce, agriculture, hunting, medicine, and theater—and gave them a place beside the liberal arts.[8] Pamela Smith has framed her detailed studies of artisanal knowledge in fifteenth- and sixteenth-century Europe as "artisanal epistemology," a term that underscores the cognitive and rational components of artisanal skill.[9]

Nevertheless, it is true that training in a craft entailed learning skills in a hands-on apprenticeship situation, while more "learned" culture involved a different kind of learning—reading and writing, what might be called book learning. In Europe after the thirteenth century, much book learning occurred in Latin in universities and was only available to male students who had been afforded a Latin education in their youth. Yet, as I have argued elsewhere, from the fifteenth and sixteenth centuries, "trading zones" developed between the skilled and the learned. In certain contexts, they each had substantive information useful to the other. Conversation and interchange (and sometimes friendship) was the result.[10] The relationships of craft practitioners to the wider culture, including learned culture, varied from one craft to another and from one region to another around the Mediterranean and Europe.[11]

Crafts and other practical endeavors from farming to irrigation were usually transmitted in the context of hands-on apprenticeship, either formal or informal. But it was also true that from ancient times and beyond, authors from both artisanal and learned backgrounds produced writings concerning practical, technological, and craft procedures. Such writings continued to be produced in Byzantine, Arab, and Western European lands in the medieval centuries and beyond. Their number and variety massively increased after the invention of printing, especially in the sixteenth century. Such writings had many authors and readers. They could be written by hand or printed; they took the form of manuals, recipe books, pamphlets and treatises; they could be illustrated (sometimes lavishly) or not. They served many purposes and should not be thought of solely as manuals of instruction for people practicing a craft.[12]

Geographic Perimeters

The geographic range of this book comprises the lands around the Mediterranean Sea—the Balkan Peninsula, including the Greek archipelago; Anatolia (present-day Turkey) and the Levant (the eastern Mediterranean, including medieval Syria and Palestine); Egypt; North Africa (called Ifriqiya after the Islamic conquest of the seventh century); the Maghreb, which refers to the northwestern region of Africa;

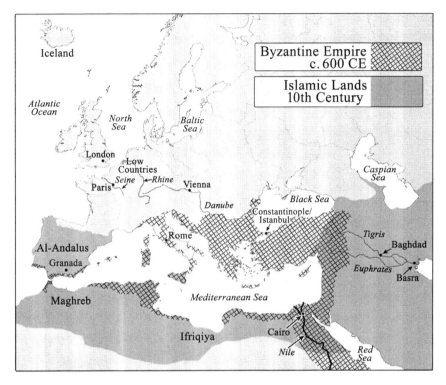

Mediterranean and European lands. Map created by Bob Korn using QGIS, open-source software.

the Italian Peninsula; and the Iberian Peninsula (al-Andalus after the Islamic conquests of the early eighth century). It also includes the British Isles, regions of central and northern Europe, and Scandinavia, many of these lands bordering the Atlantic Ocean. It is a macroregion, or what global historians have called a sub-global region, which might be called Western Eurasia / North Africa. It is bounded by ocean, steppe, and desert and features an unusual number of inland seas, from the Black and Mediterranean to the Baltic and North seas.[13]

This broad geographical compass allows different locales to be discussed side by side, not for formal comparison, but as a way of seeing the diversity and richness of solutions to common problems such as food supply and the fabrication of textiles. Borrowing from a project on the "global Middle Ages" (to be discussed below), this approach might usefully be called combinative, that is, the thoughtful juxtaposition of technological practices carried out in very different locations for similar ends.[14] Cultivating a field in the light, sandy soil of a desert-proximate region of North Africa is an entirely different affair from cultivating a field in the heavy, alluvial soils of central France. Yet it is useful to think about the two at the same time, providing a view of particular technologies, social forms, and solutions to environmental problems, each of which is contingent and dependent on its own particular social, cultural, environmental, and technological characteristics.

Most of the region below the Rhine and Danube rivers (which flow through central Europe and present-day Germany) in antiquity had been part of the Roman Empire. As that empire disintegrated, its capital moved to Constantinople in the east in 324 CE, becoming the eastern Roman Empire that eventually would be called the Byzantine Empire—an empire with changing boundaries which ended with the conquest of Constantinople by the Ottoman Turks in 1453. While the Byzantine Empire remained a unified entity, western European lands came to be divided into separate kingdoms, such as the Kingdom of the Franks in parts of present-day France and Germany, the Visigoths in the area of what is now southern France and Spain, and the Lombards in Italy.

The Islamic states developed after Muhammad (c. 570–632) founded the new religion of Islam in the early seventh century. After consolidating their power in Arabia, the early Muslims turned toward the lands of the Byzantine Empire in both the eastern Mediterranean and North Africa. They had conquered Egypt by 642, moved across North Africa, and by 711 had conquered the Iberian Peninsula. To the north, in 800 Charlemagne (ruled 768–814) was crowned by the pope, making him the first "Holy Roman Emperor," ruling a vast area of western Europe known as the Carolingian Empire, which broke into separate kingdoms after his death. By the mid-ninth century, the Islamic powers had also broken up into separate kingdoms, the Muslim successor states. The Norsemen (also called Vikings) expanded their activities and territories in the ninth and tenth centuries. These geopolitical areas and events provided essential geographic, military, and political foundations for the complex developments of subsequent centuries.

Those developments include the Crusades, a series of religious wars that occurred between 1095 and 1291 in which western Christians attempted to recapture eastern Mediterranean territories, especially Jerusalem, from Islamic rulers. Notable was the Fourth Crusade which led to the western capture and rule of Constantinople between 1204 and 1261.[15] Developments also include the flourishing of city states, especially in Italian and German lands; the rise of territorial states such as France, England, and Spain; the end of Islamic power in Spain with the defeat of the Muslims at Granada in 1492; and the beginning of Atlantic Ocean crossings and then conquests in the so-called New World.[16]

Technology's Histories

Historians of technology working in the eras and geographical range of this book stand on the shoulders of some of the great historians of the twentieth century. One of the first of these was the Belgian historian Henri Pirenne (1862–1935), who wrote a foundational historical study of Belgium, and studies of medieval towns and trade. Pirenne was imprisoned during World War I for defying the German invaders. After his release in 1918, he turned from his locally focused studies of towns and regions to broader themes, developing what has become known as the Pirenne thesis. The most important statement of this thesis appears in a short work, translated as *Mohammed and Charlemagne*, the first draft of which he finished shortly before his death in 1935. In it, he argued that the Germanic tribes that penetrated the Roman Empire between the fourth and sixth centuries preserved much of Roman culture and political institutions, and that trade remained active. In Pirenne's view, it was the Islamic conquests of the seventh and early eighth centuries that caused the collapse of the Roman Empire and effectively split the eastern Mediterranean from the west, isolating the Frankish kings in northwestern Europe. This enabled Charlemagne to create a novel Western form of government. Without Muhammad, Pirenne argued, Charlemagne was inconceivable.[17]

The Pirenne thesis has been debated from the time of its publication. Scholars have shown that Pirenne underestimated the decline of the Roman Empire even before the Germanic incursions, overestimated the negative effect of the Islamic expansion on trade, and failed to recognize the importance of regional economies. Pirenne's thesis, as Chris Wickham has noted, has been so successful because "it fits the long-standing metanarrative of medieval economic history which seeks to explain the secular economic triumph of north-west Europe." The narrative he refers to has the goal of connecting the economic developments of medieval Europe to modern capitalism and industrialization. It is teleological, "assigning brownie points as it does to developments which produce our own world economy, and marginalizing those which do not." Focusing on the countries of origin of the most influential historians of the last century (to the neglect of other regions), the narrative also profoundly overvalues long-distance exchange over local and regional trade. In addition to these criticisms, the thesis has been refuted by additional extensive empirical research, including Michael McCormick's monumental study of communications and commerce in the early European economy.[18]

The regional and local studies that were at the core of Pirenne's earlier scholarship influenced the Annales school of historiography that developed in the 1920s.

The Annalistes explicitly rejected "event-oriented" nationalist history, which they believed failed to grasp the richness of human reality. Led by Marc Bloch (1886–1944) and Lucien Febvre (1878–1956), a small group of European historians set out to expand the range of historical studies beyond their traditional focus on war, politics, diplomacy, and great leaders. These historians worked to create stronger analytical frameworks and to incorporate the insights of sociology and economic history. In 1929, Bloch and Febvre created a new journal, the *Annales d'Histoire Économique et Sociale*. Bloch advocated a comparative method, drawing on economic and social history. He considered technology as a fundamental concern along with rural and agrarian history. In a special issue of the *Annales*, published in November 1935 and devoted to the history of technology, Febvre suggested that the new discipline "incorporate all the uneven, accidental, and human elements of science and invention." In the same issue, Bloch traced the history of water-power from ancient times through the medieval period, and tied the adoption of the overshot waterwheel to the power of feudal lords on the medieval manor.[19]

A member of the second generation of the Annales school, Fernand Braudel (1902–1985), while being held in a prisoner-of-war camp during World War II, wrote his famous work, translated into English as *The Mediterranean and the Mediterranean World in the Age of Philip II*. Braudel emphasized the unity and coherence of Mediterranean lands. He divided his large tome into three parts, the first (the *longue durée*) comprised a history of humans and their environment, emphasizing the importance of geography for the history of the Mediterranean. The second (*destins collectifs et mouvements d'ensemble*) referred to the history of structures such as states, economic systems, societies, and changing forms of war; and the third (*l'histoire événémentielle*), which was least important for Braudel, comprised the history of particular events and individuals. Braudel's work has had enormous influence on Mediterranean history as it has also been criticized, especially for its geographic determinism. Important successors of Braudel are Peregrine Horden and Nicholas Purcell. In *The Corrupting Sea*, they described the Mediterranean basin as a region of easy communication among fragmented microregions, coastlands, and islands, which are highly interdependent and liable to reciprocal influence through the sea itself. More recently, Jessica Goldberg has modified this microregional view in her studies of the Geniza documents—a huge cache of letters and other documents from Fustat (old Cairo), most written in Judeo-Arabic (Arabic written in Hebrew characters). Goldberg suggests that the Mediterranean possessed "zones of macro-ecology" such as the Nile Valley, the plain of Ifriqiya, and parts of Italy—regions that regularly produced agricultural surpluses.[20]

Although not focused on the Mediterranean per se, the American medievalist Lynn White Jr. (1907–1987) dedicated his widely influential *Medieval Technology and Social Change* to Marc Bloch. White's early scholarship focused on Latin monasticism in twelfth-century Norman Sicily. However, when he lost access to European archives in the late 1930s on the brink of World War II, he turned to the history of technology. He was deeply influenced by Bloch's scholarship, especially his work on French rural history. White's argument that the adoption of the stirrup in Carolingian France led to feudalism (discussed further in chapter 6) has been rejected by historians.[21]

In another area, environmental history, White argued in a famous essay, "The Historical Roots of our Ecologic Crisis," that medieval Christianity was responsible for our present-day environmental crisis. This because in Christian belief, God gave humans "dominion" over nature (Genesis 1:28), giving humans permission to exploit the natural world regardless of consequences. Christianity also vested all spiritual power in the creator God, thus removing spirituality from nature (and supposedly ending the animism that was fundamental to pagan ways of thinking about the world). Finally, investigating nature was the medieval way of investigating the mind of God (who had created it), leading to "natural theology," and leading in turn to modern science—the results of which have often been destructive of the environment. This thesis has been criticized in detail by historians, while it continues to be repeated by others, including environmentalists.[22] Despite these specific criticisms, White's broad interests and range of scholarship made him an important and positive force for the development of the history of technology in North America.

While thinking of the Mediterranean as a coherent geographical unit has been useful, it is also true that there were extensive and complex interrelationships among Mediterranean lands, central and northern European regions, and the British Isles. The lands of the Mediterranean Sea were in no way cut off from these other regions. Further, while some European lands bordered the Mediterranean, others bordered the Atlantic Ocean, which became increasingly significant in the fifteenth and sixteenth centuries, in an age of oceanic voyages. Finally, as the editors of a recent four-volume history of the rural economies and societies of northwestern Europe have pointed out, "like the Mediterranean, the North Sea can be considered a unity."[23]

The present book focuses on specific (although vast) geographic areas. These areas were also tied to the rest of the world in complex ways—true for the earlier centuries of this study as well as for the fifteenth and sixteenth centuries. As the work of scholars of the "global Middle Ages" have shown, the assumption that globalization began with the European Atlantic voyages, and that it is necessarily connected to capitalism and modernism, is false. Rather the connections of the Western Eurasian / North African region with the rest of the world were based on the connectivity of individuals and groups and the often-changing networks of which they were a part. This complex global world was multicentered. It had shifting nodes of exchange in the forms of ports, emporia, and cities. It was connected to other regions by land and sea transport such as the silk roads and Indian Ocean shipping.[24] The Mediterranean and European worlds were embedded in networks in which knowledge, materials, and even cultural norms were circulated—not only within their own boundaries but globally as well.

Sources and Approaches

A historical synthesis such as this is based on the best scholarship available which in turn is based on a variety of sources, including those produced by diverse disciplines. The work of archaeologists is important especially for the periods and locales in which textual sources are scarce or nonexistent. Archaeologists have produced crucial knowledge about particular sites, and more recently have undertaken landscape archaeology, combining techniques such as aerial photography and computer processing of large quantities of data to understand large tracts of rural landscape over broad extents of chronological time. They have also excavated hundreds of physical objects or their fragments, such as tools and vessels, many now in museums around the world. The turn of historians to an emphasis on materiality has promoted more intense and comprehensive study of such objects, as well as physical environments more generally. Archaeobotany has provided pollen and DNA analysis of site-specific grains and seeds, as well as soil and tree-ring analysis, enhancing understanding of the relationships of climate, cultivation, and food supply. This detailed scientific work combined with traditional historical scholarship has resulted in important revisions of long-held assumptions (such as, for example in southern France, the view that agriculture declined in the early medieval period).[25] Another kind of source, especially significant for historians of premodern technology, are visual images—on frescos, mosaics, illuminated or printed books, painted altarpieces, and painted panels.[26] And finally, written evidence in whatever form, is crucial.

The history of technology has proved fertile ground for the concept of technological revolutions within specific spheres, such as agriculture or commerce. For the period encompassed by this book and for specific regions, there have been proposed, for example, the medieval agricultural revolution, the industrial revolution of the Middle Ages, the print revolution, the commercial revolution of the Middle Ages, the green revolution (indicating the Arab introduction of new plants and cultivation techniques to al-Andalus), and the military revolution. "Revolution" in its most usual modern senses can mean short-term, sudden (often violent) change, or longer-term structural change. While the latter is more appropriate applied to these technological "revolutions," I question the usefulness of the term itself. For one thing, as will be seen in the following chapters, recent historians of various subspecialties have argued against such "revolutions." Heuristically, casting an invention or group of technological processes as revolutionary tends to obscure their complex context, as well as prior developments, and to exaggerate their effects, creating more a distorting lens than a useful tool for interpretation.[27] Here, I explore technological change—both incremental and more rapid—within the broader context of traditional practices.

Profound long-term changes that affect technologies from agriculture to navigation include environmental and climate change, topics that increasingly have become a focus of historical research. Major environmental events included catastrophes such as the Great Famine in northern Europe (1315–1322), which itself was brought about by crop failures in the wake of extreme weather events, all part of a crisis in the fourteenth century. More generally, the ways in which human practices, such as agriculture and metallurgy, were influenced by environmental factors and themselves had profound environmental impacts are a focus of recent investigations.[28]

During these centuries (and in all preindustrial societies before the age of fossil fuels), several forms of energy use predominated. The first is biochemical energy from photosynthesis (and the resultant muscle power of humans and animals). The primary source for this energy is solar radiation (i.e., electromagnetic energy). In photosynthesis, plants use solar energy to create new stores of chemical energy. Animals, including humans, consume plants (or other animals that have consumed plants), metabolism reorganizes nutrients into living tissue and maintains bodily functions including a constant temperature. Digestion generates the mechanical energy of working muscles. Muscles of humans and animals were fundamentally important for work in premodern times—for carrying loads, powering grain mills and other machinery, and plowing fields, for example.[29]

A second form of biochemical energy was obtained by combustion—burning plant materials, primarily wood, for fuel, which was used for cooking, heating, and processes such as metal smelting. For certain processes, especially in ceramics and metallurgy, workers made charcoal from wood. The process involved burning wood with a limited supply of air. The charcoal maker would cover a pile of wood with earth and burn it slowly. What remained was almost pure carbon, what we now call charcoal, but what, before the nineteenth century, was called coal. It was lighter than wood (thus easier to transport), possessed greater energy density, and burned without smoke or other contaminants. The use of wood for fuel had profound environmental effects on forests.[30]

A third form of energy that workers used was mechanical energy from the natural movement of wind and water, and the use of machinery to transform that energy into a variety of uses. Water mills and windmills were particularly important, as well as various kinds of sailboats powered by wind—all to be discussed in later chapters of this book.[31]

The uses of energy and the environment as a whole were influenced by weather and climate. In these centuries, two major periods of change have been identified. The first was a period of several centuries of warming known as the "Medieval Warm Period," now called the "Medieval Climate Anomaly" that extended from about 800 to 1300. The change in terminology is apt because, as we know from our own climate crisis, the changes affected different regions differently. After the Medieval Climate Anomaly came the "Little Ice Age," the chronological perimeters of which are debated but may have ranged from about 1450 until about 1850. The transition from the Warm Period (in western Europe) to the Little Ice Age was a time of turbulent weather, called by Bruce Campbell the "Great Transition." Climate changes affected agriculture, animal husbandry, fishing, navigation, and much else. Correlating climate change with historical events and practices in particular regions is a complex, interdisciplinary task. Historians and scholars from a variety of scientific disciplines are collaborating to understand better the effect of climate on historical developments and practices. These scholars examine textual sources, such as harvest dates in particular years, fishing reports, weather records, and reports of extreme weather events, in conjunction with physical evidence, such as isotope readings of ice cores, tree rings, evidence for glacial receding and expansion in specific years, sunspots, volcanoes, lake sedimentation layers, insect remains, and pollen analysis.[32]

The bubonic plague (*Yersinia pestis*) brought two major pandemics during the centuries included in this book. The Justinianic plague ravaged Mediterranean lands from the sixth to the eighth century. The Black Death—the worst pandemic humans have ever known—entered Europe circa 1346 and is thought to have killed as much as half the population in European and Mediterranean lands. As Monica

Green has argued, the renowned Black Death of Europe was brought inadvertently by the Mongols (whose empire emerged in 1206) during a number of sieges in their westward expansion to Mediterranean and European lands. This is not to demonize the Mongols who, as Green has pointed out, were also afflicted by the disease. Paleogenetics—the study of ancient DNA (aDNA), samples of which have been obtained from the teeth of plague victims found in premodern burial sites—in the last two decades have transformed the study of premodern diseases including the Black Death. Gene sequencing combined with more traditional historical analysis (with a global orientation) has shown that the bacterium split into four new branches in the thirteenth century (called by geneticists the "Big Bang") and spread both east and west.[33]

Finally, it is important to remember that arts, crafts, and technological practices were (and are) carried out by people. Individuals always interacted with others but also possessed diverse social and legal positions within the larger societies in which they lived. One division was that of gender. Nowhere in the lands and centuries discussed here did women enjoy social or legal equality with men. Within the patriarchal and misogynist cultures in which they lived, however, they performed tasks as vitally important as those of men. Further, although work was often divided by gender, it is a mistake to anachronistically compare modern women—say women in North America in the 1950s—with premodern women. As in all historical issues, the nature of their work, their status, and the agency that they did possess, can only be understood through empirical investigation. Recent work has underscored that the nature of women's and men's work varied in diverse locations. In specific times and places, women were farmers who harvested but also sometimes plowed, blacksmiths, miners, merchants, combatants in war, textile workers, and heads of households.[34]

The same caution is needed in considering other social categories. Elites, including powerful caliphs and kings, noble aristocrats, and wealthy merchants, possessed varying degrees of power and wealth. Many elite individuals exerted great efforts to call attention to their power and authority by means of conspicuous consumption—a crucial motor for artisanal production. Similarly, farmers or peasants possessed widely varying degrees of autonomy and wealth. Some were relatively autonomous, some relatively prosperous, others lived on the edge of starvation. Some were tied to the land in various forms of serfdom, others possessed their own land.[35]

Another issue pertaining to the relationships between groups has to do with creating "others" in premodern societies, that is, groups of people stereotyped as different from the dominant group and, therefore, racialized. Recent scholarship, especially that of Geraldine Heng, has overturned the view that racism is a modern phenomenon. Once race was understood not to be a biological classification but a social construction, it can be seen that racism is pertinent to premodern as well as modern societies. "Race" is a way of marking humans as different in fundamental ways, essentializing them, and thereby justifying their bad treatment. Heng uses Jews in medieval England as an extended example. After stereotyping Jews as a kind of inhuman "other," the English herded, tagged, imprisoned, murdered, and finally expelled English Jews—activities, Heng argues, that were intrinsic to the development of the group identity known as "Englishness." Similarly, Muslims were turned into nonhumans by Christians, as were the Irish during the English invasion of Ireland in the twelfth century. Such attitudes can be found in many cultures and groups.[36]

A particular group of people in these centuries were those who were enslaved. The centuries and lands that are the focus of this book did not contain "slave societies"—defined as having 30–35 percent of the population living in slavery—as did ancient Roman society and parts of the Americas in the sixteenth to the eighteenth century. Nevertheless, slavery in medieval and early modern European lands was a potent force. Michael McCormick posited that enslaved humans were of crucial importance as a commodity for the rise of European commerce in the early medieval centuries, enslaved people being traded to eastern Mediterranean lands in exchange for luxuries such as spices and silks. Beyond this, slavery across the Mediterranean was commonplace. People could become enslaved in a variety of ways, such as by becoming booty in warfare, or being captured by pirates. As Sally McKee has argued, the gender of enslaved people was important. In Europe, for example, the price of enslaved women rose, as they were purchased for household service that often included sexual service coerced by the master.[37]

* * *

Behind all technological processes—whether farming or fishing, mining or metallurgy, building houses or aqueducts, weaving or painting—were individuals who lived and worked in the context of communities and groups of people. Only a few become visible through writings, documents, signatures, or reputation. Most remain invisible. Yet the results of their labors and skill are with us still. Although I would reject the Ikhwān's hierarchical ordering of the crafts (with tillage and planting at the top), as mentioned at the beginning of this chapter, I could not dispute the group's emphasis on the importance of food production—to which we now turn.

CHAPTER ONE

Food Production

The best ewes for breeding are those that have plentiful soft wool, with hanks growing thickly over the whole body but particularly around the throat and neck, and with the whole belly thick with plenty of soft wool, and all the same colour; they should have good eyes, long legs and long tails: these are the best for lambing. The rams should be sturdy, good looking, with blue-grey eyes, a woolly face, good horns but short, ears covered with thick wool, flat back, large testicles, and the same colour over the whole body.

"SELECTING SHEEP, JUDGING MALES AND FEMALES,"
GEOPONIKA (FARM WORK), C. TENTH CENTURY

Make tryall of your sheepe every yeare. . . . And suche as are not to be kepte cause theim to be shorne betymes and to bee marked from the other and put theim in a wood that is inclosed or in other pasture wheare they may be fatted.
And about mydsomer selle theim, for then is the fleshe of mutton in season. WALTER OF HENLEY, *HUSBANDRY*, C. 1286.

These two texts were written three centuries apart in different languages and in different parts of the world—one in the eastern Mediterranean, the other in England. The *Geoponika* is a tenth-century Byzantine agricultural encyclopedia, written in Greek, which was compiled during the reign of the emperor Constantine VII Porphyrogenitus (ruled 913–959) and was dedicated to him. Its anonymous author(s) took material from a variety of ancient and early medieval writings on agriculture and horticulture.[1] The second is a thirteenth-century agricultural treatise, which Walter of Henley (c. 1240–c. 1300) wrote around 1286 in the style of a sermon. Walter may have been a bailiff (overseer) of a large estate. His writing style shows that he had received a Dominican friar's training in writing sermons.[2] Viewed as a whole, the two works are quite different. The *Geoponika* is primarily focused on the cultivation of olives and grapes, as would be expected from its eastern Mediterranean origins, while the *Husbandry* concerns the cultivation of grain and the care of animals. Yet both give advice to farmers. The two quotes about the selection of sheep for breeding and their preparation for slaughter have striking similarities—they both treat specific empirical practices and give detailed advice for particular agricultural tasks.

Choices such as these concerning animal husbandry and other aspects of food production affected the well-being of farmers but also affected the environment. Food production (including animal husbandry) has to do with the human manipulation of the world of nature and is an important aspect of environmental history. The natural world is itself highly variable, depending on both location and time. Nature brings various types of soils, good weather and bad (heavy rainfall, floods, snow, and ice jams), locusts, mosquitoes, and diseases of humans and of other animals. The environment is not an inert, static entity but has a history, which is intertwined with human history and influenced by human interventions.[3]

Methods of producing food, including the kinds of crops that were cultivated and the kinds of animals that were husbanded, depended on climate, including rainfall and soil conditions, which could vary greatly from one region to another. The organization of land and of labor depended on environmental conditions but also on social and political contingencies. Various technologies, including irrigation, could modify the environment. Food production entailed agriculture and included various forms of pastoralism such as transhumance—the seasonal driving of animals, usually sheep, to upland and lowland pasture—as well as hunting, fishing, and the cultivation of fruits and vegetables in kitchen gardens for household use. The seasonal nature of agriculture made food preservation, storage, and exchange crucial activities.

Arrangements of land could extend from large estates to small plots. Basically, four kinds of laboring people worked this land—enslaved persons, tenants who were free or unfree to varying degrees, independent peasants, and paid laborers. These forms of labor could coexist. As Chris Wickham has noted, by far the largest group in the early medieval period were peasant tenants who worked the land within a variety of obligations owed to those who owned or controlled the land. These peasant laborers often owed a certain amount of work to the landowner, and they had to pay rents, usually in kind—such as a percentage of the crops. The common belief that the slavery of the ancient Roman world changed immediately into medieval serfdom (characterized by unfree peasants tied to the land) does not represent the reality. Ancient slavery was in decline by 200 CE (although slavery never entirely disappeared), and various kinds of peasant tenancies took its place. It is notable that except in cases where the landowners took an unusual interest, the ways in which peasant tenants carried out their work was left up to them.[4] They labored within a panoply of handed-down traditional practices, and they could innovate as well.

Most preindustrial people, including those discussed in this book, lived on the razor's edge of hunger. The great variety of foods discussed here should not suggest that they were always abundant. The grain crop was crucial. When conditions were good, crops could be plentiful, and a surplus obtained for sale or storage. But the crop could be diminished or destroyed—by bad weather, war, attacks of insects, or by plant diseases. On such occasions, famine loomed. Dearth was often accompanied by grain hording on the part of the wealthy to maximize profits.[5]

The dire consequences of food scarcity meant, as Peregrine Horden and Nicholas Purcell have cogently argued, that risk avoidance was more important than intensification of production. Risk avoidance brought with it three imperatives. The first was to diversify, that is, to plant many different crops rather than a single crop year after year (called monocropping), and to raise a variety of animals, rather than just, say, sheep. The second was to store, which was important for foods such as grains, wine, olives, and fish. The third was to redistribute, that is, to maintain regular, often small-scale markets and networks of exchange with neighboring and perhaps more long-distance locales in order to be able to acquire scarce foods by purchase or exchange.[6]

This chapter is divided into discussions of Byzantine, Arab, and western European lands. Yet it should be kept in mind that these are overlapping entities. Egypt and parts of North Africa were Byzantine and then Islamic. Most of the Iberian Peninsula was controlled by the Visigoths, then by Muslims, and then by Christians (composed of various regional and ethnic groups). Yet changing forms and degrees of military control and political governance did not necessarily mean a change in techniques of food production. More important were differences in climatic and soil conditions. Much of what is said about Byzantine food production can also apply to food production in Arab lands and in western and northern Europe. Here the differences will be a particular focus.

Food Production in the Byzantine Heartland and the Levant

In the Byzantine Empire, food production was carried out primarily by peasants, the general word for them being *geōrgoi* (soil tillers). Most peasants in the vast empire lived in small settlements—villages or even smaller hamlets. Yet living in a village did not determine the nature of tenancy—villages could range from semi-autonomous to being part of large estates owned by wealthy individuals, or by the imperial government, or by a monastery. The average village possessed common holdings, including forests; pasture; groves of walnut, chestnut, and other trees; shores of the sea or of lakes; and streams. However, most of the land of a village was divided among its households. The household and its land was called a *stasis* by the central government for tax purposes. A *stasis* could include the house, vineyard, kitchen garden, trees, fields including pasture, and a spring or well. These elements could be scattered in different places around the village. Usually, irrigated vegetable gardens and orchards were situated near the village nucleus with the fields further out.[7]

The territories of Byzantium contained highly diversified climate and soil, but most common was a scarcity of water, the predominance of rocky soil, and hot summers. These conditions led to relatively small fields for crops and grapevines, and they encouraged a form of livestock management characterized by transhumance. Among the many crops produced, the most important was grain. Archaeobotanical research, including DNA analysis of seeds found in archaeological sites, is producing a detailed picture of the kinds of grains that were planted. Several nonindigenous grain crops, such as hard durum wheat (now used mostly in pasta), as well as rice, have been claimed as new crops and as part of a medieval agricultural "revolution," but in fact these crops were firmly in place in antiquity. Cultivated grains included rye, millet, and oats.[8]

Farmers planted both summer and winter crops—winter crops in November, which were often helped by abundant autumn rain. Beans as a crop ranked second to wheat in importance. Farmers cultivated other vegetables and fruits such as olives, grapes, cucumber, cabbage, leeks, onions, garlic, turnips, carrots, squash, and melons. Orchards of fruit trees included apple, quince, pear, cherry, plum, and peach. It was, as Angeliki Laiou emphasized, a polycultural agriculture.[9]

Grafting (joining two plants together) was an important technique for both trees and grapevines. The *Geoponika*, quoted at the beginning of this chapter, describes many different methods, which often involved experimentation. Grafting instructions are provided for grapevines and trees such as olive, apple, pomegranate, fig, almond, chestnut, and walnut. Grafting reduced the time between first planting and a fruit crop, provided a remedy for sick trees and vines, allowed hybridization, and increased opportunities to adopt varieties from one region to another.[10]

Peasants cultivated the fields with a plow without wheels, inherited from the ancient Romans, the so-called scratch plow or ard (Latin: *aratrum*), which was usually pulled by a pair of oxen. The parts of this plow included a share beam, the essential element that was narrowed to a point and often strengthened by a piece of pointed iron. The plow was drawn through the top layer of soil by draft animals. The peasant farmer controlled the depth of the furrow by the amount of pressure he exerted on the plow as the animal(s) moved forward. Soil was thereby loosened and deposited on both sides of the share, making a furrow. The cultivation disturbed only the top layer of the soil, leaving the soil underneath intact and retaining its moisture, important for the dry conditions of Greece and Anatolia. Because the soil was only "scratched," it had to be worked two or three times. Peasants used harrows, made from brush and wood planking and drawn by oxen to break up the clods of earth created by the plow and to even out the seed bed. In some cases, where poor soil quality prevented plowing, peasants tilled by hand with hoes and other tools.[11]

The heavy plow, to be discussed below, was long argued to be a great advancement in agricultural technology when it came into use in northern Europe. Yet, it would have been of little use in most of the Mediterranean. Overly generalized assertions concerning the superiority of the heavy plow to the scratch plow fail to consider the appropriateness of the scratch plow to most eastern Mediterranean and North African soil conditions. And besides, there were many variations of scratch plows, including some which could also efficiently plow heavy, wet soils. Peasants also used a variety of mattocks (tools similar to pick axes with a broad blade on one end of the head and an ax on the other), spades, and hoes on their fields as well as in their gardens and vineyards. Many of these tools had iron cladding or parts, suggesting that most villages included a blacksmith.[12]

Peasants practiced crop rotation. They allowed the land to lay fallow for periods of time during which they frequently plowed it in order to maintain soil moisture and destroy the weeds. But not all land was suitable for oxen-drawn plowing. Villagers also constructed terraces on hills and mountains. For most terracing they used drystone walling, collecting the stones from the hillside being terraced and often carrying earth from elsewhere to fill the cavities inside the terrace walls. Thus could cultivators extend productive land—for trees, grains, and garden crops. The construction of terraces was laborious as was their cultivation, which was accomplished by two-prong digging forks (*lisgarion*) and drag hoes (*dikella*) made from wood or metal. These two common tools were also used for trenching around grape vines and olive trees, recommended by the *Geoponika*.[13]

Workers harvested the grain between June and August. They used the sickle (*drepanon*)—a type that is often depicted as serrated. They held the curved iron sickle in the right hand and collected the cut-off grain with the left. They left stubble

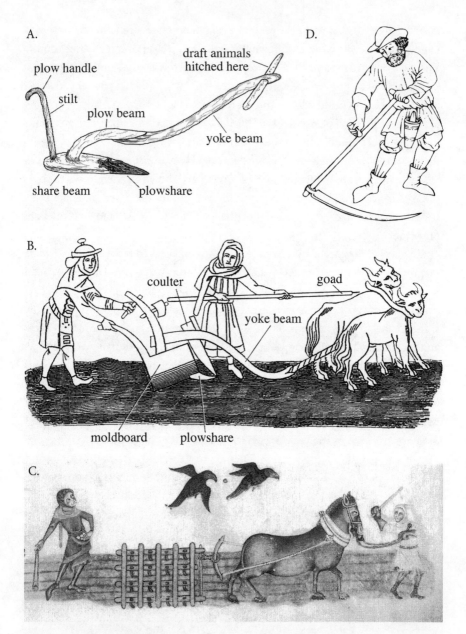

Figure 1.1. Tools and machines for field cultivation and harvesting. A. Scratch plow or ard. Photoshop image by Bob Korn. B. Heavy plow. Wikimedia Commons. C. Harrow used to break up the soil after it is plowed. Luttrell Psalter, 1325–1335 (detail). British Library, London. ©British Library Board / Art Resource. D. Peasant using a scythe. Wikimedia Commons.

and then put out cattle to graze on the field, contributing fertilizer to the soil for the next crop. Sometimes after the reaping, impoverished people, usually women and children, were permitted to glean—the backbreaking task of collecting the fallen grain left behind. For threshing (the separation of the grain heads from the straw), the workers spread sheaves of grain on a threshing floor and had oxen or donkeys drag a threshing sled (*tribolos*) over it. Then they separated the grain from the chaff with a shovel or winnowing fork, or winnowing baskets. They tossed the mixture in the air so that the lighter husks and straw would blow off. They stored the grain in either pits dug in the earth (underground silos), or in large earthenware vessels, many of which have been discovered on archaeological sites.[14]

Mills to grind grain into flour included hand mills or querns—probably the most common type—which were used in households. Querns consisted of two flat, circular stones with a hole in the top stone into which the grain was trickled, and a handle used to turn the stone. Flour came out from the bottom stone. Another kind of mill was the hourglass or Pompeian mill, which could be powered by a human or an animal—donkeys, oxen, or horses. Gradually water mills and later (in the twelfth century) windmills proliferated, but they never fully replaced human- and animal-powered mills.[15]

Olives were a staple food and were also used for other purposes such as lamp oil, a lubricant for machines, an emollient for the body, and a substrate for medicines. Syria and North Africa were major areas of olive production. The Byzantines, after losing these territories to the Arabs in the seventh century, maintained olive groves in Anatolia, southern Italy, and Greece. The production of olive oil was a complex process. Workers harvested the olives that had ripened on the trees in November and December, allowed them to ripen in storage vats for two more weeks, and then placed them in a crushing basin. They crushed the olives by pushing a millstone over them, creating a mushy pulp, which they placed in baskets and then stacked one on top of the other in a raised, circular press bed. Placing a wooden lid on the stack, they applied pressure by lowering a horizontal beam either with weights or with a windlass or by means of a screw anchored into a stone. This operation crushed the olive paste, allowing the oil to flow from the baskets into a channel at the bottom of the press bed into a settling tank below. Workers then transferred the oil into another vat, filled with water, whereupon the impurities fell to the bottom, and they drew the oil off. Oil was produced for local consumption and often for long-distance markets as well.[16]

Peasants planted vineyards everywhere in the territory of the Byzantine Empire, creating new plants from cuttings or grafts. Of great importance was the choice of

Figure 1.2. Photo from c. 1900 of Palestinian women using a quern to grind wheat into flour—a traditional technology used into modern times. Stereographic photograph by B. L. Singley. Reproduction Number LC-USZ62-69083. Courtesy the Library of Congress.

the correct variety of grape for the specific soil. The basic tool for work on vines was the pruning knife or *klaudeutērion*. Workers cut off grape clusters and put them on staves or in baskets and carried them to a wine vat. After dumping the grapes into the vat, men with bare feet treaded on the grapes to extract the juice, after which they often pressed the skins and stems further with a beam or a screw press. They placed the must (freshly crushed grapes including skins, seeds, and stems) in casks to ferment. Wine was a basic food of the Mediterranean and, like olive oil, was produced in large quantities in many places, both for local consumption and for the market.[17]

Livestock was central to Byzantine food production. This is apparent in the *Farmer's Law*, an intriguing document of unknown date and provenance but pos-

Figure 1.3. Olive oil screw press. The baskets under the press contain crushed olives that are subjected to further pressure, causing the olive oil to run down the channel into the container below the floor level. This is a photo of a seventeenth-century press in Trsteno, Croatia, but is representative of one kind of traditional oil press used for centuries in the Mediterranean. Photo by Miomir Magdevski, May 2, 2017. Wikimedia Commons.

sibly from the Anatolian plain before 850 CE, and found in numerous manuscript copies after the tenth century. The *Farmer's Law* seems to present livestock as more important than grain production. Of eighty-five articles, forty concern donkeys, sheep, pigs, and cattle, but only sixteen concern land cultivation. Other topics include vineyards and gardens, houses and barns, and agricultural tools. An example of the importance of animals: "If a man cuts a bell from an ox or a sheep and is recognized as the thief, let him be whipped; and if the animal disappears, let him make it good who stole the bell."[18]

Figure 1.4. Noah's sons picking grapes. From the Story of the Flood (detail), Byzantine mosaic, twelfth century, Cappella Palatina, Palazzo dei Normanni, Palermo, Italy. Erich Lessing / Art Resource, New York.

Livestock included pigs, goats, cattle, sheep, mules, donkeys, camels, and water buffalo. There were regional variations—camels were typical of North Africa, Syria, and Egypt, but were also to be found in Thrace on the Greek mainland. They were used as pack animals in Byzantine armies. Horses, much rarer, were the animals of wealthy elites. By far the most common animals were sheep and goats. Farmers could raise them close to home but often practiced transhumance. Animals were used for food—meat but also milk and cheese, and some, such as oxen and camels, were used for hauling carts and plows and for carrying loads. Animal hides supplied leather, and animals provided manure for the fields. Poultry was also an important part of the household food economy, as were pigeons. Columbaria (pigeon-houses) are ubiquitous in archaeological sites in the Levant, and remain today. Pigeons were prolific and easy to raise; they produced essential fertilizer and were an important source of protein in the villager's diet.[19]

The work of the peasant household was divided by gender. In Byzantine lands (and in all the lands discussed in this book), society was patriarchal and misogynist. Women in general were the subjects of their fathers and husbands. In peasant households, the head of the household was usually a male—but could sometimes be a female (usually a widow). Women could also own property. In terms of work, there was a gendered division of labor. Men usually did the plowing and most of the shepherding that was distant from the home. Women took part in the harvest, took care of kitchen gardens and vineyards, and raised barnyard animals. They performed the labor-intensive maintenance of their households, including all food preparation, and often they ground grain in addition to baking bread. They bore and raised children (often with many pregnancies and in the context of a high rate of infant and child mortality). They fabricated cloth for the family and sometimes for the market, and they made clothes (to be discussed in chapter 5).[20]

Egypt

Egypt is a special case. It has little rainfall and sits on the edge of an inhospitable desert, making the Nile River its primary source of water. Yet the idea of Egyptian agricultural workers and Nile River agriculture as an eternal and unchanging entity without history is a myth. As Alan Bowman and Eugene Rogan have emphasized, the rural history of Egypt is one of dynamism and change. The extent of the Nile River annual flood depended on the summer rainfall in the Ethiopian highlands, which varied from one year to the next. Floodwater carried mineral-rich silt to the fields along the river. A low flood failed to irrigate all available land, prevented sufficient cultivation, and could bring famine in its wake. Overly high floodwaters produced lower yields. Agricultural productivity was vastly increased by irrigation and water-lifting technologies, to be discussed in chapter 2.[21]

Within Egypt, agriculture ranged from large estates farmed by forced labor or sharecropping, to small peasant holdings. Yet most medieval Egyptian farming was village-based. Village farmers managed irrigation and tilled the land and were taxed by the central authorities. (Tax records—account books of amounts each peasant holding owed and then paid—are a primary source of information.) Peasants cultivated wheat and barley in winter; rice, cotton, and sesame in the summer; and maize and sorghum after the annual flood. Crops included flax and sugar. Farmers used fertilizers that were crop-specific and included animal dung, bird droppings, and human waste, as well as composted grass and leaves. Usual tools and machines included the mattock, hoe, plow, levelers (to even out the terrain for irrigation), and harvesting tools such as the sickle. As everywhere, animals were as important as machines. In Egypt, the most valuable and expensive animal was the jāmusā, the Egyptian buffalo cow.[22]

Egyptian village-based agriculture was hugely productive and fed many mouths both within and beyond Egypt. As Alan Mikhail has shown, after the Ottoman conquest of Mamluk Egypt in 1517, the Ottomans harnessed that productivity by organizing a legal and bureaucratic superstructure. However, they left the villagers who actually did the work great autonomy to carry out their cultivation and irrigation practices. They recognized that it was the villagers who possessed the necessary local expertise, in the face of a constantly changing environment, to carry out productive farming.[23]

North Africa (Ifriqiya and the Maghreb)

Traditionally North African history as it pertains to food production has been dominated by an overly stark contrast between settled agriculture of the coastal regions and the nomadism of the Sahara Desert. In reality, there are large areas of "pre-desert" in which farmers and pastoralists coexisted and in which irrigation agriculture and animal husbandry provided the essential prerequisites for human settlement. In addition, pastoralists themselves sometimes grew crops in the very short rainy season available to them.[24]

Peoples and geographies of the Maghreb varied greatly in the kinds of foods they produced. Coastal areas and oases of the northern Sahara produced olive oil, wine, honey, saffron, fruits, dates, sesame oil, and barley, all centered on farming villages. Wheat was an important crop in the Wadi Soffegin (in present-day Libya). The assumption that was often made about the radical decline in agricultural productivity of North Africa after the Arab conquests has been shown to be a myth, based on flawed assumptions about, first, the immense productivity of the Roman territories beforehand and, second, the lack of agricultural skill on the part of "native Berbers." In fact, it has been shown that a decline in productivity in the region began well before the Arab conquest.[25]

The conquest impacted landholding in that land grants in Ifriqiya were granted to Arab soldiers and the Berber allies, including large estates in northern and central Tunisia, and smaller estates in Sahel and the province of Tripoli—formerly owned by individuals but also by tribal groups. Some of these estates were quite large and may have been worked by enslaved labor. A well-researched area, the island of Jerba (the largest island of North Africa, off the east coast of present-day Tunisia) was dominated by coastal and inland villages in the seventh century, and later was dominated instead by new, single-family dwellings on high ground, with lower ground left for cultivation. The southeast of the island was a well-watered area with fertile soils that grew grains, olives, grapes, and fruit trees. There may have been an influx of new settlers into the island in the ninth century at a time when parts of the island were given over to large estates, perhaps to grow olives.[26]

In the western Maghreb and Morocco, there was an upsurge of oasis cultivation in the early Middle Ages. Between the seventh and tenth centuries, earlier hilltop sites were abandoned, and substantial canal networks were built to support agriculture on the oasis floor. Settlement forms included both open farms and fortified farms (*gsur*) protected by walls. In much of North Africa, agriculture focused on olives, grapes, and cereals, supplemented by pulses (beans, chickpeas, and lentils). Figs and dates were also common. A key cash crop for Tunisia was olive oil, which was exported across the Mediterranean.[27]

Food crops throughout North Africa were supplemented by animal husbandry— especially cattle, goats, and sheep for milk and meat, wool, leather, and bone. The scrub landscape was ideal for raising sheep and goats, with sheep predominating. Camels were sometimes used to pull plows. In the wetter areas of Algeria and Morocco, cattle dominated, being used primarily for milk, for pulling plows, and for powering mills. Chickens were raised for eggs and meat. In coastal regions, mollusks and fish formed an important part of the diet. There is declining evidence for pigs in medieval North Africa, probably due to Islamic dietary rules.[28]

The Arabs, with the significant help of the Berbers of the western Maghreb, conquered the Iberian Peninsula in 711 CE, after which it was called al-Andalus. After the Islamic conquests, migration and travel were relatively easy throughout a vast region, and this had important consequences for the transmission of new technologies and new knowledge. Andrew M. Watson in particular wrote classic studies which posited the resulting "green revolution." He underscored new crops moving across Africa and to the Iberian Peninsula including rice, sorghum (a cereal grass used as animal fodder and for making molasses), durum wheat, sugarcane, cotton, watermelons, eggplants, spinach, artichokes, colocasia (the tubers and leaves of which are cooked and eaten), sour oranges, lemons, limes, bananas, plantains, mangoes, and coconut palms. Many of these crops were originally from India or farther east. Most were summer crops, Watson emphasized, and farmers increased production by planting formerly fallow ground.[29]

Without minimizing Islamic agricultural contributions, Michael Decker and others have convincingly shown that this thesis of an Arab "green revolution" completely neglects pre-Islamic Mediterranean agriculture. Discussing several of the crops crucial to the thesis, Decker shows that durum wheat, rice, cotton, and artichokes were ancient crops, established well before the Islamic centuries. Further, practices such as crop rotation throughout the year were common procedures in the pre-Islamic Mediterranean. Finally, some crops, supposedly part of a "green revolution," such as mango, banana, and coconut palms, had no importance in Mediterranean agriculture.[30]

Western Europe, the British Isles, and the Scandinavian Peninsula

Unlike the cohesive villages of the eastern Mediterranean, in some areas of the early medieval west, settlements were based on estates or villas. Where the soil was rich enough to support larger concentrations of people, substantial villages developed. In regions of relatively poor soil such as Wales, Scotland, and the highlands of France, dispersed settlements were the rule, with peasants living on isolated farms or in small hamlets. They practiced what is called in-field, out-field agriculture. Each household had a small plot of land close to the dwelling (in-field) that was cultivated continuously and fertilized with the waste of humans and animals. The household also worked a plot of land farther afield (out-field) until it became depleted of nutrients after a year or so. Then the field was allowed to lay fallow and used for grazing animals, which also provided fertilizer in the form of manure. Meanwhile, the household worked another (out-field) plot.[31]

Peasants in larger villages usually practiced open field farming. The village was surrounded by a tract of land that was divided roughly in half. One half was planted each year while the other lay fallow. The field was divided into long narrow strips. Each household had a number of strips scattered in each of the two fields. The characteristic wet, heavy soils led to the development of the heavy plow. This plow, which needed to be drawn by teams of four to eight oxen, had a colter that cut the soil vertically, a plowshare that sliced it horizontally, and a mold-board that flung the slice aside. The plow cut deep furrows and turned the soil, facilitating drainage. Yet field organization varied with region, and the heavy plow did not lead to the demise of the scratch plow. Their respective uses depended on local circumstances, the heavy plow usually being used for larger, seignorial holdings (i.e., land over which a lord or an institution, such as a monastery, held tenure or title).[32]

From the tenth century on, Europe experienced a great increase in agricultural productivity. Yet the thesis proposed by earlier historians, of a technologically based "agricultural revolution," has been dismantled on many scholarly fronts on the basis of further empirical research. A classic statement of this "revolution" was made by

Lynn White Jr. in *Medieval Technology and Social Change*. White argued that increased productivity was a result of a series of technological innovations—including the invention of the heavy, wheeled plow; a traction revolution entailing the invention of the rigid horse collar; the use of horseshoes which allowed horses (which are faster and more efficient than oxen) to pull plows—and finally, a shift from a two-field to a three-field system of crop rotation.[33]

Such a view of an agricultural revolution is perhaps inadvertently supported by a parallel view of the "backwardness" of prior agricultural technologies. More recently, investigations of early medieval agriculture make such a view no longer tenable. In some areas, the three-field system was already in use in the early medieval centuries. Field organization and the rotation of crops was far more varied and complex than the idea of a linear progression from a two-field to three-field system.[34]

Historian of agriculture Karl Brunner noted that the renewed study of archaeological finds (particularly an important find of fifth-century tools and implements at Osterburken, Germany) and the reexamination of objects in museums support a view of gradual (rather than revolutionary) technological change in agricultural implements from Roman to early medieval times. For example, transitional plowshares that anticipated the heavy medieval plow have been found in late imperial Rome. Larger, animal-powered machines included harrows—heavy, wood-framed machines set with teeth that were dragged over the soil to break up clods and cover seed. Farmers also utilized numerous hand tools. François Sigaut suggests that most of the technologies of field cultivation in the Mediterranean and Europe were in place in the first centuries CE and that technical developments were evolutionary and included numerous, small, local innovations. For England, John Langdon has listed, among others, shovels, mattocks, spades, pitch forks, axes and hatchets, knives, wheelbarrows, saws, hammers, the short-handled sickle, and the long-handled scythe. Most of these tools were made of wood, some sheathed with iron. Langdon notes that revolutionary changes or innovations in toolmaking are not in evidence. Michael Toch has argued that agricultural productivity increased in the early medieval period not by virtue of technology per se, but by "the more intensive application of human work" and by "diffusion, organizational adaptation and elaboration." Georges Raepsaet and Georges Comet emphasized the complexity of technological change in agriculture and the importance of social context to such a change, while Comet elucidated the many variations of both light and heavy plows and the historical evidence for them.[35]

Horden and Purcell have conceptualized the issue as one of abatement and in-

Figure 1.5. Woman harvesting grain with a sickle, as man binds it into a sheath. Stained-glass panel. Labours of the Months, August, c. 1450–1475. Victoria and Albert Museum. Photo by David Jackson, February 2010. Wikimedia Commons.

tensification within microregions, rather than a simplistic idea of progress, or the notion of raising productivity. Intensification of cultivation may be motivated by such things as strong demand for a certain crop. In this case, farmers might bring marginal lands into cultivation or cultivate existing fields more intensely (by leaving less land fallow, for example). When that demand goes away for whatever reason, abatement occurs, in which the extra effort required for intensification is abandoned. For Carolingian lands, Paolo Squatriti notes that "farmers were not motivated by calculations of yield, instead measuring productivity in terms of labor inputs in relationship to household needs." Similarly for England, David Stone, focusing on thirteenth- and fourteenth-century agriculture, cogently argued that medieval demesnes (i.e., land held and managed by the lord of the manor) were far more competently and rationally run than has been assumed, and that decisions did not always favor higher yields and the higher costs and labor needed to produce them.[36]

Changes were not "revolutionary," since they developed slowly over centuries, nor were they entirely attributable to technological causes. Yet change in the form of a gradually increasing food supply did occur for complex reasons—technology but also a decrease in violence after the worst of the Viking raids of the ninth century were over, and a small but significant warming of the climate. Improvement in agricultural techniques played a part. Very gradually, beginning in the eighth century, peasants changed from a two-field to a three-field system of crop rotation. The new system was instituted mostly in the north because it required spring planting, successful only in regions of wet soil. Available land would be divided into three parts. One third lay fallow, one third was planted in the fall with winter crops such as wheat and rye that would be harvested in the summer. The third part was planted in the spring with oats, barley, or nitrogen-fixing legumes. This system could increase crop productivity by a third. But the changeover seems to have occurred far later than had once been assumed. And rotations always varied according to region, soil, climate, demand, and custom.[37]

The subject of animal traction and particularly the use of the horse collar has undergone similar revision. The traditional view that the Romans had used a harness that put pressure on the horse's windpipe, choking the animal when it tried to pull heavy loads, has been challenged as overly simplistic. The detailed studies of Georges Raepsaet on ancient and medieval harnessing have been particularly important in showing the complexity and relative sophistication of ancient harnessing. Medieval peoples did develop a more rigid and efficacious horse collar, but the consequent adoption of the horse as a draft animal was by no means a uniform or universal phenomenon. Horses are faster than oxen, and when shod with iron horseshoes to prevent lameness, they can be more efficient. Nevertheless, unlike oxen, which can live on grazing in pastures, horses require grain such as oats. Oxen remained the preferred draft animal in many parts of Europe, but the twelfth and thirteenth centuries in northern Europe saw the use of horses as draft animals for wheeled plows. Their employment in this way was gradual, never universal, and varied from one region to another.[38]

As in the Byzantine and Arab lands discussed above, most peasants were tied to various kinds of tenancy, free and unfree throughout the European continent. The system, variously called manorialism, demesne, or seignorial farming, often constituted an important part of the mix. As Chris Wickham noted, it was a northern development that began in the Carolingian Empire, especially with the reign of Charlemagne (ruled 768–814) in the Frankish heartland between the Seine and Rhine rivers and into the Danube Valley and Bavaria, and as a separate development in Lombardy and other parts of Italy. And although a typical manor or villa can be described, in fact there were numerous variations, in part depending on variation of soil types and climate. Further, property and tenancy arrangements were by no means static. They varied and changed over time, partly because of the influence of markets and the growth of urban centers.[39]

In manorialism, a so-called bipartite structure of land division and work was introduced. In one part, the peasants cultivated land for the lord. The second part comprised the tenement or holdings, which peasants cultivated for themselves in exchange for goods and services that they rendered to the lord and his demesne. Conditions and terms varied widely. The lord and lady of the manor might live in the manor house with their children and supervise the peasant workers themselves, or they could hire a bailiff or overseer who would supervise the work and collect the lord's revenues. The lord took about a third to a half of the produce, as well as some livestock and fish caught in the streams. The lord's animals grazed on the common pasture. Peasant men provided labor for building and digging ditches, while peasant women, supervised by the lady of the manor, often worked in the manor house, spinning, weaving, and doing other chores. Peasants labored under various conditions of freedom or unfreedom, and some villas included the labor of enslaved persons.[40]

Most of the tools, equipment, and food used by both lord and peasant were made on the manor or in the village. Peasants ground the flour from grain, baked bread, and made beer, wine, cheese, and butter. They slaughtered animals and cured meat. If the laws of the manor allowed, they fished. They gathered fruits, berries, and firewood from the forest, as well as lumber for making tools, barrels, and furniture. Peasants worked cooperatively but also according to gender. Men did most of the heavy plowing, while women and children goaded the oxen. Women bore and cared for children; raised poultry and livestock; milked cows, sheep, and goats; sheared sheep; tended kitchen gardens; fetched water from wells; took grain to the mill; gathered firewood; and tended the fire. They spun yarn, made cloth from wool and from fibrous plants, made clothing, did laundry, and prepared food and drink such as ale. All able-bodied family members, including children who were old enough, participated in the harvest. Both men and women were engaged in marketing surplus produce. In some regions, such as the Netherlands, many peasant families were also involved in "proto-industrialization" and produced products such as cloth for urban markets.[41]

As the population increased, more land was cleared for cultivation, a process that medieval documents call *assarting*. First brush and light woods were cleared by fire and/or axe, and then heavier forests were cut, and the troublesome tree roots removed. Marshland, fens, and peat bogs were drained as well—difficult tasks, which could yield rich, arable land for cultivation. Land management and agricultural practices were not only highly variable from one region to another, but changed over time.[42]

In England, important innovations occurred at the beginning of the sixteenth century. They included convertible husbandry, the draining of fens, seasonal inundation of meadows called "floating meadows" (to be discussed in chapter 2), and marling—fertilizing with marl, a mixture including clays and shells, effective in lime-deficient soils. Convertible husbandry abolished the distinction between permanent, arable (cultivated) land and permanent pasture. The same land was alternated between pasture and planting. This new practice increased the productivity of both crops and livestock. Crop yields were higher on land that had rested under grass while fertilized by the manure of grazing animals. Likewise, pasture grown on recently cultivated fields provided richer nourishment, thereby increasing the health, weight, and productivity of animals.[43]

An important change in field use both in England and on the continent involved the "enclosure" movement. Traditional medieval agricultural practices divided "commons" into parcels of land that villagers cultivated cooperatively or used for common pasture. This practice continued in many regions of Europe through the eighteenth century. However, "enclosure" also developed in which land in common use would be fenced or cordoned off for private use. In early sixteenth-century England, landlords often enclosed land and then leased it to tenants. They did this in response to high wool prices, fencing off arable land and then using it as pasture for sheep. Then, from the mid-sixteenth century, they enclosed land and cultivated it in response to high grain prices. When enclosure occurred, it ended common usage, including village strip farming, and created individual holdings. Enclosure disrupted communal regulations, usually to the detriment of the peasantry and to the benefit of private owners and tenants. It also helped to solidify divisions of wealth and property ownership already evident in medieval villages. Historians have interpreted apparent increases in productivity as resulting not from enclosure per se, but from the ability of landlords to acquire a larger share of the products. Conversely, they have evaluated traditional village strip farming as more productive than previously thought.[44]

Moving from local contexts to a longer view, the European environment was profoundly changed in the medieval centuries by "cerealization," that is, increasingly from the eighth to the fourteenth century, converting the land from whatever it was—woods, scrub, marsh and fens, peat bogs—into fields suitable for growing crops. Population pressure and cultural preference for eating grains produced large-scale intensification of agriculture toward cereal production. This process resulted in unintended environmental consequences. Land clearing led to a reduction in the diversity of plants and animals. Converting open woods to permanent fields destabilized relations between plants, soil, and water. Large-scale soil erosion and deposition or alluviation (the deposit of sediments by flowing water) occurred, changing shorelines and riverbeds. Changes in water chemistry undoubtedly also occurred. To give one possible example, around the coastal area of Pomerania in the southern Baltic, huge herring schools, which had spawned and been harvested for several centuries, disappeared around 1300.[45]

Woodlands

Trees and woodland also played a crucial role in food production, especially in regions where they were abundant, including the British Isles as well as European and Scandinavian lands. Interest in environmental history has brought new scholarly attention to trees. (In addition to being an important resource for food production, woodlands were also essential for constructing buildings and ships, as a source of fuel for household cooking and heating, and for metallurgical operations such as smelting—to be discussed in later chapters.) In terms of food production, forests played several roles. Assarting, or clearing an area for cultivation, often meant cutting down a forest or at least a group of trees, an activity that was extremely widespread, especially as the population of Europe expanded in the eleventh and twelfth centuries. It was a practice that had profound environmental consequences, but one which was by no means original to the medieval centuries. Indeed, the thesis of the "great clearances"—that a massive number of woodlands were cut down between the years 1000 and 1300—has been questioned. For one thing, the word "forest" in its several forms found in legal documents in various European languages did not mean what it means today. Often "forest" referred to a legal entity, usually land set aside for noble and royal hunting, usually land with trees but sometimes not, basically a managed game preserve. And much deforestation had already occurred in the ancient world.[46]

Forest set aside for hunting, of course, did provide food in the form of game (mostly deer and wild boar) for the tables of elite people and, in some cases, food for others as well. Hunting had immense social and political importance, being key to elite leisure and to demonstrable social superiority. Hunting included the rabbit, an animal introduced by the ancient Romans to Iberia, which had become "invasive" by the 1200s throughout Europe, causing a decline in the population of the native European hare. Other introduced animals, such as the fallow deer and the pheasant, served the hunting agendas of elite people. But elite prerogatives were not exercised everywhere and were not completely enforced where they did exist. Peasants and other non-elite persons in fact used forests extensively. For example, the Weald, a large expanse of uncleared woodland in southeast England was heavily used as pasture, mainly for pigs. Woodland could also be used to feed goats (which will climb trees to reach edible material), as well as cattle and sheep (which will not). Humans also gathered green branches that were dried and used like hay for feed and bedding (called "leafy hay"). Trees were harvested for nuts and fruit.[47]

Traditional forestry practices, called "woodmanship," created what we would call sustainable forests with a panoply of techniques. Most important was coppicing, which took advantage of the fact that cutting down a hardwood tree does not kill it. Most hardwood species such as oak, beach, ash, and walnut, will sprout from a stump—called a stool after the cutting. After the trunk of the tree has been cut, shoots eventually will grow to provide a continuous supply of poles, rods, and logs. Coppicing could not be combined with wood pasturing, since the animals would eat all the young sprouts (and yes, then the tree would die). Another technique, pollarding, involved using ladders to cut the tree farther up the trunk, out of reach of grazing animals. Workers usually carried out such pollarding with an ax.[48]

In the fifteenth and sixteenth centuries especially, forests could be used by rulers and states to augment their political authority as well as their finances. For example, in Poland/Lithuania, a region of abundant forests accompanied by extensive river networks that made forest products accessible, King Sigismund Augustus (ruled 1548–1572) enacted new, restrictive forest legislation. As Mateusz Falkowski has shown, his monopolization and exploitation of forest products, such as timber and charcoal, as commodities delivered to the Atlantic maritime states, gave him needed financial resources and an upper hand in regional political and military struggles. In other regions, playing on the fear of scarcity (as Karl Appuhn has shown for Venice) became a powerful tool in the hands of authorities to enhance their political and financial power. For the Iberian Peninsula, John Wing has shown how royal forest policy transformed between the fifteenth and sixteenth centuries—

from conserving forest resources for the common good to exploiting them for the needs of the state in shipbuilding.[49]

Fishing

The emphasis that most historians have placed on cultivation and animal husbandry for food production should not obscure the importance of fishing throughout the regions encompassed by this book—along rivers and shorelines, in the Mediterranean and Black seas, and in the Atlantic Ocean. Fishing provided nutrition in all areas adjacent to rivers, marshes, and sea coasts. People in such areas ate fish, eel, and mollusks, as did people living in the growing number of places that had access to commercial fish markets. For Christian areas, it provided a substitute for prohibited meat on particular days such as saints' days and Fridays. The technologies of fishing developed according to the particular environment.[50]

In addition, from the eleventh century, fish became available through aquaculture, the creation of artificial ponds in which to raise usually carp, a fish originally native to the Balkans in eastern Europe. As Richard Hoffmann has explicated, aquaculture entails human manipulation of all the same variables as agriculture. One method was live storage, herding fish into confined aquatic areas until time for consumption. In addition, fish farms proliferated—they have been especially studied in Poland and other areas of eastern Europe. Fish farming required management of multiple ponds—one for spawning, one for fry (fish of one to two years old), and one for final growth of two to five years. Each required management of inflow, outflow, and water level, using dams, dikes, and bypass channels. Some fishponds were surprisingly large—one built in southern Bohemia in 1492 was over five square kilometers (two square miles). These ponds, created and controlled by elites, were managed by specialists called "pond masters" or "fish masters." Fish aquaculture exerted profound environmental changes by changing the land itself and (often, depending on the region) introducing an exotic species (carp) to the detriment of native species.[51]

For the Byzantine and Arab worlds in the eastern Mediterranean, interdisciplinary studies have identified a huge number of species of fish that were caught and eaten, both marine and freshwater fish, as well as cephalopods such as octopus, squid, and cuttlefish, and crustaceans including crabs, lobster, and shrimp. Fresh fish was sold in fish markets and then consumed, but fish were also preserved (usually with salt) for longer-term storage and/or long-distance commerce. Our knowledge of fish culture comes from documentary evidence, including laws concerning fish markets and documents recording commercial transactions. It also comes from archaeological evidence, such as fish bones found on dryland sites and fishing equipment discovered on sunken ships. These ships often contain weights or sinkers used for both net and line fishing, hooks, needles for repairing nets, nets themselves (or pieces of nets), and implements for spearing fish.[52]

In medieval European lands, the technologies of fishing ranged from catching fish and other aquatic animals with the bare hands to carefully managed ponds. Fishers sometimes stunned fish with narcotic substances (examples include extracts of plants such as yew, great mullein, or nettle) or with underwater explosions (using caustic quicklime—anhydrous calcium oxide), and then gathered them up by hand. They also practiced hook-and-line fishing, using poles of various lengths, lines made of hemp or linen, and hooks made of iron, bronze, or wood in a variety of shapes often designed to catch specific kinds of fish. Sometimes fishers attached numerous hooks to short dropper lines that were tied to a main line or leader.[53]

Other methods of catching fish employed various kinds of nets. In one kind, either baskets or nets were set up so that the fish swam into them and could not swim out. Fishers frequently positioned such nets at the openings of weirs, sluices, and other structures, often in association with mills. They used other nets by moving them actively through the water to catch the fish. One kind of net, called a sein, possessed two wings and was set in an area thought to have fish, which were then hauled in to either a boat or to the shore. Another kind of net called a trawl, had rigid frames of weights or floats to keep the mouth open while fishers dragged them through the water behind their boats.[54]

Another way of obtaining fish was to create a barrier to capture and hold migrating fish. Most widespread were mill dams which captured and held the downstream migration of eels and fish, often using weirs to funnel them to a pond or tank where they could be scooped up. Tanks and sometimes ponds were created to keep the fish alive until the time came to eat them, eliminating the problem of preservation.[55]

In tenth-century Italy in the middle Po Valley in the north and in other centers such as Lucca and Ravenna, fishing was a complex, organized activity. The fishers of Pavia for example, maintained a fleet of at least sixty boats. Most fishing was inland, freshwater (as opposed to marine). After the year 750, great landowners began building inland fisheries (*piscaria*) for the rearing of fish. They were built on rivers or river deltas and featured water intake systems with sluices and gateways. At times ponds were dug for raising fish, the water being controlled by sluices, from which fish could be harvested by draining the pond. Most *piscaria* depended on the work of specialized fishers, many of whom in the eighth and ninth centuries were bonded men or serfs obliged to fish for overlords. In contrast to ancient Roman law, in which waters including those for fishing were considered public, in medieval Italy fishing waters became increasingly privatized.[56]

Nearby in the eastern Adriatic Sea along the Dalmatian coast, hundreds of islands contributed to an abundant marine fauna. Fishing techniques included attracting fish at night with the light of torches and then driving them into gill nets, in which the correct gauge of mesh entrapped fish of a certain size: the fish's head could fit through the net, but the fish would get stuck at its gills. In the second half of the fifteenth century, the new technique of drift-net fishing emerged. Drift nets are long nets with floats on top and weights on the bottom, attached to boats which are allowed to drift with the currents and winds, enmeshing large numbers of fish in the process.[57]

In medieval England, fishers practiced both riverine and coastal fishing. In York in northern England archaeologists have studied fish consumption by carefully sieving deposits to recover fish bone fragments that they have then identified by species. They have found that in the tenth and eleventh centuries there was a shift from freshwater fish and eels to marine fish, such as herring and gadids (haddock, cod, whiting) caught in estuaries or in the open sea. After the twelfth century, some freshwater fish such as perch were raised in fishponds. However, there was a decline in the abundance of all freshwater fish over time, possibly due to human pollution of the aquatic ecosystem (although this is debated). To give one example, eel—the most abundant species between the seventh and ninth centuries—declined to trivial numbers over the next several centuries. By the tenth century the marine herring had overtaken the eel, and by the late fourteenth century, eel consumption had ended.[58]

English fish technologies included the construction of weirs. A weir could be a barrier usually made of timber or wattle (poles intertwined with twigs or vines) set across the flow of a river or stream. This created an eddy in which fishers caught their prey with a net. A weir could also be V-shaped, ending in a tapered basket with a wide mouth facing upstream and closed end into which fish would swim. Fishers constructed coastal sea weirs that used the tides to trap fish, and they also fixed traps in rivers, went out in boats to capture fish with nets, and fished with baited hooks and lines.[59]

Marine fishing, especially offshore, usually developed later than freshwater fishing and often in response to a rising demand for fish, a result of a rising population, and increased urbanism. Scandinavia was a region where marine fishing was especially important. Cod, herring, and related species were major catches in Norway, especially in the Lofoten Islands, as they were in the Danish and Swedish Islands

Figure 1.6. Possible reconstruction of an Anglo-Saxon fishweir. From J. M. Steane and M. Foreman, "Medieval Fishing Tackle," in *Medieval Fish, Fisheries and Fishponds in England*, ed. Michael Aston, pt. 1 (Oxford: BAR, 1986), p. 171, fig. 21. Reproduced with permission of BAR Publishing.

of the western Baltic. On one site on the island of Bornholm, archaeologists found thirteen thousand herring bones dating from the sixth to the seventh century. The herring, which has been called the ant of the ocean, flourished in the North Sea as well as in the Baltic. In addition to Scandinavia, herring became important for northern England, Germany, and Poland. From the thirteenth century, herring processors near Copenhagen removed the gills behind the head and then immediately salted the fish; the blood still fluid in the fish allowed the salt to penetrate more thoroughly for better preservation. In another new method, barreling, workers tightly packed gutted herring between layers of salt in wooden casks. The salt pulled the moisture out of the fish, which were then repacked in fresh brine, preserving them for up to two years. This method was the key to the long-distance herring trade. Herring became available all over Europe from the eleventh and twelfth centuries and was an important long-distance commodity, only behind grain and textiles.[60]

In Norway, fishing functioned as part of the traditional household economy, especially in the north where the climate made crop harvests uncertain. In Finnmark (northern Norway), commercial fishing developed in the twelfth century. Farmers who fished part-time owned small fishing boats that could be powered by both oars and sails, and carried three to four men who used fishing lines to catch cod. In coastal regions fishing stations were constructed that served as bases for seasonal fishing. Fishermen resided there during the cod-fishing season (the winter), which lasted almost half a year, while the women managed the farms back home, including the tasks of animal husbandry. From the twelfth century a large trade in stockfish—cod dried without salt in the dry, cold, windy climate of the north—developed along with smoked and/or salted herring. (Herring is an oily fish that does not preserve as well as cod does using the stockfish method.)[61]

In the same era, the economy of Borgund in southern Norway was also based on the production and trade of stockfish. Cod fishing was accomplished with two types of line—the handline (also known as the deep bait line) and the troll line (dragged behind the boat). Both consisted of a line, a sinker (usually made of soapstone), a hook, and a snood (the cord between the sinker and the hook). It has been suggested that fishers used trolling in coastal fishing for domestic consumption, while they used handlines for the commercial fishing of migratory cod in the winter and spring.[62]

In the fifteenth century, shipwrights developed a new deep-fishing vessel called a buss (usually referring to Dutch and English vessels). Used for fishing herring, the boat could accommodate heavy loads of fish and could stay at sea for months with a crew of ten to fourteen. Buss fleets began the season in the Shetland Islands off northern Scotland for the large schools of fish that arrived in February and March. They followed the fish southward into the North Sea. By November and December, they congregated off Great Yarmouth in East Anglia (in southern England). Buyers from all over Europe came to the Great Yarmouth fish fair. Numerous temporary fish camps were established where men went out to fish, and women and children on shore labored at gutting tables, gutting the fish and then packing them into barrels. Fish guts were used as fertilizer. It was dangerous work—a slipped knife could cut and maim a worker, perhaps destroying her livelihood.[63]

* * *

In sum, fishing can be thought of as a group of aquatic technologies that include equipment, vessels, methods of capturing fish (sometimes by manipulating sea or river water), and methods of preservation and shipping. Fishing in all its manifestations constitutes a kind of food production, but it also can be thought of as a group of hydraulic technologies. Such technologies also include irrigation, drainage, and the powering of machinery, all treated in the next chapter.

CHAPTER TWO

Hydraulic Technologies

Since the "bit" [the closure of the dam] at al-Lāhūn and its installation
have been mentioned, it is appropriate that I give an account of it. . . .
The above-mentioned "bit" is a long palm log upon which straw and
rags are affixed. These are tied up with ropes, so that it becomes very
thick. There are strong ropes at its edges, and the ends of these ropes are
in the hands of large groups of men on the bank adjacent to the village
. . . called al-Lāhūn, and on the opposite bank. They release the ropes
little by little, while the water carries the "bit" and pulls it toward the gap
located in the dam of al-Lāhūn, in the midst of the structure. . . . They
release it little by little until it comes to the mouth of the gap and blocks
it, and thereby prevents the water from escaping. The men pile up soil
and clay upon it so that it resembles the bank adjacent to the structure.
. . . The purpose of blocking the gap is that the water which [otherwise
would have] escaped through it, will be available for the villages of the
Fayyum.

'UTHMĀN AL-NĀBULUSĪ, *VILLAGES OF THE FAYYUM*, 1243–1244

The above citation is a rare and fascinating eyewitness report from the thirteenth
century. It describes a routine irrigation operation carried out by villagers in
Fayyum, a province in middle Egypt located in a large depression of the Libyan
Desert about 25 kilometers (about 15 miles) west of the Nile. Its author was a government official, 'Uthmān al-Nābulusī (1192–c. 1262), who traveled to Fayyum to
survey the villages at the request of the Sultan as part of an attempt to understand
(and then revive) the region's agriculture, which they believed was in decline. Al-
Nābulusī's report, which describes the tax revenues from the year 1243–1244, is
precious for its detailed description of each of the approximately 125 villages in the
province. Here he describes his observation of a maneuver basic to basin irrigation. A reservoir had been built near the village into which water flowed (from a
canal built to the Nile River) at the time of Nile flooding. When the annual flood
began to recede, the men created a dam to keep the water in, saving it for the use
of the village. Workers fixed straw and rags to a long palm log and tied them up

with ropes. Men on each end of the log (on opposite banks of the canal) held the ropes and gradually let them out so that the water carried the log to the mouth of the opening of the embankment and blocked it, preventing the water from escaping. Then they piled on clay and soil, making an augmented embankment or dam that held the water inside.[1]

Hydraulic technologies are the focus of this chapter, including discussion of hydraulic methodologies such as the *qanat* (an underground water collection system), water-lifting machines, and machines powered by water. The chapter also treats crop irrigation, drainage, and flooding. (Urban water supplies and sanitation are discussed in the next chapter.) Hydraulic technologies are dependent on geography and climate as well as on specific technological and cultural practices. They can be brought to bear on profoundly different situations, such as the irrigation of semiarid land as in the Levant or parts of North Africa, and the draining of excess water as in the Netherlands.

Studies of irrigation systems in premodern times often have been shaped by the long shadow of Karl Wittfogel's widely influential—and more recently—severely criticized thesis that associated large-scale irrigation systems with autocratic governments. His thesis, formulated in the 1950s and infused with orientalist assumptions, is no longer accepted. It has been rejected in part because subsequent empirical studies in many parts of the world, including eastern Mediterranean, northern African, and European lands, show that large-scale irrigation systems have often been managed by small groups of people (for example in villages) without the direct interference of larger authoritarian entities. Irrigation communities may have had powerful lords over them. They may have been required to pay taxes and were constrained by other obligations, but often they managed their irrigation works without supervision from above. The citation at the beginning of this chapter describes one episode in the irrigation of an Egyptian province to be discussed in greater detail below. The point here is that this description (as well as the remainder of al-Nābulusī's lengthy report) contains not the slightest indication that a higher authority is overseeing the operation. It is a description of villagers carrying out an irrigation task using their own traditional methods and materials.[2]

Hydraulic Technologies and Machines

Irrigation technologies used a variety of water-lifting machines that raised water and deposited it into canals or cisterns (large tanks or basins often built into the ground to catch rainwater or collect water from irrigated sources). Water was also used to power machines for other purposes, most importantly in the medieval centuries, mills for grinding grain into flour.

The Qanat *and Other Drainage Galleries*

A particular hydraulic technology, the *qanat* (with many alternative names depending on location), originated in ancient times. The *qanat* is built into a hill and consists of an underground canal that begins at an upland aquifer (the underground water table). Because the flow of water is gravity-powered, the canal must slope gradually downward until it exits from the ground at the bottom of the hill, providing water for irrigation as well as for human and animal consumption. The *qanat* allows new field cultivation and settlement in arid lands. Although at one time it was believed to have originated in the Persian Sasanian Empire (224–651 CE), archaeologists now think that the technology developed much earlier, possibly independently in various locales. Yet the Arab conquest of the Persian Sasanian Empire (present-day Iran) in 651 triggered the spread of *qanat* technology through the Levant, across North Africa, and into the Iberian Peninsula. The *qanat* is the focus of an eleventh-century treatise by Moḥammed Karajī (fl. early eleventh century), who was a scholar and mathematician.[3]

To construct a *qanat*, workers first surveyed the land to find an appropriately located aquifer. Expert diggers then dug a trial vertical shaft about one meter (about 3.3 feet) in diameter, pulling up the soil by means of a windlass and piling it around

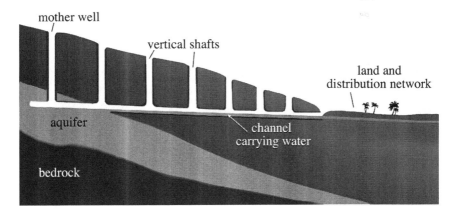

Figure 2.1. Cross section of a *qanat*. Created by Samuel Bailey, December 2, 2009. Wikimedia Commons, edited by Bob Korn.

the mouth of the shaft. This first shaft, called the motherwell, extended from the surface to the aquifer and could be from 20 to 80 meters (about 66 to 262 feet) deep. Once the aquifer was reached, workers undertook several days of testing to see if the water supply and flow were sufficient. If so, they began construction. A suitable route for the structure was determined, with the exit point usually several kilometers away. Then they constructed a series of shafts 10–20 meters (about 33–66 feet) apart along the line of the planned underground canal, and from these they built the canal itself with its necessary gradient. Workers used the shafts to enter and exit to construct the canal and later for cleaning and maintenance. As long as the *qanat* was in use, annual or biannual maintenance was necessary both to remove accumulated sedimentation and to repair broken canal walls that could be damaged by floods and earthquakes. The *qanat* was a sustainable water extraction system, which did not overexploit the resource because it drew only on the upper part of the aquifer, allowing it to recharge.[4]

There are three advantages to the *qanat*. It doesn't require a water-lifting machine or other elevation system. The water in the system is mostly underground so that it avoids evaporation, and finally, it arrives at the surface clean. There are many other terms for the *qanat* and other kinds of underground catchment systems that do not share all of its features. In a discussion of the subterranean water catchment systems in al-Andalus, Antonio Rotolo argues for the more general term "drainage gallery," referring to any excavated, underground tunnel that intercepts the aquifer and carries the water to the surface by exploiting gravity with a gradual gradient.[5]

Water-Lifting Machines

Fundamental to both irrigation and other kinds of human and animal water supply was the ability to lift water from lower to higher elevations. Two of the simplest and most prevalent machines were the shaduf and the Archimedean screw. The shaduf (also called the balance bucket or swape) was used particularly in Egypt. It consists of a long beam that rests on a horizontal crossbar in a lever arrangement. From one end of the beam hangs another long beam or a rope to which is attached a bucket. On the other end is a counterweight. The worker pulls down the vertical beam or rope until the bucket goes into the water and is filled, then the worker pushes it up and turns it with the help of the counterweight, and dumps the water into an irrigation canal.

The Archimedean screw was used throughout the Islamic world. It consisted of a screw with a helical surface on a pipe inside a hollow shaft. As the shaft was turned, the bottom scooped up the water which was pushed up by the rotating screw until

Figure 2.2. Man working a shaduf. From Amelia B. Edwards, *A Thousand Miles Up the Nile*, 2nd ed. (London: George Routledge and Sons, 1891), 73.

it poured out of the top into a canal or receptacle. It required an operator to turn it with a crank.[6]

More complex were the several kinds of wheels used for lifting water. One type was called a *saqiya*, a geared machine with a bucket chain. An animal was hitched to a shaft and, goaded by a human, walked in a circle turning the shaft. The shaft turned a horizontal lantern gear which engaged the cogs of a vertical wheel. Pots were attached to a chain (made of rope), called a pot garland, which was turned by the wheel. As the pots went around, they reached the water upside down, scooped up the water and then, arriving at the top, dumped it into a canal or trough.[7]

Another kind of water-lifting wheel was called in Spanish, the *noria*, and in Portuguese, the *nora*, words derived from the Arabic *na'ura*. (This is not to suggest that the *noria* was invented by the Arabs, as traditionally has been assumed; it was an ancient technology.) In general, the *noria* was a machine made of wood with a large wheel turned by moving water or by animals. Pots or wooden compartments were attached to its rim. The vessels on the wheel filled with water as they dipped below the water line. At the top, they dumped it into an aqueduct or canal. Historians have recently emphasized the complexity of terminology when referring to this machine (within the diverse linguistic groups that used it). There were innumerable local variations not only in terminology but in the machine itself, deriving from geographic necessity and need. Rather than focusing on where or by whom

Figure 2.3. A. Archimedean screw. Drawing from Thomas Davidson, ed., *Chambers's Twentieth Century Dictionary of the English Language* (London: W. & R. Chambers, 1907), 47. Wikimedia Commons. B. An Egyptian irrigating a field using an Archimedean screw. Museum of African Art, Belgrade. Photo by Zdravko Pečar, 1955. Wikimedia Commons.

Figure 2.4. Saqiya powered by a donkey. Notice the buckets on top of the wheel pouring water into the canal. From Eugène-Oscar Lami, ed., *Dictionnaire ency-clopédique et biographique de l'industrie et des arts industriels* (Paris: Librairie des Dictionnaires, 1886), 6: 840.

the *noria* was invented, historians have pointed to the great creativity involved in "technology of use"—the many ways that workers modified the machine to make it work in particular settings.[8]

Machines Powered by Water

Unlike the machines discussed above, which functioned solely as water-lifting devices, machines powered by water performed other tasks. Millers used moving water to power various kinds of mills that ground grain into flour. Water-powered mills were also designed for functions such as fulling cloth and sawing wood. In a famous essay published in 1935, the Annales historian Marc Bloch emphasized the innovative and extensive medieval adoption of heavy, vertical overshot or under-shot waterwheels. These were large, heavy machines in which a vertically posi-tioned wheel, placed in a course of running water was turned either by water catch-ing it below (the undershot wheel) or falling on it from above (the overshot wheel). The wheel was attached to a shaft and gearing that changed its vertical motion to horizontal, thus allowing such horizontal motions as grinding stones in a gristmill.

Figure 2.5. Three *norias* in Hama, Syria, on the Orantes River. Photo by Erik Albers, August 1, 2005. Wikimedia Commons.

Bloch suggested that the ancient Romans, although they possessed water-mill technology, failed to exploit it fully because they could as easily order enslaved people to grind their grain by hand. He also connected the rise of the vertical watermill to the rise of feudalism, specifically with seignorial lords who had the means to build heavy overshot waterwheel mills and the power to enforce their use on peasants. Bloch's view of the ancient Roman failure to fully exploit available technologies was reinforced by M. I. Finley's influential thesis (originally published in 1973) that the ancient Romans were motivated more by social than economic values. Historians such as Jean Gimpel and Lynn White Jr. argued for an "industrial revolution" of the

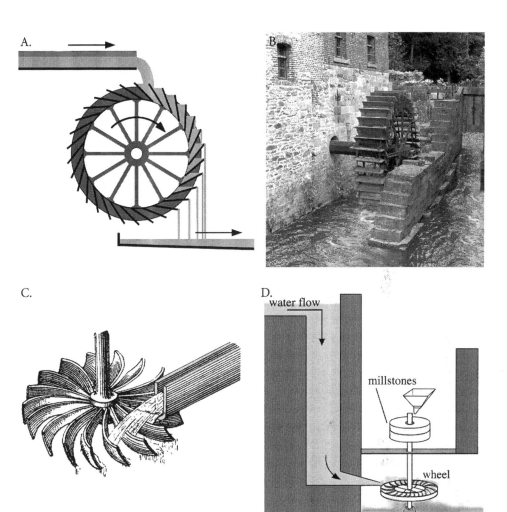

Figure 2.6. Waterwheels. A. Diagram of overshot waterwheel. Drawing by Malcolm Boura, April 26, 2017. Wikimedia Commons. B. Undershot waterwheel, Braine-le-Château, Belgium, twelfth century. Photo by Jean-Pol Grandmont, June 11, 2005. Wikimedia Commons, edited by Bob Korn. C. Horizontal waterwheel. From MM. Larive et Fleury [*sic*], *Dictionnaire Français illustré des Mots et des Choses*, vol. 3 (Paris: Georges Chamerot, 1889), 196. D. Arubah-style mill. Photoshop drawing by Bob Korn.

Middle Ages in part through the acceptance of the view that the ancient Romans failed to exploit watermill technologies, whereas medieval peoples embraced them.[9]

These ideas concerning the water mill and the medieval "industrial revolution" have been challenged—on the one hand by the recognition that the ancient Romans exploited waterpower to a greater extent than was formerly believed, and on the other, by the understanding that the medieval exploitation of the waterwheel developed very slowly in some areas of the medieval world, and more rapidly in others.[10] These mills required significant capital investment and were usually available only to the elite. The lord often operated the mill as a monopoly, forbidding peasant households the use of querns or hand mills or the cheaper, more flexible horizontal mill (to be discussed below) and forcing them to use (and pay for) his mill.

Wherever it was in use, the water-powered mill had many variations, depending on geography as well as social organization and local custom. Mills usually included weirs (walls or dams placed in a flowing stream, river, or canal to direct the water to the mill) or other features of water management, a millhouse, and other buildings nearby to store grain as well as tools for the mill's repair. All mills had two millstones (heavy, horizontally placed stones for grinding the grain), the top one constructed for circular movement. Because the stones wore down, the miller's skill set had to include the ability to dress the stone and recut the groves.[11]

The simplest water-powered mill had a horizontal wheel that lay in a stream or river, or was turned by water flowing from a millrace or funnel. It had no gearing. At the center of the wheel (fitted with angled or spoon-shaped paddles) was a vertical shaft that went through a fixed millstone and turned a second millstone. Millers poured grain into a hopper on the top, which funneled the grain between the two stones to be ground into flour. As the waterwheel turned, the shaft and upper millstone also turned in the same direction. There were many regional differences and variations of the horizontal mill, such as shape and positioning of the paddles, and the way in which the water to power the mill was delivered—such as in a penstock type of delivery system, where the water was delivered under pressure. For example, on the Iberian Peninsula, one such mill, was a tank or *arubah* mill.[12]

Vertical waterwheels were more powerful than most horizontal wheels, but their gearing, which changed horizontal to vertical motion (when a toothed cog gear jointed a lantern gear), made them more complicated and expensive. They also required reliable water flows, making them more prevalent in regions with fast, continuously flowing rivers and streams, such as in French and German lands. Some vertical mills, called float mills, were mounted on barges in rivers or on the piers of bridges, both using the current of the river to turn the wheels. The archaeological study of mills ideally entails investigation not only of the machine itself but of the watercourses, weirs, dams, and millponds necessary for the working of the mill, as well as the surrounding landscape.[13]

Heavy, vertical overshot watermills required capital investment. They were placed near urban centers or were controlled by powerful lords, who may have held a monopoly in the area for grinding grain. Millers were essential skilled workers for many communities. But mills were often the focus of controversy. Water rights (giving the mill the necessary water to power the machine) could harm downstream users, and the water structures of mills could interfere with navigation on rivers. Monopolies for grinding grain were often given to millers (or their lords) or to churches and monasteries. The millers of such establishments were sometimes accused of charging too much for grinding the grain or of cheating in weighing it.[14]

Irrigation of Crops

Irrigation and the varied forms that it took depended on topography and also on such issues such as rising populations and the resulting need for more crops. Many areas of the eastern Mediterranean, North Africa, and the Iberian Peninsula were too dry for farming without irrigation. The forms that such hydrological works took depended on conditions of climate and soil, as well as local traditions and practices.

The Eastern Mediterranean

The Euphrates River, originating in the mountains of southeastern Anatolia in present-day Iraq, presented particular problems because the river valley was restricted by rocky hills, preventing the construction of long canals. Instead of canal construction, water was lifted from the river by *norias* and *saqiyas* (described above). Farther south, canals were possible. For example, at the site of Barbalissus (modern-day Meskene [or Maskanah] in northern Syria), a fortress city on the left bank of the Euphrates, a canal system, approximately 30 kilometers (19 miles) long, was constructed. As with many such systems in the eastern Mediterranean, the chronology of construction is not entirely clear. It is thought in this case that the Umayyad dynasty (Muslim rule established in 661 based in Damascus) improved a system that already existed but needed repair, rather than building an entirely new structure. On one of the tributaries of the Euphrates, the Khabur River, workers built multiple canal systems over the entire length of the river, irrigating grain crops along the way.[15]

Figure 2.7. Wadi Degla canyon, fifteen miles from Cairo, Egypt. Photo by Bassant Meligy, March 11, 2021. Wikimedia Commons.

Farther south in eastern Lebanon, the Biqāᶜ, "a mosaic of microenvironments" forms an upland valley. The Romans had left their mark on the region with the drainage of large tracts of marshland and with increased settlement—colonies of Roman veterans who had been rewarded with land. At some time during Roman/Byzantine times, agricultural production increased. This was made possible by the construction of a long canal (the Lebwe) originating in an oasis in the Massyas valley of Phoenicia. The canal's water supply was supplemented by a series of *qanats*. The water supplied a large palace complex and much else. It was stored in the massive Harbaqa Dam, which had a capacity of five million cubic meters (about 180 million cubic feet), enough to irrigate about 3,200 hectares (about 7,900 acres) of grain. The dam was constructed of large limestone blocks set over a core of unworked stones. Gates probably controlled outlets at the bottom. The two major branches of the Lebwe watered a major area that greatly extended the area of settled farming. Workers maintained these canals for centuries.[16]

In the Negev (today spanning the southern part of Israel and half of Palestine west of the Jordan River), there was much human activity from the fourth to the seventh century with the growth of small settlements and isolated farms. Agriculture was based on the successful management of surface-water resources and of wadis. (A wadi is a dry riverbed with high sidewalls carved into the landscape by rushing water during infrequent but torrential rainfalls.) Wadis were farmed by building stone cross-dams and leveling the resultant fields. The dams prevented flood devastation. Rain was captured in channels that flowed into storage cisterns— water used for animals and for irrigating fields. For irrigation, water was at times lifted from cisterns with a *saqiya*-driven, water-lifting wheel.[17]

Egypt

Egyptian agriculture, as discussed in chapter 1, was based on the annual flooding of the Nile River. Yet it was never a matter of simply waiting for the flood. Egyptian peasants, most living in villages, constructed and managed thousands of irrigation works to regulate and conserve the floodwaters. Agriculture was made possible by a complex system of canals, water-lifting devices, weirs, and other mechanisms of controlling and directing the water—the irrigation infrastructure essential to the growing of crops. This infrastructure was organized around the annual flood. The summer rains in the highlands of Ethiopia caused the river to rise in upper Egypt by June and in Cairo by July. The river reached its highest level in late August and early September, fell to half that level by November, and reached its lowest level in May. The summer flooding was the basis of what was called the winter crop, including wheat, barley, lentils, clover, and flax, which was harvested in September and October. Water was stored in basins and canals that then enabled the production of the second crop between January and May—wheat, barley, cotton, melons, sugarcane, and sesame.[18]

The irrigation and other hydraulic works needed constant attention. Canals, aqueducts, dams, bridges, waterwheels, and sluice gates had to be constructed and maintained. Canal embankments broke, villages and individuals excessively siphoned off water, canals and basins needed dredging. These water technologies were all in the service of agriculture—crops had to be planted, cared for, and harvested at the same time that complex and ever-changing irrigation infrastructures had to be managed.[19]

In part because of al-Nābulusī's fiscal report (cited at the beginning of this chapter) we know more about Fayyum than any other region. A long canal, the Baḥr Yūsuf, connected the Nile to the area—part of a system of canals and embankments that allowed sufficient water to reach the depression without flooding. Recent studies show that in the medieval period from the thirteenth and fourteenth centuries, there was extensive investment in irrigation under the Mamluk sultans.

However, there was no centralized irrigation bureaucracy, few irrigation taxes (although many taxes on crops and produce), and little direct interference. Rather, local communities, organized by tribal groups, were in charge of their own irrigation. This decentralized organization supported a fully functioning irrigation system that continued to develop.[20]

The problem that the Fayyum peasants faced was that the supply of water from the Nile flowing through the Baḥr Yūsuf canal and then through the Grand Canal (the arm of the Baḥr Yūsuf that entered the Fayyum depression) over several centuries had been providing less and less water. The complex reasons for this diminution included silting of the canal and water being drawn off by upstream villages. As Brendan Haug has emphasized, different regions of the Fayyum dealt with different environmental situations—some possessed much more reliable water supplies than others. In the irrigation system described by al-Nābulusī, each Fayyum village that was irrigated by the gravity-fed canals possessed a water right that was measured by the width of the weir at the head of their local feeder canal. These water allocations were meant to ensure that each village along the gravity-fed canals received a fair share of the water and that downstream villages were not disadvantaged. In pre-Islamic Fayyum (both pharaonic and Greco-Roman) these allocations were determined by central-government authorities. However, in the thirteenth-century situation reported by al-Nābulusī, state involvement seems to have been minimal. Local land-grant holders, as well as tribes and clans, negotiated water rights and coordinated labor contributions among themselves.[21]

When the Ottoman Turks conquered Egypt in 1515, they harnessed the great productivity of Egyptian irrigation agriculture to feed their empire, and especially Constantinople (eventually called Istanbul). Alan Mikhail's investigation of extant court and other documents shows that the Ottoman sultan based in Istanbul and the Ottoman bureaucracy in Egypt working under the sultan collected taxes, provided money when needed, and exercised general administrative oversight. However, the Ottoman rulers recognized and relied on the peasants who had traditionally carried out irrigation and agriculture in thousands of villages across Egypt. Their skill, technical knowledge and expertise, and their local knowledge of complex and constantly changing Nile River environments based on long experience over generations proved indispensable. As a result, Egyptian peasants under the Ottomans enjoyed great local autonomy with respect to water resources and irrigation infrastructure, and the empire often deferred to their knowledge and expertise, while also benefitting from the enormous supply of grain that they produced.[22]

Floodwater Farming in North Africa

The coastal zones of North Africa have sufficient rain for "dry farming," that is, farming without irrigation. Below these coastal areas is a vast area of "pre-desert"—arid lands that extend south to the Sahara Desert. These pre-desert lands have supported both pastoralists and agriculturalists with the aid of water management practices. Our knowledge of these hydraulic technologies is primarily based on archaeological work, including the UNESCO Libyan Valleys archaeological survey. The researchers surveyed a region measuring 300 by 250 kilometers (about 186 by 155 miles) that consists of an undulating plateau or hamada composed of limestone and basalt, cut through by the two great wadi systems—Sofeggin and ZemZem.[23]

Archaeologists have investigated numerous small constructed walls near the wadis or across them that are evidence for floodwater farming. Detailed studies of the walls in tributary wadis show that most served to capture, direct, and store the water available from the intermittent rainfall. Walls for sluices were designed with specific functions. Some served to trap surface runoff from large catchments on the plateaus and lead it to the wadi floor. There it was impeded by walls crossing the wadi horizontally so that small lakes were created. The water then sank into the wadi floor, providing a moist soil for crops and pasture. Walls were also built to direct water to cisterns used for water storage for humans and animals. Some walls directed water into a feeder and conduit, which ended in cisterns used by settlements, or into particular buildings. This floodwater irrigation was based on intricate knowledge of local hydrologic, geographic, and topographic conditions. It maximized the capture of run-off water both on the hamadas (plateaus) and in the wadis, and it formed the basis of rich agricultural productivity, including grain crops and olive trees.[24]

This floodwater farming in the Libyan Valleys arose in the early centuries CE of the Roman Empire and continued through the seventh and eighth centuries and beyond. Intense floodwater farming appears to have developed as a response to the Roman conquest of North Africa and the resulting growth of coastal cities. This farming was not undertaken by the Roman colonists but by indigenous Libyans, using traditional practices. Key to intensification was the emergence of incentives to create surpluses (especially of olive oil) for sale to the growing populations on the coast. A decline in these intensive agricultural practices is evident from the fifth and sixth centuries CE, as the more marginal wadis began to be abandoned. There is no evidence that this decline was caused by environmental or climatic changes. Rather the Vandal invasions into North Africa beginning in 430 CE and subsequent complex political and military struggles that continued for several centuries began to limit the demand for surplus olive oil in the declining coastal cities. Gradually, the region changed to a cereal-based farming of subsistence pastoralists.[25]

European Lands and the British Isles

For Valencia on the Iberian Peninsula (al-Andalus during the Islamic centuries), Thomas Glick has shown that agriculture was based on customary practices that sometimes were centuries old. Fields surrounded by irrigation canals called *huertas* were controlled by irrigation communities made up of farmers irrigating from a single main canal. The Iberian Peninsula as a whole was influenced by both Muslim and Christian cultures. Arabs and Berbers (often organized by clans) were particularly active in importing irrigation technologies, facilitated by the similar environments of the eastern Mediterranean and North Africa. Traditional techniques included weirs, dams to raise water into irrigation canals, cisterns and tanks, as well as *norias* (both water and animal powered), and various forms of the *qanat*. In addition to irrigation, hydraulic systems were used to power mills and to supply water to population settlements.[26]

In northeast Spain—the arid Huecha valley in Aragón, Christopher Gerrard has underscored the complexity of irrigation networks. In this region there were eleven major gravity-flow canals (*acequias*). Along the network, water was captured, stored, and distributed in a variety of ways. These included the *azud* or diversion dam, which diverted part of the flow to an adjacent canal, and the *presa* or gravity dam, which blocked the water course to create a reservoir pond. Intricate cooperative arrangements functioned among the relevant communities, but there were also conflicts. For example, the powerful Cistercian monastery, newly established in 1146 at Veruela, aggressively pursued its own water rights and purchased land that included springs and other features, which ensured more abundant water for itself. In general, however, water rights in al-Andalus were linked to the land and its crop not to particular land owners. No one was ever due a specific quantity of water but rather an amount proportional to the specific field—a share of the available water at any particular time (which varied).[27]

On much of the Italian Peninsula, grain crops were planted in winter and harvested before the dry summer—sufficient rainfall meant that irrigation normally was unnecessary. Irrigation systems that did exist were small-scale, local networks of storage tanks and ditches, used only seasonally when needed. Such irrigation was applied to vegetable gardens (a crucial food source for urban and rural households alike). These local systems included wells, cisterns, and water channels.[28]

In England and northern Europe, farmers increasingly used a kind of irrigation called "water meadows" or "floating meadows," a practice dating from medieval times. In this method, a thin layer of moving water was passed over the meadow during the winter months (from late December to March). This encouraged the growth of grass. Animals such as sheep and cattle were then put to graze on the meadow until May. It was then watered again so that a large hay crop could be harvested in June. Sometimes a further irrigation allowed a second and even a third hay crop to be taken later in the summer. This system was particularly developed in the Wessex chalklands in southern England, and it was practiced also in areas of France, Germany, the Netherlands, and Scandinavia, among other locales. The procedure added nutrients to the soil and helped to control pests. The types of floating meadows included catchwork systems located on slopes, which required the construction of a canal from a water source to the meadow. The water had to flow downstream, and then when it reached the meadow, it was guided by the arrangement of parallel "gutters." A more complicated bedwork system was needed for wide, level valleys. This method was expensive and required a specialist (in England called a drowner) to manage and maintain the meadow.[29]

Drainage

In some areas the issue was not insufficient water but rather too much water, requiring drainage for field agriculture and other purposes. For example, in arable land on the Italian Peninsula, farmers often dug ditches to drain off excess water during rain. These drainage ditches were labor intensive but improved the workability of the soil. They also sometimes served as highly effective boundary markers.[30]

The situation was far more complicated in the Low Countries (approximately present-day Netherlands, Belgium, and Luxembourg). The region contained huge peat bogs, which were five to seven thousand years old. Peat develops in anaerobic conditions in which waterlogging prevents the decay of dead organic material by inhibiting its exposure to oxygen. Capillary action within the moss raises the water table of bog peat above the surrounding level, creating "pillows," with the centers higher than the peripheries. It was in this natural environment that peasants, beginning in the ninth century, began to dig drainage canals to enable denser settlements and more extensive field cultivation. Once the water was lowered by about a meter (about 3.3 feet) peat accumulation stopped, and sod developed that could support cultivation. Workers created low embankments around fields to be cultivated, thereby preventing water from flowing back into the newly drained land.[31]

Until the mid-twelfth century, this practice of making cultivatable fields by means of drainage canals worked well. However, the unintended consequence of these drainage practices was that the land subsided, which made it far more susceptible to flooding. Inhabitants responded by making an ever more complex web of dikes (raised embankments on land), dams, sluices (gates or other devices to block the flow of water), and drainage canals. Drainage units tended to become smaller—parcels of land all at the same level would be grouped together and surrounded by embankments to separate them from parcels of land at other levels. (This enclosed land at its own level and with its own drainage outlets is called a polder.) These complex systems were connected to each other and had to be coordinated. To facilitate this, inhabitants created autonomous regional water boards, which cooperated in overseeing various necessary hydraulic tasks.[32]

The subsiding of land and flood problems were exacerbated by the harvesting of peat. Inhabitants had begun harvesting peat for fuel around 800. They cut it into blocks, dried it, and used or sold it—a practice that contributed to great economic expansion and also created profound environmental changes. The subsiding land, combined with huge storms that occurred from the mid-twelfth to the mid-thirteenth century, made flooding commonplace. A loss of the gradient between higher land and the water table made gravity-based drainage impossible. In the late twelfth century, storms broke what remained of the peat ridge in the lowland zone of the Netherlands, creating the Zuider Zee or Southern Sea.[33]

By the fifteenth century a confluence of environmental circumstances created a crisis—sinking peat bogs, a rising water table, the formation of large lakes that engulfed coastal farms, and the advance of the North Sea. Barrier dams were often a part of a long chain of dikes and dams. Workers built the dams with increasingly huge gates (sluices), which let fresh water out and prevented salt water from entering. From early in the century, technologically complex windmill-powered pumps became a new and more effective way to drain the land. These machines, combined with increasingly massive sluices, required a more complex water administration. Sluice gates became ever more complex as they addressed specific hydraulic situations, and they were increasingly anchored with deeper piles and caulking. In the sixteenth century, brick and stone replaced wood as the primary materials, making them stronger and more durable. Large dykes, dams, sluices, and canals, combined with technologically complex windmill-driven pumps, averted environmental catastrophe.[34]

Other geographic areas were characterized by complex ecological systems known as fens (inland freshwater wetland composed of peat) and marshes (coastal wetland composed of marine silt). As Eric Ash especially has shown, in England, conflict developed in the late sixteenth century when those who wanted to drain the fens—a huge expanse (about 2,850 square kilometers or about 1,100 square miles) of wetlands in eastern England—pitted themselves against local fenlanders, who wanted to maintain the status quo. The latter had long lived there, pasturing small herds of livestock, fishing for eels and fish, hunting waterfowl, digging peat for fuel, and collecting sedge (a grass-like plant) and reeds for roof thatching. From the medieval centuries, fenlanders assembled local "commissions of sewers." These were complex and evolving groups of officials made up of village and crown appointees. The commissions worked primarily to control flooding and repair local drainage works. In contrast, pro-drainage advocates, connected to elite (and non-local) landholders and the crown, wanted to drain the land completely, in order to feed large herds of livestock and plant newly arable land for marketable (and taxable) crops. The latter group eventually won, and the fens were drained (primarily in the seventeenth century), leading to a fundamental change in the ecological and environmental system.[35]

Channeling Water

Medieval monasticism brought about monastic establishments often outside the cities but needing a water supply for significant numbers of people living together. Monasteries in western Europe were particularly active in creating water conduits and other kinds of water supply systems. The monks of Farfa in central Italy, for example, built themselves an aqueduct in the 760s and 770s. Also, in the eighth century at Brescia in northern Italy, Abbess Ansilperga of Santa Giulia provided lead pipes to bring the convent's aqueduct water inside, while Abbot Zacharias of Santa Sophia in Benevento channeled water off the town aqueduct to his monastery— with the local duke's approval. Monasteries also dug wells, built cisterns, and used river water where it was available. In a well-studied fourteenth-century hydraulic system in northern France (the Carthusian monastery of the charterhouse of Bourgfontaine), the system included a springhouse that collected the water at the springs and a channel leading from the springhouse to a conduit house at the monastery. The water was then distributed by pipes to each individual monastic cell. It is notable that the pipes from the springhouse to the conduit house carried the water uphill under pressure (acting as a syphon).[36]

Urban centers also used these traditional methods of obtaining water. (Urban water supplies and drainage will also be treated in the next chapter.) As European urban populations grew in the twelfth and thirteenth centuries, they needed more water as waste and pollution became greater problems. In areas controlled by the Romans in antiquity, Roman infrastructure sometimes remained and was repaired and utilized. Water could be channeled from streams (and sometimes rivers) to rural monasteries or urban centers. As Roberta Magnusson has summarized, complex water supply systems consisted of three subsystems. First was the reservoir or artificial pond that collected the water at the source. Workers often built small roofed buildings over these (or over a section) to protect them from pollution and perhaps theft. Second there was the system of pipes or channels that transported the water to the desired endpoint. Finally, there was the distribution system that brought the water to users, where the water flowed into a distribution tank and then went out via pipes, either to fountains or to individual buildings.[37]

Between source and final distribution, water was carried either in open channels, as in the ancient Roman aqueducts (some of which remained in use), or in pipes. Open systems had to be built with a gradient so that the water flowed to its destination neither too quickly nor too slowly. Other systems (especially, for example, in England) used pipes, which could be made of lead, terra-cotta (earthenware), or wood. Trained artisans created these pipes—metallurgists made lead pipes, potters made terra-cotta pipes, and carpenters made wood pipes. Whatever the material, pipe joints were vulnerable to leakage and had to be constructed with particular care. Closed pipes did not need even gradients, as long as they didn't have air pockets and as long as the beginning of the system was higher than the end. Workers dug either earth channels or stone-lined culverts in which to lay the pipes.[38]

An alternative system was used by Siena, the hill town in Tuscany. There, workers had constructed underground seepage tunnels or aqueducts, which they called *bottini*. These tunnels collected water from underground aquifers and carried it to fountains in the city.[39]

Flood Control

In some regions, as we have seen in the Netherlands, the problem was too much water, resulting in flooding. For example, the Po River Valley in northern Italy experienced periodic floods. Unlike in ancient Roman times, there was no single governmental entity in the medieval centuries to coordinate large-scale drainage projects. Rather, towns along the river were often in conflict. In any flood, the degree of devastation depended on who or what was in the flood's pathway. In the early medieval centuries, people living along the Po protected themselves by moving back from the floodplain (possible in a time of low population). Instead of houses and buildings along the river, the riverbank and floodplain reverted to its original state—a marshland. Marsh functioned as a buffer zone absorbing floods and snowmelts. Such wetlands developed in many floodplains in Italy. They were not wastelands but productive spaces where fish, wood, waterfowl, mammals, and reeds and other plants flourished, and sometimes pasture developed.[40]

However, especially in the fifteenth and sixteenth centuries, growing populations were living on the floodplains of rivers, ensuring that some floods became catastrophes. These floods were in part a result of the climate change called the "Little Ice Age" (1300–1850). Based on analysis of hundreds of tree-ring records and drilled ice cores in locales around the world, we know that cooling in European lands began around 1300. It involved not a steady decline in temperature but a zigzag of climate shifts that brought extreme weather events.[41]

In Rome, the Tiber River had flooded periodically since antiquity. However, flooding became more destructive in the sixteenth century. Three floods—in 1530, 1557, and 1598—caused catastrophic devastation. In these floods, hundreds of people drowned as did numerous animals, sewers overflowed into the streets, buildings collapsed, and river mills were destroyed leaving a dearth of flour. Pestilence and hunger followed. People from many backgrounds, both practical and learned, wrote tracts analyzing the causes of flooding in general and suggested solutions to Tiber River flooding in particular. Suggested remedies included straightening the river that wound through the city, cleaning the bottom of the river, eliminating the arches of some of the river's bridges, building embankment walls along the river, and constructing immensely long run-off canals. Although Pope Pius IV (ruled 1560–1565) did have deep moats built around the Castel Sant'Angelo (a fortified castle near St. Peter's Basilica), which probably helped somewhat with flood runoff, the problem was not solved until the late nineteenth century when engineers constructed the high river walls through the center of the city that still exist today.[42]

The Danube River, as it flowed past Vienna (in present day Austria), presented even more problems in the sixteenth century than the Tiber. Because of the severe storms, ice jams, and floods in 1548 and in 1572, the river changed course, abandoning its main branch (the Tabor arm, which had flowed by the city) in favor of the more northern Wolf arm. In addition, on a regular basis, river islands were flooded and sometimes disappeared under water, while new islands appeared. Bridges crossing the river (sometimes over islands) were repeatedly destroyed and then had to be rebuilt, often in new locations. For a variety of reasons, defensive and commercial, the city needed the river to flow by it—the northeast drift of the river caused serious problems. Commissions were formed, engineers employed, institutions created to repair damaged bridges and rebuild destroyed ones. Official bridgemasters repeatedly inspected and saw to the repair or rebuilding of the bridges. For example, new bridges built in 1566 were all destroyed in an ice jam of 1570 and had to be rebuilt. New bridges meant new roads. Innumerable conflicts arose as these new roads had to be cut through various properties. Concerning the river itself, weirs, embankments, and fascines (bundles of brushwood bound together and used to strengthen embankments) were repeatedly constructed, destroyed, and constructed again. Ongoing litigations among a monastery, a hospital, and the citizens of Vienna went on for decades over conflicting claims to the same island. In the end it was difficult or impossible for the judges to know which island was being referred to, since so many had disappeared as others appeared. The Danube visited similar disruptions on other cities on its course, such as the city of Krems, 80 kilometers (50 miles) west of Vienna on the Wachau, a narrow passage of the Danube. There in January 1573, following the highly destructive summer flood of 1572 from which the city had not recovered, an ice jam built up and then broke to flood the city, ruining houses, vineyards, wine and grain storage cellars, and part of the city walls.[43]

Finally, in lands that bordered the North Sea—England, the Low Countries, northern Germany, and Scandinavia—repeated storm and flood disasters occurred in the later medieval centuries. Thousands of acres and hundreds of villages were permanently lost to the sea. Such disasters should not be seen simply as nature wreaking havoc on innocent victims (although there were surely plenty of innocent victims). We have seen how in the case of the Netherlands, human actions brought about enormous changes in the land—changes that increased flood risk. It is also true that responses to that risk varied greatly in diverse locales and over the centuries. As Tim Soens has emphasized, times of greatest flooding did not necessarily correlate with the greatest efforts to protect against that flooding. In many North Sea locales, initially individual peasants cooperating with their neighbors kept their own dykes in repair. Increasingly, however, water boards and other authorities gained control of the administration of such hydraulic works and "bought protection" by hiring teams of laborers and "experts" to take care of dyke repair. Flood prevention was thus monetized all around the North Sea by the sixteenth century. War, plague, famine, and scarcity of labor sometimes prevented the hiring of such workers. In addition, those with the authority and means to undertake flood prevention, sometimes opted not to. With less prevention, floods did more damage. Eventually, the people who actually worked the land had little say in what happened to it. Flooding could lead to changes in land use—from cultivated fields, for example, to pasture (which was less vulnerable to damage from the salt of floods from the sea). The human world was inexorably intertwined with the world of nature.[44]

*   *   *

The denser settlements of people became, whether those settlements were called towns or cities, the more crucial the hydraulic situation became in terms of sanitation, drinking water, and various other uses. As will be seen in the next chapter, urbanism was closely tied to hydraulic infrastructure, as well as to building construction and other technological practices.

CHAPTER THREE

Urbanism, Building Construction, and Urban Water Supplies

I arrived at length at the city of Miṣr [Cairo], mother of cities . . . mistress of broad provinces and fruitful lands, boundless in multitude of build-ings, peerless in beauty and splendour, the meeting-place of comer and goer, the stopping-place of feeble and strong. Therein is what you will of learned and simple, grave and gay, prudent and foolish, base and noble, of high estate and low estate, unknown and famous; she surges as the waves of the sea with her throngs of folk and can scarce contain them for all the capacity of her situation and sustaining power. . . . It is said that in Cairo there are twelve thousand water-carriers who transport water on camels, and thirty thousand hirers of mules and donkeys, and that on its Nile there are thirty-six thousand vessels belonging to the Sultan and his subjects, which sail upstream to Upper Egypt and downstream to Alexandria and Damietta, laden with goods and commodities of all kinds.

IBN BATTUTA, *THE TRAVELS OF IBN BATTUTA, A.D. 1325–1354*

It is my task to help you begin by giving you tough, indestructible cement which you will need to set the mighty foundations and to support the great walls that you must raise all around. These walls should have huge high towers, solid bastions surrounded by moats, and outer forts with both natural and manmade defences. This is what a powerful city must have in order to resist attack. On our advice, you will sink these foun-dations deep in order to make them as secure as possible, and you will construct such high walls that the city inside will be safe from assault.

CHRISTINE DE PIZAN, *THE BOOK OF THE CITY OF LADIES* (1405)

The two authors of these citations came from two very different worlds. Ibn Bat-tuta (1304–1369) was born in Tangier, Morocco, to a family belonging to the Berber tribe of Luwata. At the age of twenty-one, he set out on a pilgrimage to Mecca, a desideratum for all pious Muslims. On the way, he stopped at Cairo and eventually traveled far beyond to India and then China. Upon his return, he dictated an account of his travels (a *rihla*, or book of travels) from which I have taken the above description of the city of Cairo.[1] Christine de Pizan (c. 1364–1430) was born in Italy but spent most of her life in France and wrote in French. She was the first medieval woman to earn her living as a writer and the first to combat directly the traditions of misogyny that pervaded medieval culture. Her description of building the walls of a city are metaphorical—with the help of Lady Reason she is building a fortified city for ladies only. But her description of the walls and defenses reflects the concerns and structures of many medieval and early modern cities.[2]

While each city is different, these two passages point to some shared urban characteristics across the Mediterranean and in Europe. Ibn Battuta describes many buildings, often splendid ones; visitors arriving from afar; a stopping-off place of both the weak and the powerful; the necessary provision of water (water-carriers); numerous animals (used for portage and transportation); and the extensive exchange of merchandise transported in and out of the city. Unique to Egyptian cities such as Cairo was the importance of the Nile River, here noted as a conduit for commercial exchange. Although not all cities had walls and fortifications, many did. As Christine de Pizan suggests, their strength against sieges could be of crucial importance to the safety of residents and to their survival.

This chapter treats the changing forms of urbanism. It then turns to building construction as a key activity in all urban centers, and finally it considers an issue pressing in all urban centers—water supply and sanitation.

The question of what a town or city was—which settlement should be called a city and which not, in various regions (and languages)—is a topic of ongoing discussion among historians and archaeologists. Was the "city" primarily an administrative entity, an economic entity, or both? What combination of characteristics must it possess to be called a city? Must they include a relatively large and diverse population, a market, economic diversification and artisanal specialization, street planning, walls, judicial and administrative functions, certain building types? Global perspectives have undermined the idea of admitting to the category "city" only settlements of certain characteristics, which happen to have been those of many European cities.[3]

Yet many sites in the Mediterranean and northern Europe are universally agreed upon to be cities—Constantinople/Istanbul, Damascus, Cairo, Rome, Paris, London, Nuremberg. Cities were not just a collection of buildings but the form and fabric of the settled area as a whole—streets, walls, sewers, and other infrastructural features and their arrangements vis-à-vis each other. Cities were not static entities—they underwent change and development as well as decline.[4]

Recent scholarship has emphasized not only structures but people on the streets, who they were and the ways in which built spaces influenced their practices and behaviors. In his groundbreaking study of street life in Renaissance Italy, Fabrizio Nevola has conceived the street as "an ecosystem, influenced by multiple factors in its form, the way it was used and how it was perceived." Scholarship on streets aims to trace the relationships between "physical and social fabrics" of the city, including spaces for the display of power. It also investigates the often-ephemeral comings and goings of ordinary people, which were influenced by status and gender. In addition, cities drew strangers, people both from far-off places and the immediate countryside. And, as Miri Rubin has emphasized, women and Jews, even if they were long-term residents, had much less freedom and autonomy in the streets than did elite male residents.[5]

Streets themselves were by no means static and unchanging. Rather, as the history of the streets in one city, Rome, exemplifies, over the centuries, streets, plazas, and other public spaces could change radically. As Jan Gadeyne noted, in the medieval centuries, people made "short cuts" through ancient ruins, and thereby created new footpaths and new streets, changing the nature of public and private spaces. Streets also underwent varying degrees of paving (with cobblestones or bricks), and of cleaning, or lack thereof.[6]

Cities and Buildings

Cities and their structures can be studied as a whole and from various points of view. Increasingly, collective works are appearing focused on many aspects of one particular city within specific chronological frameworks.[7] Towns and cities can be investigated with a focus on features such as a city's streets, as mentioned above, or its walls.[8] Cities have been investigated as conceived in maps and urban images, as Jessica Maier has done for Rome and Ferdinand Opll has done for Vienna.[9] They can be studied from the point of view of contemporary travelers, such as Ibn Battuta, cited at the beginning of this chapter.

Other scholars have investigated single buildings or kinds of buildings. Architectural historians have written histories of specific buildings, such as Nicola Camerlenghi's study of St. Paul's Outside the Walls in Rome, or Finbarr Barry Flood's study of the Great Mosque of Damascus.[10] Some have focused on particular kinds of building such as Maureen Miller's study of bishops' palaces in the medieval towns of northern Italy and the ways in which their changing forms reflected the changing political and social positions of the bishop; Iñigo Almela's study of the role of religious architecture in the urban renewal of Marrakesh in Morocco; or

Niall Brady's study of the Gothic barn in England. These works all include consideration of the relationships of building design and construction to social and/or religious authority and power.[11] Another focus of research has been the mathematical and drawing practices of builders and architects.[12]

Cities changed profoundly over the centuries. Some cities of the Mediterranean and some parts of Europe, such as Rome and Paris, had been founded by the ancient Romans and thus possessed Roman cores and the remains of monumental buildings such as baths and amphitheaters. Those cities changed as Roman hegemony disintegrated in the fourth and fifth centuries and new cultural values and new elites emerged. Other settlements that became cities had non-Roman origins— they could begin, for example, as a trading post or as the seat of a bishopric, or as a fortified stronghold controlled by a lord.[13]

In Byzantine cities, population declined radically during the sixth and seventh centuries, a result of plague (which began in 542 and returned periodically until 767), as well as war, conquest, earthquakes, and other catastrophes. During the seventh century, the Byzantines lost much territory to the Arabs and the Slavs. The effects were devastating for the government, especially since support for the army and governmental administration came from taxing the provinces. As a result, many public services were terminated and great public buildings fell into ruin, becoming in effect quarries that provided stones for constructing more modest structures. As aqueducts disintegrated, cisterns were built everywhere. Great landlords and church officials took over the functions of defunct city councils. In 717–718 the Byzantines defended Constantinople from the last Arab siege; thereafter their fortunes improved. Between the ninth and twelfth centuries, considerable urban recovery occurred. Builders especially constructed churches and then, monastic complexes, as Constantinople continued to function as the capital and commercial center of the Byzantine Empire. The city was devastated in 1203 when it was defeated by the Venetians during the Fourth Crusade, during which it was sacked and also severely damaged by three catastrophic fires. After the return of Byzantine rule in 1261, it gradually recovered until its defeat by the Ottoman Turks in 1453. The Ottomans undertook extensive urban renovations in the city eventually called Istanbul. Similarly, Arab towns conquered by the Ottomans, such as Aleppo in Syria and Cairo in Egypt, experienced expansion and renovation.[14]

In Syria and Palestine, there exist many abandoned cities from the Roman and Islamic periods. The reasons they were abandoned and when varied from one city to the next and are debated among historians and archaeologists. What is of interest here are the changes that occurred in cities founded by the Romans that were not abandoned but rather transformed into new forms of urbanism in the early medieval centuries. Focusing on the city of Scythopolis (Bet She'an in present-day Israel), Chris Wickham has pointed to a development which occurred in many other cities as well—demonumentalization. That is, the ancient Roman monuments at the center of the city—the colonnaded agora (the public, open space used for markets and assemblies), baths, temples, and amphitheaters—were gradually transformed. As the old monumental city center lost its centrality, a commercial repurposing took place. The Roman baths became linen workshops. The theater, amphitheater, and agora turned into industrial-scale ceramic workshops. Colonnades of the agora now served as dividers for shops and workshops. Artisan workshops and storefronts occupied the ancient Roman monumental center in Scythopolis and in many other cities of the Levant as well. New buildings were also constructed, such as *sūqs* (monumental markets), mosques, and palaces.[15]

Concerning North Africa, the belief that the Arab conquest led to a radical decline of the cities has been found to be untrue. Indeed, there is a scarcity of archaeological evidence for seventh and eighth century North Africa, but this is a result of the practices of late nineteenth- and early twentieth-century archaeologists, who simply destroyed medieval accumulations, creating no records whatsoever, so that they might quickly reach the ancient Roman level. Recent scholarship shows that there was significant continuity between late Byzantine and early Arab urbanism in North Africa. Most towns were not destroyed or abandoned after the Arab conquest (Carthage being the exception). Byzantine fortifications and towns usually were simply occupied by Arab garrisons. But changes did occur. Some cities gained new administrative, military, or trade importance. In addition, there was a relative neglect of public spaces, less spending on monumental buildings, increased artisanal and craft activity in town centers, and partial abandonment of some urban areas. Formerly public or religious spaces gave way to residential or industrial uses. Fortifications became a defining feature of many cities. Christianity and churches continued but gradually came to be repurposed as more people converted to Islam. Housing became simpler, and both residential and public structures were subdivided for housing. Pottery kilns and olive presses were no longer attached to churches but were located in the environs of individual households or communities, suggesting a loosening of state control. A spolia industry, which collected ancient Roman stone blocks and architectural elements for use in new construction, developed in many cities.[16]

Wickham's meticulous itemization of the fate of cities around the Mediterranean and northern Europe in the late antique / early medieval centuries shows that cities changed in different ways. In some areas, especially Italian towns, *città da isole* occurred—islands of settlement developed, surrounded by relative emptiness. This fragmentation occurred in northern European towns as well—referred to today as polynuclear settlements. The success or failure of cities in these centuries depended on the choices made by urban elites, who were patrons of monumental buildings such as mosques and churches, and of private houses, all requiring a specialized artisanal population. In two cities of Europe to the north—Paris and Cologne—the old Roman centers continued to flourish, while the rest of the city became fragmented. In Paris the old center consisted of the island in the Seine (*Île de la cité*) and its nearby streets. In Cologne, the Roman center by the Rhine River continued to flourish, while the rest of the city fragmented. In these and other cities of Europe, churches were being built, often on the outskirts. In many cities, processions, usually led by priests or bishops and followed by pilgrims and others, tied the dispersed settlements together.[17] In sum, in the early medieval centuries, cities changed, sometimes radically. But with some exceptions, they did not disappear.

From the eleventh century, towns and cities around the Mediterranean and northern Europe again expanded and flourished in part as a result of the growth of commerce, and they became dynamic centers. In western Europe, the growth of towns and cities took a variety of forms. Occasionally they developed out of marketing centers created by powerful lords, or they could expand around a monastery or a cathedral or the palace of a bishop. Towns could grow around a lord's castle, a river crossing, a trading post, or a coastal port and trading center. In the eleventh century, kings and bishops began giving towns their own charters, guaranteeing them a significant degree of autonomy and freedom.[18]

Cities needed sufficient food for their populations. They could be densely populated, but they also contained gardens—vegetable gardens for food production and ornamental gardens for pleasure and for expressing the power and authority of rulers, as especially Caroline Goodson has shown for early medieval Italy, Anatole Tchikine for Florence and Naples, and D. Fairchild Ruggles for a broad range of Islamic sites.[19] In addition, however, specialized artisans and many other groups who did not produce food sufficiently or at all, acquired food and raw materials from the surrounding countryside. All premodern towns and cities had essential ties to the nearby countryside, but often they also needed to acquire grain from farther afield. As George Dameron has shown for the northern Italian city-states, communal governments viewed grain security as essential for political stability as they also attempted to undermine the grain security of their enemies by destroying their fields and blocking grain shipments.[20]

Cities also needed fuel—for heating in colder climates and for various industrial processes such as brickmaking and glassmaking. As has been shown for fourteenth-century London, firewood's availability depended on distance and ease of transport from the city. London used large amounts of firewood, faggots (bundles of brushwood), charcoal (lighter and thus more transportable than firewood), and coal. About ninety individuals worked as wood sellers in the city.[21]

Building Construction

Historians and archaeologists have studied monumental building far more extensively than ordinary buildings, such as modest houses, apartment buildings, and shops. Yet for the majority of non-elite people, these kinds of ordinary structures, referred to as vernacular architecture, would have been far more important than monumental structures. Of course, ordinary people may well worship in, visit, or admire monumental buildings in their own cities or as travelers to other cities. Elite individuals funded such monumental structures for many reasons including piety, familial benefits, and the augmentation of their own power and prestige. Indeed, as Hana Taragan has shown, wall paintings of builders constructing an edifice in an Umayyad Palace in Qusayr 'Amra (near present-day Amman, Jordan) served as symbols for the ruler as a great and glorious builder.[22] A building in itself, such as a great cathedral or mosque, could function as a cohesive force, augmenting city pride among the urban population.

In Constantinople standard building materials consisted of stone, brick, and lime mortars, combined with marble and other decorative elements, the latter usually taken from older buildings. Builders covered roofs with ceramic or stone tile, or with lead sheeting. Brickmaking was a major industry during all of the Byzantine centuries.[23]

Byzantine masons were trained in workshops and based their work on hands-on practice. Unlike ancient and late-antique builders, they do not appear to have used drawings. They began by marking the plan on the ground with stones and then laying it out with ropes. Design and construction went hand in hand. Masons typically constructed walls by laying alternating bands of brick and stone. They constructed vaults and domes with bricks and mortar alone. They anchored scaffolding into the walls as they constructed them. When the building was completed, they either removed the scaffolding and filled the anchor holes, or left it in place for painters and mosaicists to do their specialized work. The mason who built the structure and the painter who painted murals on the walls were at times one and the same.[24]

Figure 3.1. Section of the Hagia Sophia, Istanbul, showing dome, pendentives, and piers. Illustration by Oskar Klaussmann, *Die Hauptstädte der Welt: Reich illustriertes Prachtwerk* (Breslau: Schottlaender, [1897]). Wolfgang Dietrich / Alamy Stock Photo (labels added).

A spectacular achievement of the early Byzantine Empire was the Hagia Sophia, the largest domed building in the world for centuries thereafter. Constructed between 532 and 537, it comprised a huge square hall covered by a massive hemispherical dome made of bricks set in mortar. The dome rests on four piers from which spring innovative structures called pendentives (spandrels in the form of concave triangles that form the transition between the piers and the dome). The walls are made with thin bricks set in mortar interspersed with courses of limestone. Such brickwork formed light, thin vaults. Windows surround the base of the dome and make the dome appear to be suspended above the vast interior space. This great church, converted to a mosque in 1453, had a profound influence on subsequent Christian and Islamic architecture. In the fifteenth century and beyond, Ottoman builders observed, maintained, and adopted some of the features of the Hagia Sophia to their mosques.[25]

Turning to Egypt, there is much evidence for ordinary housing both in papyrus documents and in archaeological finds. Egyptian cities contained apartment buildings, often two or three stories high, with rooms opening into a collective (usually long, narrow) courtyard. People owned or rented a group of rooms, or a single room, or even a subdivided room. In Kōm el-Dikka, a neighborhood and archaeological site in Alexandria (with a peak population of about 200,000) on the Nile Delta, such apartment buildings, made of limestone, featured workshops for glassworkers on the ground floor and living quarters in the upper stories. In Saqqāra, ancient Memphis, now a famous archaeological site 48 kilometers (30 miles) south of modern-day Cairo, excavations have revealed that seventh- to tenth-century apartment buildings had weaving workshops and kitchens on the ground floor. Medieval Egyptian cities produced artisanal products of all kinds including cloth (usually wool and linen), glass, leatherwork, metalwork, stonework, woodwork, and ceramics. Cairo itself, founded by the Arabs in 653, had grown to a population of about 250,000 by the eleventh century.[26]

Builders used mudbrick for building construction throughout the Nile Valley, as well as in the eastern Mediterranean, and elsewhere in North Africa. The dry climate of these regions made it an ideal material. In Egypt it had been in use since the First Dynasty of the pharaohs (c. 3218–3035 BCE) when all-mudbrick towns proliferated and even huge palace complexes adopted the technique. It remained a standard construction material through the medieval centuries and beyond. The actual composition of the bricks varied, depending on local soils and their combination of clay, sand, and other materials. To make the walls of buildings, workers laid the bricks in various patterns to avoid cracking. This might involve laying alternate courses of stretchers (bricks laid horizontally along the wall's axis) and headers (horizontal bricks laid perpendicular to the axis), all bound by a mortar also made of earth mixed with straw or other materials. They then covered the wall with a plaster, which could be painted, and thus serve a decorative function. The plaster also added strength and protected the surface from weathering.[27]

As the Arabs conquered large regions of the eastern Mediterranean and North Africa, they developed particular building techniques and architectural forms. For example, as Ignacio Arce has discussed, during the Umayyad Caliphate (661–750) in Bilad al-Sham (present-day Syria, Palestine, and Jordan), new structures and cities were built that reflected the dynasty's power and authority. Builders borrowed techniques from both the Byzantine Empire and the Sasanian Empire to the east as they produced lime, mortar, and bricks; constructed walls of various materials and configurations; made arches, domes, and vaults; and created decorations with fresco paintings, stucco, and other materials. They borrowed much (sometimes using skilled artisans from those areas), but in the process they also created an original architecture and new urban forms.[28]

Islamic architecture in various regions possessed some common characteristics. There was an emphasis on enclosed spaces and courtyards. The traditional house presented the street with high, windowless walls. Larger buildings were often constructed with four vaulted halls (*iwans*) surrounding a square, central courtyard. This basic configuration was (and is) used throughout the Islamic world in mosques, palaces, madrasas (schools attached to mosques), and caravansaries (stopover lodgings for caravans that accommodated camels, people, and goods). A feature of some mosques was the hypostyle (many-columned) plan. Examples include the prayer hall of the congregational mosque of Kairouan (an intact ninth-century mosque) in present-day Tunisia and that of the Great Mosque of Córdoba on the Iberian Peninsula. In Córdoba there is a courtyard in the center and a "forest of columns" that form arcades or colonnades, all covered by a wood-beam roof. Domes projecting up from flat roofs were later added to many such buildings. Interior decoration was highly important. Artisans made surface decorations with tiles, mosaics, painted designs, modeled plaster, and open, patterned cutwork in walls and vaults. These motifs were applied with a rich repertoire of geometric and floral designs, inscriptions, and calligraphy. Textiles—both hung on walls and used as rugs—were intrinsically important to Islamic architecture.[29]

Building materials of mosques depended on location and availability. In Syria, traditional ashlar (squared stone) masonry persisted from ancient through Islamic times because readily available Syrian limestone was both durable and easy to work. Stoneworkers were divided into quarrymen and masons. Masons were specialized: some prepared rough blocks for inner walls and foundations. Others produced finished ashlar blocks, and others were skilled carvers of ornamental stonework. Woodworkers included sawyers who cut rough timber to the correct dimensions, carpenters who carried out interior woodwork and who fabricated chests and door locks, and turners who made the wooden screens for windows and other ornamental woodwork. Such elaborate specialization facilitated rapid completion of building projects. Standards of workmanship were guaranteed by craft and market laws administered by an urban magistrate called a *muhtasib*.[30]

Initially, Islamic builders constructed rounded arches adopted from Roman and Byzantine precedents. But then they developed the pointed arch and other kinds of arches in a great variety of shapes. Because of the scarcity of timber, they invented a variety of vaults and domes that required little or no wooden centering to support

Figure 3.2. A. Great Mosque of Córdoba, interior, eighth–tenth centuries. Photo by Richard Mortel, August 11, 2016. Wikimedia Commons. B. Court of the Myrtles (Patio de los Arrayanes), the Alhambra. Photo by Bob Korn, May 7, 2015.

the structure as they were building it. They also devised methods for making the transition from walls to a dome. One solution was the pendentive used in the Hagia Sophia. They also developed domes with ribbed vaulting and double-shell construction, making the dome lighter and thereby allowing a larger structure.[31]

Spectacular extant examples of Islamic architecture are found in the fortress-palace complex of Alhambra in Granada, al-Andalus (present-day southern Spain). Ibn-al-Ahmar, the first Nasrid emir and founder of the Emirate of Granada, initiated construction of the complex in 1238, and construction continued for centuries. The complex was self-contained with a mosque, public bath, palaces with courtyards, gardens, reflecting pools, fountains, and artisan workshops. It was supplied with its own water system, fed by canals built from the mountains to the east. Antonio Fernández-Puertas has shown that the buildings and ornamentation of the complex were conceived as a whole, designed with all of their elements in proportional ratios. Originally the buildings were decorated with vibrant, shimmering colors in ornamented walls, floors, tiles, carved stucco and wood, ceramic tile mosaics, glass, and carved marble. The result created the perception of "shimmering surfaces" and a "pulsating luminosity" stabilized by the geometric patterns. Olga Bush has convincingly argued that artisans created the original polychromatic colors (many of which are lost) and planned their relationships to elements such as vaults and walls following the precepts of the *Book of Optics* by Ibn al-Haytham, known in the West as Alhazen (c. 965–c. 1040). Bush has also shown that poetry inscribed on the highly decorated walls and beautifully woven textiles were intrinsic to the architecture of Alhambra.[32]

In the eastern Mediterranean in the early fifteenth century, the Ottoman Turks gained control of the Anatolian Peninsula and in 1453 conquered Constantinople and other territories to create the Ottoman Empire. In the next century a succession of Ottoman sultans, beginning with Süleyman (ruled 1520–1566), created a centralized imperial power that was augmented by architectural and infrastructure projects. The renowned court architect Mimar Sinan (1489 or 1490–1588) designed numerous mosque complexes and other buildings. He helped to carry out the imperial policy of developing urban settlements at uninhabited mountain passes, creating monumental complexes that included religious, educational, social, and economic buildings, and providing running water, often brought long distances. These complexes created the centers for new towns. He also designed many caravansaries and fortified castles along the *hajj* or pilgrimage route to Mecca.[33]

Sinan's patrons were the sultans but also their wives, daughters, and other family members including female relatives. He became famous for centralized Friday

Figure 3.3. Süleymaniye Mosque, Istanbul. Photo by A. Savin, February 9, 2020. Wikimedia Commons.

mosques (mosques hosting Friday noon prayers) with unified spaces that became the main focus of patronage in the sultanic court. His mosques varied according to the status of the patron and the location. Provincial mosques consisted of domed cubes of average size with lead-covered, hemispherical domes and pointed cylindrical minarets (typically shaped towers on or adjacent to mosques, usually used to issue the call to prayer). These unpretentious buildings contrasted with the elaborate domed superstructures of prestige mosques built in Istanbul, the most famous of which was the Süleymaniye complex, built on a hill of the city overlooking the Golden Horn (the inlet to the Bosphorus Strait that flows through the city). Designed as a center of higher learning, it included five madrasas dedicated to theological studies, and a sixth that was a medical school.[34]

In Western Europe, understanding of the development of rural settlements has been influenced by "the Tuscan Model." Based on archaeological studies of rural areas around Siena in Italy, it shows that the ancient Roman villas declined in the fifth and sixth centuries and that nucleated settlements or villages formed in the seventh century. By the tenth century, they had transformed into estates and then castles with accompanying social hierarchies. Archaeological traces show that

Figure 3.4. Reconstructed longhouse in the Vikingmuseum, Borg, Lofoten, Norway. Photo by Jörg Hempel, May 28, 2008. Wikimedia Commons.

humble dwellings of the early villages included cave dwellings and huts dug partially underground and made of materials such as earth and wood with roofs of straw—which eventually became roofs of tile. Housing also included timber-framed huts built at ground level with walls constructed of wickerwork and clay plaster. Archaeologists have discovered structures used for food-storage, as well as granaries, and have determined the nature of the foods stored through archaeobotanical investigations. They have also discovered fences, palisades (walls made from wooden poles or stakes), roads, ditches, walls, courtyards, pottery kilns, and forges.[35]

In early medieval northwest Europe (to the tenth century) a significant kind of structure was the timber longhouse, the focal building of most farmsteads in southern Scandinavian and northern German lands. These buildings were usually oriented east-west with the living quarters containing a hearth at the west end, a central entrance hall, and a byre (barn) at the east end. They possessed a complex arrangement of interior roof-supporting posts. The most important structural development occurred in the later seventh and eighth centuries with the development of an open hall from which animals were excluded. In these new structures, the load-bearing posts were placed in the walls, freeing the interior space. Gradually the longhouse was replaced by smaller houses with detached barns, possibly because field cultivation became more important than cattle raising.[36]

One method of discovering how different structures and parts of structures were used is by phosphate analysis. High phosphate levels in the soil indicate animal and human waste; low levels suggest storage or work areas. Other indicators include loom weights (weights used to keep the warp threads taut on vertical looms), indicating a work area for fabricating textiles, and seeds in postholes, suggesting a grain storage area.[37]

Another kind of structure in northern Europe and in England was a protective enclosure, known as motte and bailey castle, made of earth and wood. William the Conqueror first built such structures in Normandy and then, after his conquest of England in 1066, used them to receive supplies and house the soldiers who were brought to put down revolts by the Anglo-Saxons. The motte consisted of an earthen mound with a wooden fortification on top, the whole surrounded by a deep ditch. The bailey consisted of an enclosed yard built outside the ditch and encircling the motte. During construction, earth from the ditch would be thrown into the center to create the mound. Then the wooden structure was built on the top, sometimes just as a tower to house soldiers, but as time went on, more elaborate structures were constructed. Larger residential mottes housed the family and its retainers. The bailey was surrounded by a fence or palisade, usually constructed out of logs. Sometimes a protective ditch was dug around the bailey as well.[38]

Gradually, stone castles—often defined as the defended residence of a lord—replaced motte and bailey structures. Although the specific origins are controversial, by the late eleventh century they dotted the European countryside, many constructed on motte and bailey sites. Originally, they consisted of a stone tower, meant to garrison troops, which was surrounded by massive stone walls. As uses multiplied beyond mere fortification, they became increasingly elaborate, eventually becoming residences for noble families as well as administrative centers. The most important part of these massive structures were their walls—as much as 6 meters (about 20 feet) thick at the base—constructed by making two parallel walls out of ashlar blocks and filling the space between with stone rubble. Castles usually featured immense towers as well as courtyards, and rooms that housed family members, visitors, troops, and supplies.[39]

Churches and cathedrals (the church of a bishop) comprised another important type of building that spread across both Europe and the Byzantine Empire. The Gothic cathedral emerged in the twelfth century and entailed specific building

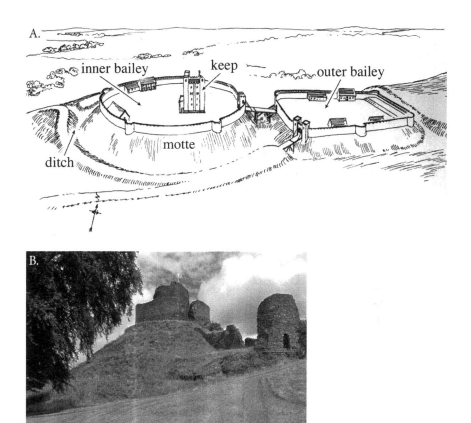

Figure 3.5. A. Motte and bailey castle. Bird's-eye view of Castle Hedingham, Essex (partial reconstruction, 1931), originally built c. 1140. HIP / Art Resource, New York. B. Remains of a motte and bailey castle in Launceston, Cornwall, England. Photo by Colin Park, August 12, 2014. Wikimedia Commons. C. Krak des Chevaliers, Crusader castle in Talkalakh, Syria, twelfth century. Photo by Vyacheslav Argenberg, August 17, 2008. Wikimedia Commons.

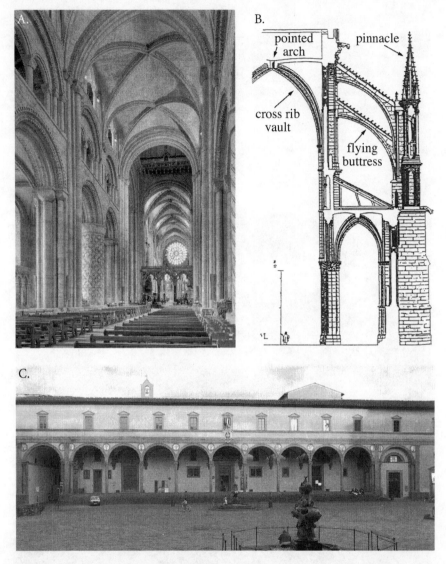

Figure 3.6. A. Nave of Durham Cathedral, Durham, England, showing rib vaulting. Photo by Michael D. Beckwith, June 21, 2019. Wikimedia Commons. B. Pinnacle, buttresses, and cross-rib vault of Reims Cathedral, Reims, France. From Eugène-Emmanuel Violet-Le-Duc, "cathédrale," *Dictionnaire raisonné de l'architecture française du XI e au XVIe siècle*, 10 vols. (Paris: B Bance, 1858–1868), 2: 317, fig. 14. C. Front of the Ospedale degli Innocenti, designed by Filippo Brunelleschi. Photo by Warburg, June 1, 2008. Wikimedia Commons.

techniques, each of which initially developed separately. One of them, the ribbed vault, was first constructed in Durham Cathedral in England. The ribbed vault is an arched ceiling with sections separated by projecting ribs, over the wide nave (center aisle) of the church. A second feature of the Gothic style was the pointed arch, adopted from Islamic northern Africa, and the third was the flying buttress. The flying buttress on the outside took the weight of the vaulting and the roof, enabling masons to build thin, high walls filled with stained glass. Other features included church spires and pinnacles. As a result, Gothic buildings seem to soar and on sunny days are flooded with light.[40]

Medieval masons learned their craft by apprenticeship. They built cathedrals bay-by-bay, experimenting and watching to see if cracks developed in the tension-sensitive lime mortar used to cement the cut stones. If they observed cracking, they made changes. In designing the building as a whole, they used "constructive geometry," in which they manipulated geometric forms using drawing tools such as the straightedge and dividers to determine the overall dimensions of ground plans, facades, and the other elements of the building. Once the dimension of one element had been decided, the dimensions of all the others were determined. The unit of measure became a module for the entire building. It is notable that the Gothic style entailed the use of drawing as a tool of design.[41]

The master mason was in charge of both design and construction and usually stayed on site during the construction process, working closely with both the patrons and the masons, journeymen, and apprentices. He supervised all construction involving stone and brick. He oversaw three sites—the quarry where the stone was first cut; the lodge, a wooden building that served as the workshop; and the construction site itself. Stonecutters cut the stone from the quarry, carters carried it to the lodge either in carts pulled by oxen or, if there was a nearby river, by barge. At the lodge stonemasons cut each stone according to where it was to be placed in the construction. To shape stones exactly, masons used templates for piers, columns, vaults, doorways, and windows. Medieval masons and other workers in construction could be highly mobile, moving from one site to another as they finished the work or for other reasons.[42]

A very different style of architecture, the classical, developed in the late medieval centuries. The classical style was strongly influenced by the principles of architecture displayed in ancient Greek and Roman buildings and in the treatise *De architectura* by the Roman architect Vitruvius. It emphasized proportionate symmetry and often featured columns and arches. Around 1500 classical architecture began to replace Gothic architecture as a favored style in Italy and elsewhere. An early example is the Ospedale degli Innocenti (Hospital of the Innocents) in Florence, begun in 1419 and designed by the Florentine architect Filippo Brunelleschi (1377–1446). The flourishing of classical architecture accompanied the development of the cultural movement known as Renaissance humanism, involving the rediscovery and intense study of ancient texts; the reform of the Latin language, replacing medieval Latin with Ciceronian Latin; and an interest in ancient structures including both ruins and intact buildings such as the famous Pantheon in Rome. This classical architecture underwent a broad development throughout Europe, its principles applied to structures as diverse as churches, loggias, great public buildings, and the palaces of merchants, princes, and emperors. It was accompanied by the use of drawings to make plans and elevations both for patrons and as instruments to aid in design.[43]

While most classically styled buildings were made of stone, from the early medieval centuries, wood remained centrally important for building construction. This is especially the case in regions that were sufficiently forested to provide a supply of timber, including England and areas of France, Germany, and Scandinavia. Early medieval structures there included small buildings with walls built up from a sunken floor. Some of these structures were dwellings but could also be used as workspaces (e.g., for women weaving cloth) and for grain storage. In addition to these small buildings with sunken floors, most timber houses before 1100 were small, single-story buildings stabilized by posts rammed into the earth or into a wooden beam lying on the ground. They usually had thatched roofs. Vulnerable to fire and rot, these structures were often repaired and rebuilt. Eventually builders used fire-prevention materials such as tile for floors, walls, and roofs, and they placed timbers on stone foundations to prevent rot.[44]

From the twelfth century, building construction was transformed with more elaborate and more substantial structures. As Niall Brady has elucidated, in England between 1100 and 1550, numerous lords built large, indeed magnificent, barns on their estates for the storage and processing of grain but also to enhance their prestige. Some were built of stone with timber roofing, others entirely of timber. Timber construction also included larger houses.[45]

In England and Europe, the technique of timber framing developed in the late twelfth century. A timber frame building is a load-bearing timber structure with a frame consisting of vertical, horizontal, and obliquely placed timbers. In English wooden churches, the vertical wooden wall supports or studs were placed closer together than in houses. In the spaces between, builders put an infill such as brick or wattle and daub (in which a lattice of wood or reeds is daubed with mud or a

Figure 3.7. A. Reconstruction of a medieval, Anglo-Saxon, timber, thatched-roof house, Jarrow Hall Museum, Jarrow, South Tyneside, England. Photo by Neddyseagoon at English Wikipedia. Wikimedia Commons. B. Timber-framed barn, Coggeshall Grange Barn, Coggeshall Abbey, twelfth century. Photo by Brian Snelson, April 7, 2009. Wikimedia Commons.

Figure 3.8. A. Half-timbered house with exposed timber construction and infill of both wattle and daub, and mudbrick, Bad Langensalza, Germany. Photo by Sebastian Wallrothy, September 30, 2004. Wikimedia Commons. B. Borgund Stave Church, Lærdalen, Norway. Photo by Simo Räsänen (Ximonic), June 18, 2013. Wikimedia Commons.

combination of mud, dung, straw, and other materials). This infill was not load bearing. Carpenters joined the heavy timber beams (in England usually oak) with various kinds of joints, such as scarf joints, and then pegs with mortise and tenon joints and trusses (load-bearing triangular forms) to stabilize them. The technique allowed larger, more durable buildings. In England, timber frame houses often contained a large hall heated by an open hearth, around which were smaller rooms used as living areas and workshops. In addition to these large dwellings, builders constructed smaller timber frame houses—the differences between larger and smaller houses reflecting social hierarchies in the city.[46]

Because wood is vulnerable to fire and weather events, most wooden structures have not survived but some have. Surviving wooden buildings include twenty-eight stave churches of Norway—named after the load-bearing posts called stave in Modern Norwegian—built between the twelfth and fourteenth centuries. More than two thousand wooden churches are estimated to have been built in Viking (medieval) times in Norway. One kind of footing for these buildings is called earth-fast, meaning that the roof-bearing posts along the perimeter of the structure were set into pits in the ground. A second kind of footing was ground-set, in which a pit was dug and flat stones placed in it, providing a raised base for the weight-bearing posts and for a lumber floor (with space underneath). Gound-set footing gradually replaced earth-fast footing in churches after 1100.[47]

The building trades flourished in fifteenth- and sixteenth-century western Europe and England fueled by the desire for more spacious and comfortable living quarters, by a growing culture of consumerism, and by a new ethos of conspicuous consumption by elites. In England, commercial farmers and landowners enlarged and improved their houses. Country farmhouses became more comfortable as they acquired second stories, staircases, glazed windows, and exterior chimneys. Numerous parish churches and other buildings were constructed and renovated in Elizabethan England, where Elizabethans created a consumer society as they increasingly purchased household furnishings and objects of various kinds. In many cities of England and Europe, the construction of palaces, churches, and public buildings created building booms that also included the creation of elegant fireplaces, elaborate furnishings, rugs, curtains, paintings, and other ornamentation of the completed structures.[48]

Urban Water Supply and Sanitation

The denser a settlement, the more pressing water supply became as more people needed water for drinking, cooking, washing, and other purposes, including the care of animals. Sanitation became a serious issue. The traditional methods of obtaining water in medieval and late medieval cities include digging wells and building cisterns. Although some elite, wealthy, or fortunate individuals could have fresh water piped to their palaces or houses (usually fed by gravity-powered aqueducts), such houses were always in the minority—in no premodern city were the majority of houses supplied by fresh water. Both public and private fountains (connected to a piped water supply) were features of some urban areas. In addition, many cities and towns were built next to rivers, and some possessed freshwater springs. Beyond readily available natural supplies, workers could build structures such as *qanats*, fountains connected to a water source, and aqueducts to bring available water to urban populations. Cities situated within the area of the ancient Roman Empire often used aqueducts built by the Romans. However, such massive structures needed ongoing maintenance to function properly or at all. Both in Italy and in other areas, such as Constantinople, Spain, and Portugal, the maintenance and/or building of aqueducts was associated with water supply but also the expression of political authority and power. Without regular, ongoing maintenance, ancient aqueducts fell into ruin. On the other hand, as Anatole Tchikine has emphasized, the technology required for creating and maintaining such aqueducts never entirely died out during the medieval centuries.[49]

Here I look at only a small number of the hundreds of towns and cities that flourished in the medieval and early modern centuries. These examples demonstrate the relevance of the geophysical site for issues of water supply and sanitation, the common problems shared by all urban settlements, and the relevance of social and political circumstances to the specific practices followed.

Constantinople was the largest city in medieval Europe until the thirteenth century. The central city had no rivers to supply fresh water, and its underlying geology mitigated against wells. It was a relatively small settlement until the emperor Constantine (ruled 306–337 CE) made it the capital of the Roman Empire. As the city grew, a long aqueduct was built, completed during the reign of the emperor Valens (ruled 364–378 CE). Usually referred to as the Aqueduct of Valens, it is now understood as a complex system of channels and tributaries with two main lines, the longest reaching to a water source in Vize in Thrace, about 118 kilometers (about 73 miles) from Constantinople. The complete sinuous system extends for more than 550 kilometers (about 340 miles). As in all functioning aqueducts, it was built with a descending gradient to enable the flow of water to the city. Another aqueduct supplied by closer springs, known as the aqueduct of Hadrian, also functioned. Aqueduct water in Constantinople was channeled into over 150 cisterns.

Figure 3.9. A. Section of the Aqueduct of Valens, Istanbul. Photo by Mond079, May 2, 2019. Wikimedia Commons. B. Interior view of the Basilica Cistern. Photo by Metuboy, July 27, 2022. Wikimedia Commons.

They included the famous Yerebatan Saray or Basilica Cistern, which was covered by vaulted roofs supported by free-standing columns. The water from the Valens aqueduct also supplied three massive, open reservoirs outside the walls of the city.[50]

Medieval Constantinople inherited the physical structures of this massive, late-antique water system, which included not only many cisterns (which continued to be built through the centuries) but also fountains, baths, and pipes to certain houses. How all of these elements were connected is not fully understood. The medieval city also inherited the laws that provided for maintenance and regulation for agricultural uses, for distribution and allocation of the water within the city, and for maintaining and financing the system. However, over the centuries, the far-flung Valens aqueduct could not be maintained. Disruptions such as the siege of the Avars in 626 cut off the aqueduct (restored only a hundred years later). Varying degrees of decline and repair ensued, but by around 1170 it was deemed beyond repair. Ultimately, the medieval city's water supply would come from the same sources as Hadrian's aqueduct, fed by springs located to the north of the city. With the Ottoman conquest of 1453, the Turks used these closer sources to further develop the (mostly underground) urban aquatic infrastructure.[51]

The Ottoman sultans in the late fifteenth century were centrally concerned with building waterways and increasing the city's water supply. Hundreds of skilled and unskilled workers built and maintained the complex water structures of Istanbul. Key workers were the highly skilled water-channel builders (ṣuyoclu), who built multiple waterways, each managed separately and having its own endowment (or source of funding). These water channels needed constant maintenance and frequent repairs due to damage caused by earthquakes, fires, and floods. Workers on these extensive projects included diggers, pumpers (using pumps and waterwheels to remove water from underground excavations), masons, carpenters (who built scaffolds), porters who carried stone from the quarries, and laborers including enslaved workers.[52]

Cairo, located on the banks of the Nile River, greatly expanded in population after its founding by the Arabs in 653. The burgeoning city needed increasing water supplies but could not use direct canalization of Nile waters because of the dangers to densely populated areas of uncontrolled flooding. Both government and private elite patrons oversaw the construction of intermediate structures, including great reservoirs and artificial lakes. Egyptians used cisterns and wells for all purposes except drinking water. Water venders obtained water for human drinking directly from the river (drawing water towards the center to avoid the sewage and garbage near the banks). Water carriers also drew water from the reservoirs. As Ibn Battuta makes clear in the quote at the beginning of this chapter, Cairo featured numerous water venders (*saqqa' ūn*), who sold the water throughout the city. By the ninth century, five- to six-story apartment buildings became the usual habitations for the nonwealthy. The sanitation system consisted of drainage canals beneath the floors and courtyards leading to cesspools, which were cleaned from the outside. The courtyards of these buildings usually contained two cisterns carved from the rock for water storage. One, lined with bricks, stored cleaning and cooking water; the second, lined with smooth plaster and vaulted, stored drinking water. Both cisterns were located near the street, accessible to the water carriers. Eventually in an improved hydraulic system, water was pumped to flow into fountains and basins in the courtyards, while buckets on waterwheels raised the water from the Nile to the reservoirs. Later in the fourteenth century, the reservoirs received water by a canal from the river, controlled by dykes with an aqueduct built on top.[53]

In Rome, in central Italy, the ancient Romans had built eleven aqueducts that drew massive amounts of water to their enormous baths and to public and private fountains. In medieval Rome, with declining population and without significant aqueduct maintenance outside the walls of the city, many gradually fell into ruin. However, there were also medieval repairs. For example, in the 770s Pope Hadrian I (ruled 772–795) repaired four of the ancient aqueducts including the Acqua Vergine, an aid to the city as well as the ecclesiastical complexes of St. Peter's and the Lateran Basilicas. Later popes undertook repairs intermittently. In the fifteenth century, Pope Sixtus IV (ruled 1471–1484) achieved a major repair of the Acqua Vergine to the springs near the city walls. A century later, in 1570, workers restored the entire channel out to the Salone Springs about 13 kilometers (about 8 miles) from Rome. Because the aqueduct was sourced by many streams including those near the Roman walls, it had always produced some water (albeit sometimes only a trickle), but now abundant spring water became available to the low-lying areas of the city. More than a decade later, in 1586, Pope Sixtus V (ruled 1585–1590) had a new aqueduct built, the Acqua Felice, serving higher areas of the city. To construct it, workers used part of the conduit and many of the materials of an ancient aqueduct. Another aqueduct, the Acqua Paola was completed in 1612.[54]

Yet the main supplier of water for most of medieval and late medieval Rome was not the aqueducts, but wells, cisterns, and the water of the bountiful Tiber River that flowed through the center of the city. Until the successful repair of the Acqua Vergine, in 1570, *aquaroli* or water sellers, carrying their barrels of water on donkeys, were a common sight in the street. Although these water sellers drew some of their water from springs, most they obtained from the Tiber River, also the depository for massive amounts of sewage and other refuse of the city. Decades of papal decrees and other regulations on street cleaning and related topics provide enough description of the streets to make it certain that the streets were often filled with mud, sewage, animal dung, garbage, and construction materials.[55]

Venice, built on a lagoon, was awash in water with canals serving as streets, but that Adriatic Sea water was salt water and thus undrinkable. Nor did the city possess any of the three sources of fresh water available to most European cities— rivers, spring-fed aqueducts, and groundwater wells. Instead, Venice built an immense network of cisterns to capture rainwater. Every square and many private palaces and houses owned by elite Venetians possessed a "well" (*pozzo alla veneziana*), which in fact was an underground cistern, of which there would eventually be more than five thousand in the city. Specialized workers constructed the cisterns and maintained them. The cistern received rainwater channeled into it from a nearby source such as a rooftop. Only the well-head was visible, surrounded by a stone paved area interspersed with gulley grates made of Istrian stone (*pilelle*), through which surface water could drain. The underground tanks were constructed of brick and lined with a thick layer of impermeable clay to prevent rainwater from flowing out and salt water from flowing in. The cistern occupied most of the area under the square. Yet Venice's supply of rainwater was not sufficient. Watermen in barges also carried water to the city obtained from the nearby Brenta River and transferred it to the cisterns.[56]

Toledo on the Iberian Peninsula was situated high above a deep gorge, at the bottom of which flowed the Tagus River. Its height and distance from the river meant that the city suffered from chronic water shortages. When the Muslims gained control of the city in 711 CE, they built many *norias*—both for irrigation of the surrounding fields and perhaps also for the urban water supply. After the reconquest by Alfonso VI (ruled 1065–1109) in 1085, people bought water from water carriers who obtained water both from the river and from nearby springs, transporting it to the city with donkeys. Residents also obtained water from underground cisterns and deep wells. Centuries later, in the 1540s, the Holy Roman Emperor Charles V (ruled 1519–1556) and his son Philip II (ruled 1556–1598) rebuilt their enormous palace (*Alcázar*) overlooking the gorge. They needed abundant water for its gardens. The well-known clockmaker Janello Torriani (1500–1585) designed and built (at his own expense) the famous water-lifting machinery between 1565 and 1569. It consisted of two large water wheels and (going up the hill) a system of oscillating machines that lifted the water. Torriani's ingenious machine (the technical details of which historians have long debated) succeeded in delivering much-needed water. The machine worked, and Torriani built another one to increase the supply. However, the city, which had been promised some of the water, did not get any and thus did not pay Torriani at all, and Philip II paid only a small portion of what he had promised. The machine, which needed constant maintenance and repair, soon fell into ruin. Torriani died bankrupt.[57]

Lisbon, on the Atlantic coast of the Iberian Peninsula, is also situated on the Tagus River. Urban dwellers obtained water from underground springs (some hot) near the river. From the late thirteenth century, two public fountains were constructed by the river. Residents whose dwellings were far from the river used wells and cisterns. In the late fifteenth century, a monumental fountain was constructed at the river port to provide water for ships. The city transformed in the sixteenth century as it became both a capital and the first center of a transatlantic empire. The waterfront became the monumental and governmental center of the city. By the end of the century, the population had reached 120,000, putting strains on both water supply and sanitation. Attempts at improvement included street piping constructed to drain water from the streets. Enslaved people (more than a tenth of the population in the sixteenth century) transported water between the fountains and houses.[58]

In Lisbon, sewage accumulation and the lack of infrastructure were chronic problems. Fifteenth-century documents associate epidemics with streets full of sewage and garbage. Regulations ordered street-cleaning by residents, the paving of streets, the installation of sewer pipes, and the covering of open ditches. A hierarchy of sewage conduits began with two very large conduits, about 2 meters in diameter (about 6.5 feet), and then smaller ones, down to small private conduits. The large conduits were made of masonry, lined inside with limestone ashlar masonry. They were vaulted and had an earthen floor.[59]

London was also a city with a river running through it—the Thames, along with its tributary stream, the Walbrook. The medieval city featured hundreds of tenement houses, each of which functioned as a hydrological microsystem. The houses had narrow street frontage with an open yard or garden behind that included privies and sheds. Some of the houses possessed their own wells sunk into the water-bearing gravel level (the aquifer). In the late thirteenth century, most well linings were wooden, some made by stacking barrels one on top of the other into the pit. Gradually stone-lined wells became more common. Other houses possessed rainwater cisterns. Some residents carried water from a public well or from the river, or they could purchase water from a professional water seller. These tenement buildings were vulnerable to water damage during rain and flooding. Privies were built over pits that residents also used for garbage disposal. Although often dug shallow, they sometimes extended to the porous gravel level and could endanger the purity of the groundwater. As the population grew, the tenements became denser. Buildings featured roof gutters with spouts that emptied unwanted water into the yards, onto the property of neighbors, and into the streets. Open water courses down the center of streets directed water to a river or stream and could be used to dispose of waste. Thousands of court cases attest to the numerous conflicts between neighbors concerning water and waste disposal. Riverside tenements positioned their privies over riverbanks. Dyers, tanners, and butchers also dumped their noxious by-products into the Thames or into its tributaries.[60]

In the early thirteenth century, London citizens, supported by the king, acquired a plot of land and created a reservoir into which water from surrounding springs was piped. A conduit made of lead and timber followed a circuitous route to a building (called the "Great Conduit") in the center of the city containing a reservoir lined with lead. This was the first of several public conduits piped into the city. Yet such water remained a small fraction of water used by Londoners who continued also to use all of the other methods of water collection and waste disposal mentioned above.[61]

During the sixteenth century, a growing population made increased water availability ever more urgent. In 1582, the city permitted and supported financially a German (or perhaps Dutch) engineer, Peter Morris (d. 1588), in his installation of a complex machine under London Bridge. Called "the London Bridge Water Works," the machine was powered by an undershot waterwheel driven by the Thames tidal flow, which stopped (for about an hour) and then reversed directions when the tide turned. The wheel powered two piston pumps, which forced water up to wooden pipes that led to directly to dwellings, as well as to cisterns in the city. As Leslie Tomory has emphasized, the innovative technology combined with a "for-profit" business structure became foundational for the development in later centuries of a networked city of water delivery based on private companies.[62]

For towns in central Europe along the Danube River, including Vienna, much documentation exists for hydraulic measures. Water management in Vienna is first recorded with reference to fire prevention. Fire was an ever-present danger in medieval and early modern cities. A fifteenth-century Viennese regulation required all people to store vats of water in their courtyards and attics in case of fire. For drinking and other uses, the city possessed numerous wells, the earliest from the thirteenth century. In addition, account books attest to the construction and repair of sewage drains (for example under butcher vending tables). Other documents record complaints about blocked and foul-smelling sewers. In 1565, workers constructed a pipeline from springs 8 kilometers (5 miles) distance from the city, increasing the availability of fresh water for all purposes. The Danube River town of Krems, 61 kilometers (38 miles) west of Vienna, possessed similar hydraulic structures, especially wells. At Krems, pipes were installed at the well in the main market. Workers frequently repaired and replaced wooden pipes. They sealed pipe sections together with hemp rope.[63]

To the west and north, German towns often were situated on fast-running rivers and streams. However, the tenth-century settlement of Gutingi, later part of Göttingen, was near a stream but not on it. Archaeological evidence shows that the settlers diverted the stream to make it flow through the village. Builders of Göttingen itself, founded in 1180, surrounded the town with walls and a moat. Small streams flowing through the western part of the town served as the water supply. In some north German towns, the marketplace was situated directly by a stream. In addition, the workshops of tanners and dyers, which required plentiful water (and also emitted pollutants), often were situated near the river. Wells in Göttingen and other German towns were ubiquitous. Town dwellers often built wells behind their houses. Most were dug about 8.5 meters (about 28 feet) deep to reach the groundwater. They were usually built using limestone, a readily available material in the area. Some were draw wells with a windlass, in which buckets of water were pulled up with a chain or rope. Only in the fifteenth century did towns construct public wells in the streets. Houses often possessed cesspits—at some distance from the wells. Many north German towns constructed waterworks (*Wasserkünste*). That is, they created reservoirs of water that they then lifted with waterwheels to wooden pipes that brought the water to fountains and other locations in the city.[64]

As Chaim Shulman has chronicled, central European towns created water-raising machinery and distribution systems as the need developed. For example, the north German city of Lübeck authorized the brewers to make a *noria*-type device to lift water from the Wakenitz River. The water flowed into an elevated tank and was then distributed to buildings used to brew beer. In Zurich, scooped waterwheels were installed at the bridges over the Rathausbrücke. Pots attached to the rim of the wheel poured water into a tank and then distributed it. In late fourteenth-century Bremen, private shareholders established a water-lifting waterwheel on the bridge crossing the Weser River to supply water to themselves. Similar water-raising and distributing machines were constructed in other central European towns and cities, such as Berlin, Gdańsk (Danzig), and Augsburg.[65]

* * *

The towns and cities discussed in this chapter were spread over a large geographic area. Yet merchants, pilgrims, and other kinds of travelers traversed that area in surprisingly great numbers. They also communicated with people in distant lands by means of letters, books, and other kinds of writing. The eastern and western Mediterranean cities and towns were connected to each other in many ways, and were also connected to the British Isles and northern Europe. Such connections were possible because of technologies of transportation and communication, the focus of the next chapter.

CHAPTER FOUR

Transportation and Communication

In the name of God. I, Michael of Rhodes, shall write below about the
time I came to Venice. It was on June 5, 1401.

And first, I signed on in Manfredonia [a port in Apulia in southeastern
Italy] as an oarsman with the nobleman Pietro Loredan. . . .

I signed on as an oarsman in 1404 on a voyage to Flanders. . . .

I signed on as a *nochier* [seaman] in the guard fleet of 1410. . . .

I signed on as *paron* [officer] of a small galley in the guard fleet of
Albania in 1419. . . .

I signed on as *comito* [commander of the ship] on the voyage to
Trebizond of 1426. . . .

I signed on as *comito* on the voyage to Tana of 1427. . . .

I signed on as *comito* to Alexandria . . . in 1433. . . .

I signed on as *armiraio* [commander of the convoy] in 1436 on the
voyage to Flanders. . . .

I signed on in 1443 as *homo de conseio* [officer] on the voyage to
London. . . .

SELECTIONS FROM THE ANNUAL LOG (FROM 1401–1443)
OF THE VOYAGES OF MICHAEL OF RHODES

Michael of Rhodes (Michalli da Ruodo), as he reports in his remarkable book,
joined a Venetian galley (a ship propelled by both oars and sails) in 1401 as an
oarsman. Since the galley oarsmen of Rhodes were enslaved and those of Venice
were free, we can hypothesize that Michael was an enslaved Rhodian oarsman who
jumped ship in Manfredonia (the Italian port in Apulia closest to the Island of
Rhodes) to become a free Venetian oarsman. (This is a hypothesis only; there are
no records of Michael on the island of Rhodes itself.) In Venice, Michael performed
the arduous labor of rowing for four years, and then, year after year until 1443, he
hired on to Venetian galleys, being awarded various higher positions, including
(twice) *armiraio*—commander in charge of an entire convoy.[1]

While Michael was working his way up on Venetian galleys, he was also writing
a book. The first half concerned problems in practical mathematics (e.g., if a load
of pepper weighing 400 pounds costs 49 ducats, how much is 315 pounds worth).

Michael's mathematical section included fractions, algebra, and square and cubic roots. Historian of mathematics Raffaella Franci concluded that he was a good mathematician, worked the problems himself (as opposed to copying the solutions from another book), and that he loved mathematics. Michael also recorded each of his voyages with Venetian convoys (a selection cited above), which included trips to the Black Sea, Constantinople, Alexandria, London, and Bruges. His book contains the first extant tract on shipbuilding, instructions for making sails, calendrical material, and portolans (sailing directions). Michael created his own illustrations, including illustrations of ships, signs of the zodiac, and of St. Christopher (patron saint of sailors). Finally, he designed his own coat of arms (something usually reserved for nobility) emblazoned with an "M" (for "Michael"), flanked by two turnips, and topped by a mouse eating a cat! He died in 1446. His book also contains the 1473 testament of a dying sailor on a ship in the eastern Mediterranean, showing that it was carried around by others after his death.[2] Michael's book is one of many examples that show that transportation and communication went hand in hand.

Transportation: Vehicles and Infrastructure

Traveling from one place to another requires more than the conveyance itself, such as a ship or a camel. Transportation needs infrastructure such as ports, caravansaries, roads, and bridges; and accompanying technologies such as shipbuilding and leathermaking (the latter for such things as harnesses and saddles).

As the life of Michael of Rhodes exemplifies and as historians such as Michael McCormick have shown, from the early medieval centuries, travelers crisscrossed the Mediterranean Sea and used overland routes as well. Pilgrims sought to visit the Holy Land and Jerusalem. Many travelers formed part of diplomatic missions from one ruler to another, and merchants traveled on both land and sea. Enslaved persons suffered involuntary transport, often to unknown lands.[3] Travel on water was often far easier than travel by land, making the craft of shipbuilding and sailing fundamental to many economies.

Yet interregional commerce was by no means conducted at the same intensity during all the centuries with which this book is concerned. In the early medieval centuries from 600 to 900, especially around the Mediterranean, the transportation infrastructure of the ancient Roman Empire had broken down, and the many routes traveled by both Imperial Roman state cargo ships (mostly carrying grain supplies—the *annona*), and private cargo ships, had dissolved. Yet travel continued. It gradually increased from the tenth to the fourteenth century with commercial activity, which included the flourishing of long-distance trade, the development of banking and credit systems to support that trade, and the growth of industries that manufactured products such as fine wool cloth that could be traded for spices, silks, flax, and other products from the East. By the fourteenth century, cities in the West were thriving and both interregional and long-distance trade flourished across the Mediterranean, in the British Isles, in northern Europe, and around the Baltic Sea. Despite the crises of the fourteenth century, including famines and the Black Death, commerce continued and by the fifteenth century was moving toward full recovery and expansion. However, it is anachronistic to consider the flourishing commerce of the late medieval centuries simply as the prelude to modern capitalism. Rather, as Martha Howell has argued, it should be studied on its own terms that include concepts concerning the ownership of property and the practice of gift-giving that profoundly differed from the modern.[4]

Ships and Shipbuilding

Transport by water was the most common way to travel in premodern Mediterranean and European lands. The forms of ships varied tremendously, depending on time period and region, and for this period are usually classified by their hulls. (The hull is the body of the ship without its rigging, sails, rudders, anchors, or other accoutrements.) Historical shipbuilding is the focus of a subspecialty that in the past fifty years has been greatly enriched by nautical archaeology, both on land and underwater, the latter focused on shipwrecks and their contents. Archaeologists have discovered underwater a few intact ships and many fragments, and have studied them in detail—and this is an ongoing process. Such archaeological investigations, which take years of specialized excavation and study, are to be distinguished from the activities of treasure hunters, who often pull up objects from shipwrecks, aiming to make a profit without sufficient (or any) investigation of the finding context.[5]

Changes in ship design, as John Pryor has emphasized, were evolutionary and cannot always be deduced from terminology found in literary sources. One important change in the early medieval centuries in the Mediterranean concerned hull construction. Ancient and early medieval shipwrights used mortise and tenon construction, constructing the hulls of the ships by butting the planks up against one another. They secured them at the edges with wooden stubs called tenons that fitted into recesses called mortises, which they joined together with a dowel. By the tenth century shipwrights had abandoned this labor-intensive method of building a hull. Now they constructed an internal skeleton frame to which they nailed planks edge-

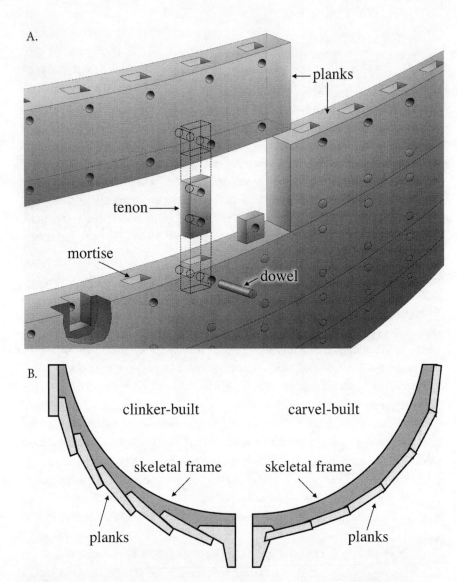

A.

planks

tenon

mortise

dowel

B.

clinker-built   carvel-built

skeletal frame   skeletal frame

planks   planks

Figure 4.1. A. Mortise and tenon joint for hull of ancient / early medieval Mediterra-
nean ship. Illustration by Eric Gaba (Sting), January 2008. Wikimedia Commons.
B. Clinker-built and carvel-built hulls. Illustration by Willhig, English Wikipedia,
December 13, 2007. Wikimedia Commons.

to-edge (carvel built) and then caulked. The strength of this hull now relied not on the skin, as before, but on the skeletal frame. Even though ships built with this method—the frame built first and then the planks attached—tended to leak more, they were far cheaper to fabricate, and ultimately opened up new possibilities in ship design.[6]

An example of a ship built with this "frame-first" method is the eleventh-century shipwreck at Serçe Limani found at the bottom of a small harbor on the coast of southern Turkey. It is called the "glass ship" because it was carrying a huge shipment of glass evidenced by nearly a million glass fragments or shards. These glass fragments were (probably) being carried in wicker baskets from an Islamic glass factory on the Syrian coast. They served as ballast on the ship, and upon arrival, would have been for sale to Greek glass factories, to be recycled and made into new glass vessels. (Making glass objects from old glass is far easier than making them from scratch.) The merchants to whom the ship belonged were Greek Christians, while the ship's contents suggest extensive Byzantine Greek and Islamic cross-cultural ties.[7]

Beyond differences in hull construction, two basic kinds of ships prevailed in the Mediterranean—the round ship and the galley. The round ship was a basic cargo ship powered by a sail. In Venice, for example, where most were privately owned, it was called a *cocca*. It was outfitted with a single, square sail. There were many variations, classifications, and names of round ships, but in general they were wide and tall cargo ships and sailed unarmed with a relatively small crew.[8]

The galley derived its primary power from oarsmen, but it also had sails. In the early medieval centuries, the Byzantines utilized a type of vessel called the *dromōn*, a large, oared ship with a single-mast, square sail. During the tenth and eleventh centuries, larger and larger war galleys were built in response to the threat posed by the Islamic naval fleet. Shipwrights in Italy (Venice and Genoa) and in Spain (Barcelona) also built galleys, some for commercial convoys and some for battle. In Venice, for example, from the early fourteenth century, the private system of merchant shipping was supplanted by a public navy made up of merchant galleys. Regular convoys to the most important trade hubs were established. State-protected merchant convoys of galley ships departed annually on the following predetermined routes—"Romania" (to Constantinople, and Tana and Trebizond on the Black Sea), "Beirut" (Syria and Alexandria, Egypt), "Aigues Mortes" (Marseilles, Aigues Mortes, and Barcelona), "Fiandra" (Bruges and England), and from the 1460s, a route to North Africa.[9]

Over the centuries, Mediterranean sailing vessels changed in terms of the type

Figure 4.2. A. Manuscript illumination of ships with square sails. From Titus Livius, *Les Trois Decades*, ms 309/760, fol. 96v (detail), French, fifteenth century, Musée Condé, Chantilly, France. Photo by Erich Lessing / Art Resource. B. Venice, boats in background with lateen sails. From Hartmann Schedel, *Liber chronicarum* [*The Nuremberg Chronicle*] (Nuremberg: Anton Koberger for Sebald Schreyer and Sebastian Kammermeister, 1493), fol. 43v (detail).

of sails that they used. Ancient Roman and early medieval vessels used a square sail. Gradually the lateen or triangular sail developed. Julian Whitewright has shown that lateen sails were present in the Mediterranean as early as the second century CE. It was part of the development of a wide variety of sail forms and accompanying kinds of rigging. Lateen-rigged ships were not faster than square-rigged, nor was there a linear technological development from one to the other. Rather, in order to lower cost, the square sail rigging had been simplified, a process that caused it to lose some of its performance capacity. Lateen sails had fewer rigging components, were cheaper, and had greater maneuverability.[10]

From ancient times, the Arabs had been seafarers to the East on the Indian Ocean and beyond, and by the mid-ninth century, they were sailing to China. The hulls of these ships were constructed with planks placed edge to edge that were lashed or sewn together (rather than nailed) with fiber made from the husks of coconuts or from date palms. The Arabs turned to Mediterranean seafaring after their Mediterranean conquests of the seventh and eighth centuries. How their long, Indian Ocean experience influenced their Mediterranean maritime endeavors is not entirely clear. For one thing, the planks of their Mediterranean ships were nailed rather than sewn. These vessels can be divided into three general categories—deep seagoing ships, coastal boats that sailed on rivers and along sea coasts, and riverboats. Riverboats could be royal barges, gondola-type boats for river processions, or ferry and transport boats, including barges with oars. Seagoing ships included large war vessels with from 16 to 40 oars and up to three sails (which were stowed during battles). The oarsmen—who were enslaved—sat on wooden benches, four to six men to an oar, their feet chained to the benches, where they rowed for hours. Galleys could also operate as merchant/passenger vessels. One kind of ship was a commercial ship called a "rounder" because of its belly-shaped hull. Another type of Arab ship served as a vessel for the transport of horses. Islamic war galleys in the Mediterranean initially were larger and heavier than Byzantine galleys, but by the twelfth century they were very similar. Indeed, Arab vessels and those from western Europe were often indistinguishable, especially given that fleets from each usually included vessels captured, either in warfare or in acts of piracy, from the other.[11]

The complexity of early Mediterranean ship types can be glimpsed in Lawrence Mott's description of a single fleet belonging to the crown of Catalan and Aragon on the Iberian Peninsula and led by Roger of Lauria (c. 1250–1305) in the so-called War of the Sicilian Vespers, fought between 1282 and 1302 against squadrons of Angevin, French, and Genoese opponents. As Mott's study shows, Lauria's fleet contained galleys—referring to a wide range of vessels from small (light) galleys of 45 to 80 oars, to great galleys having 100 to 150 oars. Most had two lateen sails, which could be struck and stowed prior to battle. Other crucial vessels were transport ships, which included the *tarida* (oared vessels capable of carrying cargo and horses) and the *galea aperta in puppa* (a galley open at the stern), with a large stern hatch allowing direct access to the hold, but about which little else is known. Both of these vessels could unload horses with their riders directly from the stern to the beach by means of a ramp. Another ship of the fleet, the *navis,* was a standard mer-chant vessel, a round ship with two lateen sails. These round ships were privately owned, most by Roger of Lauria himself. Then, there were *galioni* and *llenys,* sim-ilar or identical oared merchant galleys with 40 to 80 oars, used to transport goods quickly. Another ship, the *veccettae* carried between 14 and 28 oars and probably a lateen sail, its main function being to carry messengers and messages, and for use against pirates. Finally, the *barcae* or barge was a vessel with numerous variations in construction, with both oars and a sail, used to transport men and supplies in coastal waters and for patrols.[12] This example of a single fleet points to the com-plexity of ship types, which was the reality in the Mediterranean and in northern seas as well.

In Europe's Baltic regions and in early medieval England, shipwrights con-structed hulls very differently from those of the Mediterranean. They built clinker (or lapstrake) hulls with overlapping horizontal planks—that is, they placed the bottom edge of each plank over the top edge of the plank below and nailed them together. Scandinavian peoples from a variety of cultures who came to be called Vikings used this method to build what are called longships. Viking ships are well-known in part because of the Viking practice of ritual ship burial. The most strik-ing of such burials was excavated in 1904 in Oseberg in eastern Norway. Behind the mast of the ship, a burial chamber had been constructed containing the corpses of two women, along with personal belongings and objects, including sledges and a wagon, fifteen sacrificed horses and a saddle. Both were high-status women; one had died of cancer at around eighty years old, the other was around fifty-five. (The Vikings practiced ritual human sacrifice, but whether the fifty-five-year-old woman was a sacrificial victim is unknown.)[13]

Early Viking shipwrights built diversely, but most ships were slim, lightly built, without a strong keel. (The keel is the "backbone" of the ship that runs along its bottom from bow to stern.) Such ships could not carry a mast and were rowed. In the early eighth century, the Vikings began to build ships with a stronger keel, which allowed both a deeper, flatter, stronger hull, and the addition of a heavy mast. These changes made their ships oceangoing, and they thereby began their highly profitable raids on more southerly European lands and the British Isles—raids that continued for three centuries.[14]

In other parts of northern Europe, shipwrights built the cog, a ship usually constructed with oak timbers. Probably first constructed by Germans in the tenth century, the cog was widely used from the twelfth. It was a flat-bottomed ship (bot-

Figure 4.3. The Oseberg ship. Viking Ship Museum, Oslo. Photo by Petter Ulleland, November 30, 2016. Wikimedia Commons.

tom planking laid edge to edge), and the steep sides were clinker-built with a single mast and a single, large, square-rigged sail (unrelated to the ancient Mediterranean square sail). Sailors steered them with a rudder attached to the sternpost. Larger ships reached a capacity of 250 tons. Their builders added superstructures or "castles" for purposes of defense, as well as a top castle to the mast for use as a lookout and for throwing missiles. It was an ideal cargo ship, its flat bottom allowing it to be beached and unloaded at low tide in the (many) areas where jetties were lacking. The cog became the workhorse of the Hanseatic League—the confederation of merchant guilds and towns that dominated northern European trade from the twelfth to the fifteenth century.[15]

Influenced by Mediterranean shipbuilding traditions, northern shipwrights gradually gave up clinker hull construction and took up frame-first construction. Whatever type of hull they were building, shipwrights used similar tools. Most important was the framesaw, which two-man teams of sawyers used to transform fallen trees into timbers and planks. Shipwrights also used a variety of axes, adzes (tools with curved blades used for shaping rough timber), augers, drills, hammers, and chisels. They held planks together with clamps and used caulking irons to make the joints watertight.[16]

As Spain and Portugal turned toward the Atlantic Ocean with its huge waves, gales, and high-velocity tidal currents, they needed new kinds of ships and new methods of navigation. First Portugal and Spain and then England developed ships that could better handle Atlantic Ocean conditions. Beyond ship construction, as will be discussed further in chapter 6, oceanic voyages produced significant developments in navigational instrumentation. More broadly, conditions were created, both on ship and on land, that promoted interactions between practical navigators and men trained in mathematics and astronomy.[17]

For Atlantic sailing, shipwrights developed the caravel, a ship with lateen sails, originating from a Portuguese fishing boat. The caravel eventually acquired four masts with two lateen and two square rigged sails. Another type of ship was a trading ship called a carrack, an amalgam of northern European and Mediterranean ship construction. It was a large ship and could have one large mast, or eventually, two or three masts. The three-masted carrack was the forerunner of the galleon, which (in the seventeenth century) replaced it as the most important oceangoing vessel sailing out of Europe.[18]

In many regions, especially in northern Europe and England, river transport, often supplemented by canals, formed important transportation vectors. In the early medieval centuries and beyond, localized trade networks were linked along the river valleys of the Loire, Seine, Meuse, Moselle, and Rhine. By 700 these river routes were connected to wider markets via the Saône and Rhône rivers with the city of Marseilles, gateway port for the Mediterranean trade. River navigation often required dredging the river, modifying its water course, and building canals. In 793 Charlemagne (ruled 768–814) attempted to have a navigable canal built between the Rhine and the Danube rivers near Weissenburg in Bavaria—a huge effort, which failed. In France between the ninth and twelfth centuries, a number of river diversions were undertaken for various reasons, some defensive. The work of canalization increased in the second half of the twelfth century in France, in Flanders, and in Holland. In twelfth-century Italy, canals were built including the Naviglio Grande, constructed from the Ticino to Milan, so that boats could carry marble for the construction of the Milan Cathedral. Both irrigation and navigation canalizations continued in Lombardy and Piedmont through the fourteenth and fifteenth centuries. England also had a rich tradition of canal building and river modification both for drainage and navigation. Particularly important were the monastic and ecclesi-

astical establishments that created canals, modified rivers, and built quays (structures built parallel to the riverbank for docking boats) for the purpose of transporting heavy building materials such as stone for their construction projects. Workers
built numerous channels and canals in the coastal wetlands of England, for example at the Somerset Levels, an area of southwest England. Wetland water management included drainage channels, fishing weirs, millworks, and navigation. At times
these different usages conflicted with one another. For example, boat traffic was
frequently hindered by fishing weirs and millworks.[19]

The kinds of boats used for river and canal haulage differed from seagoing vessels and were variable in form. For example, as John Langdon has discussed, in
England between 1294 and 1348, there were many vessels called *batella* that could
refer to a variety of boats ranging from rowboats to larger boats with sails. Other
riverboats could be powered by poling or pulling by ropes from the shore. Another
kind of vessel was the "shouts" (from Latin *shoutae*), used in Thames River navigation. They were flat-bottomed, like barges, but pointed at both ends. They could
be used in shallow water and over the blockages of mill weirs (and many mills were
equipped with a cable and winch for hauling up the barges and over flash locks—
a kind of lock used in early mill weirs). Barges and other kinds of vessels varied
considerably, depending on specific river systems. These vessels carried loads such
as animal carcasses, horseshoes, nails, wheat, oats, beans, and flour.[20]

Ships required infrastructure of shipbuilding facilities or shipyards (often referred to as arsenals), ports, quays and docks, and cargo-transfer facilities. The
Arabs in the eastern Mediterranean maintained shipyards and shipbuilding facilities in many locations such as Alexandria, Acre in Syria, and Tunis in North Africa.
For the Catalan-Aragon fleet referred to above, arsenals for the fleet were maintained in three locations in Sicily—Trapani, Palermo, and Messina, of which Messina was the largest. All ships required constant maintenance, most of which was
carried out at the shipyard. The vessels also had to be stored when not in use, usually
the winter months. Before a vessel left port, it had to be rigged and outfitted—a
complex operation. Ships not in use were "disarmed" and stored. Disarming entailed removing all the rigging and sails, the oars, all other gear, and all the ship's
stores. The gear was inspected, repaired, and stored. Arsenals needed a large supply of wood to repair the hulls. Where possible, these hulls were stored under
sheds, protecting them from drying out in the sun and from collecting rainwater
and thereby rotting. During storage, workers inspected the planking, replaced broken and rotting planks, nailed loose boards, and replaced loose caulking. They covered the parts of the hull needing waterproofing with pitch. In Barcelona, thirteenth-

century ship sheds, equipped with block and tackle hoists, are extant and serve today as a maritime museum. During the rule of Philip II (ruled 1556–1598), the arsenal of Barcelona became the center of galley construction. By 1574, shipwrights had constructed a fleet of 150 galleys.[21]

Another well-known arsenal, that of Venice, was designed not only for ship-building and ship maintenance and repair, but also for the fabrication of rope and guns, as well as sails. (Many of the sailmakers were women.) From the early fifteenth century, Venetian shipbuilders experimented with a variety of new galley designs. They were supported by the Venetian Senate, which encouraged rivalry among shipwrights. The Venetian Senate also imported Greek master shipwrights from the island of Rhodes to supervise shipbuilding and compete with Venetian masters. (Michael of Rhodes may well have obtained the ship drawings that appear in his book from a fellow Rhodian.) Matteo Valleriani points to the enlargement and complete reorganization of the Venetian arsenal in the sixteenth century in response to the threat of the expanding Ottoman Empire. Instead of one ship being the finished object of a production line, each of the components—such as oars, sails, castles, hulls—had their own separate production lines. That way components of a ship could be pulled together quickly in a military emergency.[22]

A major issue everywhere in the Mediterranean involved shipworms, a soft-bodied, bivalve, marine mollusk from the family Teredinidae, sometimes known as Teredo worms. They enter submerged ship timbers while very small and then grow rapidly inside the wood. After one or two years most hulls in the Mediterranean were riddled with boreholes that filled with water, radically reducing the ship's seaworthiness. To repair such hulls, the ships had to be hauled out of the water or, in the case of larger round ships, careened (pulled over by the mast on a beach or other landing place to repair the exposed side). Then caulkers applied a thick coating of tar or resin on both the outside and inside of the hull below the waterline.[23]

In addition to shipyards, nautical infrastructure included ports and landing places—these often connected to port cities and merchant communities. Ports of the Mediterranean and Europe were connected to each other by intricate networks of transportation and information. Archaeologists have recently been using network theory (a mathematical and graphic method of tracing the way complex elements of a network interact) to better understand the complex relationships among harbors and ports, products, and social factors. Ports and port cities often changed under the influence of shifting power structures and shifting patterns of commerce. As ship design changed and developed—increasing cargo capacity and oceanic capabilities—port cities improved their harbors, ports, docks and quays, wharves and jetties, and other infrastructure for transferring and storing cargo, including the provisions of ropes, pulleys and cranes.[24]

Çağla Caner Yüksel has investigated the ways in which two western Anatolian port cities—Ephesus/Ayasuluk (today Selçuk) and Miletus (Balat)—changed with fluctuating circumstances over the centuries. In ancient times they were both thriving port cities; they experienced decline in the early Byzantine centuries, revival from the seventh to ninth centuries, deurbanization from the ninth to the eleventh, and revitalization with the rule of the Turks from the late eleventh to fourteenth. The harbors of the cities changed in tandem. In Ephesus the medieval harbor silted up and led to the building of a new harbor in a slightly different location. The latter became a port for lighter vessels in later Byzantine times. In the flourishing Turkish period, another highly equipped harbor was built nearby (in south Pamucak) to handle heavy maritime trade. Such changes became characteristic of the Mediterranean, North, and Baltic seas and of Atlantic Ocean ports.[25]

In Egypt, Alexandria was a "gateway city," that is, a major port for shipping around the Mediterranean and beyond. The city, however, was not on one of the two navigable branches of the Nile that flowed into the Mediterranean. As John Cooper has shown, those branches were notoriously difficult to navigate, making transport into and out of the Nile dependent on weather, the skill of navigators, and the condition of the canal leading from the Nile to the port city of Alexandria. Alexandria itself was located on a wide bay in the center of which was a small island with a lighthouse. The island protected the harbor from sea currents and strong winds. The port was divided into two by a long quay that stretched from the lighthouse to the shore. The eastern part, dangerous because of its narrow entrance, was reserved for Christian and other non-Muslim boats. Only Muslim ships were permitted in the western half, which was closed with iron chains at night. An official in charge of the port issued permits, and no ship could enter or leave without one. The port was highly active except during the winter, the season in which ships normally did not sail. Between 1300 and 1600 the old, walled city of Alexandria declined (eclipsed by the new capital of Cairo), but the port continued to flourish within the network of both long-distance and regional shipping.[26]

In England, numerous small ports and landing places existed on rivers. They served to accommodate barges and riverboats carrying various kinds of goods and supplies, including building supplies such as stone. Some served as ports for boats delivering goods to manors, or to trade fairs. Some served as landing places for ferryboats. Although London was on a river (the Thames), about 50 kilometers (about 31 miles) from the sea, the daily tide of up to 7 meters (about 23 feet) allowed the port of London to become a major port for North Sea and long-distance maritime trade.[27]

Another major English port was Dover Harbor—key for northern European commerce and the ships of the English Royal Navy. Eric Ash has investigated the history of the problems of the harbor and its sixteenth-century reconstruction. The harbor was locally controlled and maintained until the mid-to-late sixteenth century when the privy council of the royal government took control. The main problem of the harbor were the tidal currents coming from opposite directions that collided nearby, slowing the speed of the water. This slowdown caused the water to drop its load of sand and pebbles onto the bottom, creating numerous shoals and sandbars. A local solution put into place in the late fifteenth century—a pier built out from the foot of the nearby cliff—eventually made the problem worse. Ash shows that the controversies surrounding the reconstruction of the harbor mainly involved two issues—the materials to be used for the seawalls that were eventually built and the location of the entrance to the new harbor. After many false starts, two crucial elements ultimately led to successful reconstruction. One was the "expert mediator," Thomas Digges (c. 1546–1595), a mathematician whose social position and experience allowed him to communicate both with elite members of the privy council and with skilled workers. The second element comprised the workers of the nearby town of Ramsey, who had built seawalls with readily available chalk and mud for their own town. After the decision to construct new seawalls of these materials (as opposed to the far more expensive and less efficacious lumber or stone), these men were hired to carry out the work, which they completed with resounding success. One result of the successful reconstruction of the harbor in the 1590s was that the city of Dover flourished with the continuous arrival of numerous ships from afar.[28]

Part of the essential infrastructure of ports (and, it should be added, of roadways used by travelers) were lodgings and warehouses for storing commercial goods. Olivia Constable has written a comprehensive history of these establishments around the Mediterranean. In the medieval centuries they were called *funduqs* (in Arabic) or *fondacos* (and variants thereof). They served a variety of functions, depending on the time and place—lodgings, warehouses, commercial depots and emporia, taverns, brothels, prisons, tax-stations, and offices.[29]

Ships required many kinds of skilled and unskilled workers to rig and otherwise prepare the ship for a voyage, to propel and navigate the vessel, to keep it afloat, to organize any merchandise on it, to oversee passengers, and to organize, supervise, and feed the crew.[30] Navigation developed with adoption of an array of instruments, some of which are discussed further in chapter 6.

Roads, Wheeled Vehicles, and Beasts of Burden

Within the vast area that had been the ancient Roman Empire, Roman roads continued in use throughout the medieval period. Although they had been constructed very solidly, usually of stone, many nevertheless fell into disuse as time passed. Medieval patterns of trade were not served particularly well by this ancient road system, which had been constructed primarily for the rapid movement of Roman armies. Some Roman roads, particularly those in the Byzantine Empire, were kept in repair, while others were abandoned as they deteriorated, and others were dismantled so the materials could be used elsewhere. In the Byzantine Empire, roads were repaired by forced labor, and those that were maintained over the centuries often became narrower—suitable for foot traffic and animals (donkeys, mules, and horses for hauling loads and for riding)—but not for carts.[31]

Medieval roads and road systems are often difficult to detect and traditionally have not been a focus of archaeological investigation (which has focused instead on buildings and urban settlements). In her study of medieval roads in east-central Europe, Magdolna Szilágyi explicates some of the methodologies and source materials now being used. They include archaeological investigation, including aerial archaeology; written sources of various kinds; cartographical evidence; and toponymy (the study of place names). The most basic (and common) medieval wayfare was the footpath, created by people walking repeatedly on the same route (as between one village and another). Similarly, riders of horses created "bridleways." Other kinds of roads included cart and sled roads. Workers constructed medieval roadways variously—using packed earth, clay, gravel, and/or stone. Roads also took a variety of forms, which included "hollow ways" (roadways that had sunk into the ground with high sides due to erosion and constant use).[32]

In his study of English medieval roads, Paul Hindle points to road names as indicators of use: salt road, corpse road (used to carry corpses to authorized burial sites), pilgrimage road, and drove road (on which herds of animals were driven). Throughout Mediterranean and European lands, trains of pack animals—donkeys, horses, mules, and camels—were common ways of transporting goods, as were animals pulling carts. The donkey (or ass), a small and docile animal, was widely used as a pack animal and for riding. The versatile mule, a cross between a horse and a donkey, was used both as a pack animal and for pulling wagons and canal

Figure 4.4. A hollow way, Chemin Creux du Coté de la Meauffe, Manche, France. Photo by Romain Bréget, July 20, 2014. Wikimedia Commons.

boats. Horseback riding increased as an improved saddle and the introduction of the stirrup and the iron horseshoe made it attractive to a wide range of riders. New roads in the medieval period were often created through habitual use, whether by people walking on foot or leading pack animals or carts.[33]

Oana Toda has investigated road systems and bridges in medieval Transylvania, an eastern European region encompassing present-day Romania. Sources for locating and studying such systems include documentary evidence such as administrative and account documents, cartographic sources, and archaeological investigations. Transylvanian roads include stone roads (very infrequent, often referring to an ancient road still being used); corduroy roads, that is, roads over marshy or muddy terrain made of tree trunks; and earth road tracks, including hollow ways. River crossings here and elsewhere could be fords, ferries, or bridges. Documentation is often available for bridges because elite and/or noble individuals, or privileged groups such as monasteries, collected bridge tolls, resulting in account records as well as documents coming out of disputes and litigation. The right to collect tolls was often accompanied by a responsibility to maintain the structure.[34]

Figure 4.5. Stone bridge with pointed arch—thirteenth-century Eurymedon Bridge (Köprüpazar Köprüsü) near Aspendos, Pamphylia, Turkey. Photo by Klaus-Peter Simon, May 7, 1988. Wikimedia Commons.

Throughout the Mediterranean and Europe, workers constructed bridges of various types—reflecting both geography and culture. Medieval bridge types included wooden bridges, stone bridges, those with stone piers and wooden decks, and pontoon bridges. The latter, common in the eastern Mediterranean, consisted of a plank roadway anchored to boats that were strung across the river on chains running from one bank to another. In the tenth century there was a pontoon bridge across the Tigris River in Baghdad. From the twelfth to the sixteenth century in Anatolia, Turkish peoples built stone arched bridges, often with pointed arches. Andrew Peterson's studies of medieval bridges in coastal Lebanon and Palestine reveal the difficulties of inventorying and dating structures that still exist. Existing structures or their ruins often possess long histories of repair and rebuilding and sometimes rest on ancient foundations. Crucial dimensions of stone arched bridges included the rib width (i.e, the thickness of the stones forming the arch) compared to the width of the span of the arch. Wider spans allowed more river water to flow through, but if the rib width was very thick, the piers supporting the arch had to be thicker, reducing the clear space for water. In the early medieval West, most bridges were made of wood, but during the eleventh century, stone bridges became more common (built along with stone castles and cathedrals). Even stone bridges required a substantial amount of timber for scaffolding and for the centering (structural support during construction) needed to build arches. In both the eastern Mediterranean and northern Europe, builders sometimes constructed bridges with cutwaters (wedge-shaped piers that divide the current and protect the bridge from damage from floating debris and ice). Bridges often combined stone arches and timber roadways. The piers of the bridge were sometimes surrounded by "starlings"— made by driving piles into the riverbed side-by-side to create an enclosure that reached above the water level and that was filled with rubble.[35]

The main reason bridges fail during floods is because of "scour," the process in which the moving river waters remove material from the riverbed around the piers or other bridge supports, creating holes that undermine the structure. Because bridges narrow the water channel and tend to create eddies, scour happens around them. Bridges are also threatened by debris pushed against them (a common occurrence during flooding), and ice flows. Finally, shifting river channels (the natural tendency of rivers to move within the flood plain), threaten bridges.[36]

In premodern England, three kinds of bridges predominated: timber girder bridges (also known as trestle bridges), stone and timber bridges (stone piers with timber roadways), and stone vaulted bridges. To construct timber bridges, workers used powerful pile drivers to drive large timbers into the bed of the river. Or, they used large soleplates—horizontal timbers laid at the bottom of the bed into which the vertical posts holding up the bridge were jointed. Another kind of timber bridge was made with piers in the form of large wooden boxes filled with rubble. Stone vaulted bridges proliferated in England and Europe in the twelfth century. An important development was the creation of segmental arches (arches with an arc of less than 180 degrees, which are much stronger than semicircular [180-degree] arches). The foundations of the bridge were best laid dry if conditions allowed. Workers built a cofferdam by driving piles into the riverbed in a circle or rectangle, making them watertight and then removing the water with buckets. As in the eastern Mediterranean and northern Europe, workers sometimes also built starlings, around the piers of the bridge, as protection against scouring.[37]

As we all know from experience, roads and bridges disintegrate without ongoing maintenance. Premodern people knew this too and expended considerable effort to maintenance, repair, and rebuilding. David Harrison has shown for England that, against previous interpretations, great efforts were made to maintain English roads and bridges. In Italy in the city of Rome, maintenance of roads and bridges was a focus of ongoing conflict (about who would do the work and who would pay for it). For example, there are frequent contracts dating to the sixteenth century for paving Roman streets, but there is also evidence that many streets were

Figure 4.6. Cofferdam with water being drained with buckets. Agostino Ramelli, *Le Diverse et Artificiose Machine del Capitano Agostino Ramelli* (Paris: In casa dell'autore, 1588), plate CXI (detail). Courtesy Linda Hall Library.

unpaved or in urgent need of pavement repair. There are also records concerning the repair of bridges and damage to them during floods.[38]

In some regions, roads fell out of use altogether. Through an immense expanse of territory extending from Morocco to Afghanistan, including Syria, Palestine, and Egypt, wheeled vehicles disappeared in the medieval centuries. They were replaced by camels for both riding and cargo transport. Caravans of camels accompanied by merchants and perhaps other travelers including pilgrims traversed North Africa, the eastern Mediterranean, and central Asia. In desert and semi-desert regions, camels made better headway than other available forms of transportation, and they have a unique capacity for storing water and food. Managing a caravan was a complicated endeavor, as complicated as putting a large merchant ship to sea. Merchants from distant cities often undertook caravans as joint ventures. Caravan sizes varied from twenty-five animals to a thousand or more. Whatever size, it was necessary to have a stopping place every 20–25 miles, whether a town, or oasis, or caravansary. Workers built caravansaries with walls and gates, a large central pen for the animals, and storage areas for cargo, supplies, and provisions. They built lodges for travelers on the second floor. Islamic rulers often built caravansaries near oases and towns in order to encourage trade.[39]

Technologies of Communication

"Communication" is a term that covers many human activities, including speaking and listening, creating pictorial images, observing and reacting to those images, and writing and reading. The technologies of these forms of communication include writing implements such as paints, brushes, and reed pens; substances such as inks; writing surfaces such as ostraca (potsherds or broken pieces of ceramic); tablets covered with (erasable) wax; papyrus, parchment, vellum, and paper; and machines such as the printing press. Physical forms of written communication include single sheets, rolls, and codices.

The most common writing surface of the ancient and early medieval Mediterranean was papyrus, which is made from the stalks of an aquatic plant plentiful in the Nile Delta. Papyrus grew as tall as fifteen feet above the water with pithy stalks that could be two inches thick. Workers peeled off the outer layer and sliced the remaining material into thin strips. They placed a layer of these strips vertically on a hard surface and covered it with a second layer positioned horizontally, wet the mat, then beat or pressed it, and let it dry in the sun. The material exudes a resin that acts as a glue, fusing the crisscrossed strips into a solid but flexible sheet that they trimmed and polished with a volcanic glass called pumice. Single sheets could be used for documents, but workers often made the papyrus into rolls. They pasted the edge of one sheet to the next with a flour paste, smoothing the joint flat. The size of the roll was usually about twenty sheets. The resin that held the sheets intact also acted as a sizing—the material that prevented the ink from soaking through the surface. Scribes could correct errors by wiping off wet ink or scraping off dry ink.[40]

The codex form (plural codices)—the books with which we are familiar—is a book made with folded sheets sewn along one edge. It began to be made with papyrus and was popular with the early Christians, increasing in use from the second to the fourth century. Scholars debate the reasons for the adoption of the codex as the standard form of the book. Some argue that it appealed to Christians engaged in itinerate proselytizing because it was easier to carry than a roll and could contain more writing in a smaller object. The codex in general replaced the roll by the fourth century CE, but the roll continued to be used for special purposes, such as for the rolls of the exchequer in England (treasury records), prayer rolls, and genealogies.[41]

Early codices were usually made of parchment or vellum. Parchment usually refers to sheepskin or goatskin, while vellum refers to calfskin (although the distinction is not always clear). To produce parchment, leatherworkers soaked skins in a bath of lime or urine for ten days to remove any flesh. They scraped the surface smooth, then made it smoother by rubbing it with pumice and whiter by rubbing it with chalk. Parchment and vellum are fine, durable materials. (Many documents written on such animal skins in European archives are in pristine condition after hundreds of years.) However, they were expensive. A large parchment Bible, for example, required several hundred animal skins.[42]

Paper had its origins in ancient China, where it was made with a variety of bast fibers (the woody fibers obtained from various trees such as the mulberry). In Islamic regions of the Mediterranean, by the eighth century, artisans were fabricating paper out of cloth, usually hemp or linen. The key to its rapid spread was the decision of the Abbasid rulers (750–1258) to use paper for their official records. A paper mill established in Baghdad in 794–795 turned out sufficient quantities of paper to supersede papyrus and parchment at far less cost. Further, ink could not be scrapped off paper, making it seem more trustworthy than papyrus as a writing surface. Papermakers soaked the cloth and hammered it into a pulp, which they poured into a rectangular mold that framed a cloth sieve. There were two kinds of paper molds. One could be floated in a vat of water from which the liquid gradually drained out. The sheet was sun-dried while still in the mold. A second was a dipping mold, designed so that sheets could be removed while still moist. They were then "couched," that is, pressed between two pieces of wool felt, and then hung to dry. The mold was then reused while the sheets were drying. Papermakers then applied a size or glue, gypsum, gum, or starch to the sheets in order to make the surface impermeable to ink. Without sizing, the paper would absorb the ink like a blotter, making it impossible to write on.[43]

Workers fabricated inks according to which surface was to be used. For papyrus and paper, they prepared ink with lampblack or carbon combined with plant gum. They made carbon by heating plant products in earthen pots and lampblack by burning various waxes, oils, or resins and collecting the soot. For parchment surfaces, workers prepared a brown ink from gallnuts (swellings on the bark of oak trees caused by gall wasps laying their eggs), which contained gallic acid and tannin. They mixed a solution of this substance with ferrous sulfate (which they called copperas). This metal tannate ink penetrated parchment or vellum like a dye but could not be used on paper because the acids it contained eventually made holes.[44]

The adoption of paper had cultural components. For example, although the Byzantines in Constantinople had extensive contacts with the Abbasid court in Baghdad, they did not adopt paper on a large scale until the thirteenth century. The technology of papermaking was transmitted to Christian Europe in the eleventh and twelfth centuries (through Islamic Sicily and Spain). Europeans adopted papermaking and soon introduced new techniques, including the use of meshed wire in molds instead of cloth sieves. In Italy, papermaking was taken up in the town of Fabriano on the Adriatic Coast and then in the Veneto. By the end of the thirteenth century, Fabriano paper had begun to dominate Mediterranean markets. The Italians invented watermarks—raised designs on the mold screen—to identify their papers. Because the starch used in (often dryer) Islamic lands tended to mildew in the wetter Italian climate, they began to make size from gelatin, produced from the horns, hoofs, and hides of animals. Gelatin sizing made a harder surface, which worked better with the quill pens that Europeans had used customarily on parchment and now used on paper. In Islamic lands, writers used reed pens, suitable for writing on the softer paper sized with starch.[45]

Traditions of writing and bookmaking differed from one region to another. In Byzantine lands, most books concerned theology and the liturgy and were made of parchment. Because of its high cost and the time it took a scribe to make copies, books were expensive. The value of one ninth-century manuscript, for example, has been estimated to be the equivalent of half a year's salary for a civil servant. Private collectors rarely owned more than twenty-five books, and there was no such thing as a book market. Readers acquired books by commissioning a copy of a particular book at a scriptorium, or going to a library, or borrowing from friends.[46]

The Arabic/Islamic tradition was quite different. First, there is the issue of Arabic script. Pre-Islamic Arabic script had only a cursive form with many of the letters joined by ligatures. There were many ambiguities. After the rise of Islam, this became problematic for the transmission of authentic copies of the Qur'an. As a result, Arabic script was "regularized." Medieval Islamic writing appeared in books but also on cups, vases, textiles, and buildings. Medieval manuscripts of the Qur'an were copied on parchment sheets that were put into protective boxes, or else they were sewn and bound into covers made of wood covered with leather. The importance of the Qur'an ensured that Arabic would become the common language of most Islamic lands. It also encouraged ordinary people to read.[47]

Under the Abbasid rulers in the ninth century, secretaries exerted significant influence on the courts. Official documents were judged by their elegance, their subtle references, and the skill of the calligraphic hand as well as by their content. Thus did the art of writing move from the primary domain of Qur'anic copyists to the secular world. Paper facilitated the work of writers, as did a new simplified script called "broken cursive," as did the use of lampblack ink, all of which contributed to an outpouring of books. The arrival of paper, the production of books, and the translation of texts from Greek and other languages into Arabic were simultaneous and mutually reinforcing developments. Topics and disciplines addressed in books included theology, law, philology, grammar, biography, history, poetry, philosophy, navigation, practical arts, and engineering, astronomy, astrology, mathematics, alchemy, and cooking. There were even works of fiction.[48]

Much Islamic book production was carried out in mosques. An author would sit in front of auditor-copyists, as many as twelve of them, who would take dictation. A close associate or student might sit nearby to serve as an intermediary. The author would have written a draft of the work to follow while dictating, but he would often speak from memory. A copy of a book was not considered authentic unless "authorized," a process accomplished by "check-reading"—the scribe reading the text back to the author. Thus could an author create multiple copies at a single reading. The proliferation of books created many new professions—copyists, bookbinders, dealers in paper, and book dealers. Numerous libraries, both private and public, were established throughout Arab lands.[49]

In the early medieval West, book production was carried out for the most part in monastic scriptoria—workshops dedicated to the copying of books. A scriptorium had to be attached to an institution wealthy enough to provide it with the expensive materials of parchment and pigment. It also had to have skilled scribes with the time to copy books. A single scribe, or five or more scribes working in shifts, might copy a book. In the late eighth and early ninth centuries in the Frankish kingdom of Charlemagne (ruled 768–814), a new script was developed called the

A.

B.

Figure 4.7. A. Page of text written in Carolingian minuscule. British Library MS Add. 11848, fol. 160v. Photographer unknown, December 2010. Wikimedia Commons. B. Decorated initial letter (*U*). Evangelistar von Speyer. Badische Landesbibliothek, Karlsruhe, Cod. Bruchsal 1, fol. 68r. Wikimedia Commons.

Carolingian minuscule. It was uniform, consistent, and regular—far easier to read than most prior scripts.[50]

With the expansion of schooling in the twelfth century and the rise of universities in the thirteenth, secular scribal workshops flourished in many towns. Professional stationers contracted artisans (who prepared parchment), scribes, and illuminators who specialized in elaborate border decorations or in miniature illuminations. Stationers also supervised the *pecia* (piece) system that developed in the thirteenth century around the universities, in which they gave out sections of university-approved copies of texts to scribes for copying (to be sold to university students). The system included an elaborate system of checking copies against the original for accuracy.[51]

Manuscript illumination was common in religious books such as psalters (books of psalms and other devotional texts) and gospels, as well as in many other kinds of books. Usually, a scribe would copy the entire book, leaving a space for the first letter of the first word of every chapter. Then a rubricator would add the ornamental letters, and specialized illuminators would paint pictures, either in spaces left for them throughout the text or on separate pages. Illuminators might first lay down a layer of gesso (a thick, water-based substance made of plaster, gypsum, or chalk and bound with glue). They might gild areas of the picture by laying down gold leaf. Then they might tool the surface with decorations. They created images by adding pigments, which were derived from plants and various other substances mixed with a binding medium of glair (clarified egg white) and other substances.[52]

Into this flourishing culture of manuscript books, printing emerged. One kind of printing was woodcut printing or xylography. The technique of woodblock printing came from Asia (where it had been practiced for centuries) and was well-suited to Western paper. In Europe, block prints appeared in the fourteenth century and proliferated rapidly. They first comprised single sheets containing only images, usually of religious subjects. Eventually block printers added written legends. Then small, multipage woodcut books appeared—the first books in the West that were cheap enough to be within reach of ordinary people. Woodblock printing continued after the invention of the printing press, and in addition, printers used woodblock images in their moveable-type printed books.[53]

The printing press involved a composite invention of three separate elements— moveable type cast in metal, an oil-based ink, and the press itself. To make moveable type, for each letter the smith made a punch from a piece of hard metal with a letter engraved on the end. The printer used the punch to strike a die in a softer metal that held the impression or intaglio, and then used the die to make as many "sorts" or letters as desired. Letters were made from metals such as tin or lead that fuse at low temperatures. Goldsmiths and minters, who were familiar with using punches to make designs in coins and leather, developed the technique.[54]

Between 1430 and 1450 experiments on the invention of a viable commercial printing press using moveable type were proceeding in several locations. There is enough evidence to show that this experimental phase did occur but not enough to know exactly who was doing what. The first successful printing business finally emerged in Mainz, Germany, and is associated with three men: Johann Fust (c. 1400–1466), a rich citizen financier; Peter Schoeffer (c. 1425–c. 1503), who apprenticed to Gutenberg and who possibly worked as a scribe before becoming a printer; and Johann Gutenberg (c. 1400–1468), who is generally accepted as the inventor of moveable, lead-based type. The expenses of developing the new printing press drove Gutenberg into bankruptcy. The court cases that resulted, including Fust's lawsuit against Gutenberg in 1455, have provided much information, although who did what in terms of the various inventions that led to printing is not entirely clear. What is clear is that there were several inventors, not just one, of the different components.[55]

The influence of printing on culture and society is the focus of the thriving field of the history of the book and of printing more generally. The historian Elizabeth Eisenstein in a groundbreaking and controversial book argued that there was an unacknowledged "print revolution" that among other things led to the end of the errors inherent in scribal copying, to a "fixity" of ideas and images, and to the development of technical and scientific literature. It also led to the rise, she argued, of scientific illustrations because it allowed the reproduction of identical images through many copies and editions. Historians have criticized her thesis on many grounds, including its technological determinism, but have also used her work as a springboard to further studies. As Ann Blair has explicated, printing did lead to a huge proliferation of books, far greater than that of manuscript culture, and a resulting information overload. This led to the habit of note taking, to the creation of a variety of reference books, and to various other strategies, such as indexes, aimed to manage the huge increase of available information. Paul Dover suggests that the wide availability of paper was more important than print; that paper writings from notes, to records, to letters proliferated along with printed books; and that the separation of manuscript and print cultures is a distortion.[56]

In the mid-fifteenth century another kind of printing emerged both in Germany and northern Italy, namely engraving, a technique of incising images on a metal surface, usually copper, often first with the light lines of a drypoint needle and then the deeper grooves of a burin. After the plate was prepared, engravers placed it in a press and printed the image on sheets of paper, either to make a single sheet for sale or to make an image as part of a book. Somewhat later around 1500, artisans developed etching, a process in which they waxed the plate (initially iron but then copper) and incised an image into the wax. Then they dipped the plate into an acid bath, which cut the image in the unprotected part of the metal. Printers sold thousands of single-sheet engravings and etchings of images in the sixteenth century. They also used the plates to add illustrations to printed books. Unlike woodblocks, which could be inserted into the page with the type, engravings and etchings had to be printed separately on a special press.[57]

In many ways, printing amplified and transformed the world of manuscripts. Books of secrets, containing recipes of all kinds including recipes for household remedies and medicines; religious tracts, particularly promoting the Protestant Reformation in German lands; almanacs, popular writings often referred to as "cheap print"; and manuals and "how-to" books proliferated. With more available reading material, literacy rose among the general population, including women. Printshops proliferated and became trading zones in which authors, copyeditors (called correctors), illustrators, and other artisans worked together and communicated on the tasks at hand but ultimately discussed other substantive issues as well.[58]

Travel and Commerce

Extensive scholarship of the last twenty-five years has produced a complex picture of the many ties among the cities, regions, and locales of the Mediterranean with each other and with northern European lands. Michael McCormick provides a mass of evidence to show that the Arab conquests did not end commerce across the Mediterranean but provided a new basis fueled by the export of enslaved persons to the Islamic world in exchange for the riches purveyed by that world including silks and spices. Peregrine Horden and Nicholas Purcell emphasize the significance of trade between small settlements and microregions, including those in close proximity. They suggest that a focus only on the great trade routes by land and sea comprised a top-down approach that devalues the significance of short-distance exchange among microregions using small boats and ubiquitous donkey trails and footpaths. Numerous other studies explore commerce and communication between a great diversity of regions and microregions. Interchanges could be cooperative or hostile—piracy, including piracy sponsored by states or by elite individuals, was a given of both Mediterranean and eventually Atlantic seafaring. ("Piracy" was a fluctuating concept and only came to be used as a word of condemnation when states in the fifteenth century began to criminalize certain rivals by labeling them as pirates.)[59]

The ways in which communication, travel, and commerce were intertwined are particularly illuminated by the Geniza documents discovered in Fustat (Old Cairo). Geniza is a Hebrew word meaning storehouse or treasure. Jewish belief was that writings in which the name of God appeared (or even any document written in Hebrew) should be buried in a cemetery. As a result, a Jewish community in medieval Fustat placed in a storeroom connected to their synagogue, thousands of fragments (an estimated 400,000) written mainly on vellum and paper but also on papyrus and cloth—intended for later burial. Most were written in the Arabic spoken by the Arab-Jewish community of Fustat but written with Hebrew characters, a form now called Judeo-Arabic. The Geniza documents constitute an immense and

Figure 4.8. A scholar (Solomon Schechter, 1847–1915) studying fragments of the Cairo Geniza in Cambridge University Library (detail). Photographer unknown, c. 1898. Wikimedia Commons.

priceless collection of documents concerning work, marriage, and commerce of all kinds pertaining to this Egyptian Jewish community, especially from the eleventh to the thirteenth century.[60]

The Geniza documents include hundreds of letters exchanged by commercial associates, family members, and friends. These indicate that regular travel between Egypt and the Iberian Peninsula and around the Mediterranean was nothing unusual. Long-distance commercial transactions could occur because of networks of merchants, who helped one another, often using the *ṣuḥba*, a reciprocal agreement between two merchants to act as agents for one or more specific transactions. As Jessica Goldberg has shown, Geniza merchants worked within Islamic, not European, networks. S. D. Giotein emphasized that ties were often strengthened by means of marriages arranged between merchant families in Cairo and those in distant lands—al-Andalus, Tunisia, Morocco, Arabia, Palestine, Lebanon, Syria, and Iraq. Hundreds of Geniza documents refer to far-flung travels for the purpose of trading, with little mention of political interference. Jewish traders frequently traveled on Muslim ships. Skilled artisans seem to have been highly mobile throughout

Islamic lands. Communication was accomplished by the exchange of thousands of letters delivered by a combination of public postal systems between specific cities, courier services, and personal business associates.[61]

Economic historians have found it useful to distinguish between markets and emporia. A market is a place where local products were exchanged, whereas an emporium collects goods from outside a region and distributes them to merchants who take them elsewhere. For both Mediterranean and European lands, historians have often emphasized long-distance trade at the expense of local and regional exchange. But both were important. One result of neglecting local commerce is that women's work has thereby been neglected. In general, women were far more involved in local markets than in long-distance trade; in long-distance trade, they did not have the legal standing or credit to adequately function (and far fewer women than men traveled long-distances).[62]

Recent scholarship has begun to break down the modern national, linguistic, and disciplinary boundaries that have tended to artificially (for the eras of this book) isolate one region from another (such as the Iberian Peninsula from North Africa). This scholarship has also proposed a complex model of cultural and knowledge exchange in these diverse areas, rejecting traditional assumptions about the superiority of European culture and rejecting a linear progression towards the "Scientific Revolution" and "modern" science. The exchange of material goods in commerce and the exchange of information and knowledge went hand in hand.[63]

* * *

The many details of commerce, travel, and cultural exchange are known to historians because of archaeology, the study of material objects, and because of thousands of written documents, including letters written by travelers, travelers' accounts, and writings such as the *Book of Michael of Rhodes*. Trade involved the purchase of raw materials, such as lumber, metals, and stone; the sale of enslaved persons; the exchange of objects such as relics; and the purchase of fabricated or crafted things such as textiles, clocks, and paintings. The excavation and collection of raw materials and the fabrication of objects is the focus of the next chapter.

CHAPTER FIVE

Crafts and Industries

We therefore say that the practical craft consists of externalizing the form that the knowledgeable maker has in his mind by positing it in matter. As for the product, it is an aggregation of assembling matter and form together. The initiation of this happens under the influence of the Universal Soul as aided by the power of the Universal Intellect, which is commanded by God, may his Glory be exalted.

IKHWĀN AL-ṢAFĀ, "EPISTLE 8: ON THE PRACTICAL CRAFTS"
(TENTH CENTURY)

Therefore, most beloved son, you should not doubt but should believe in full faith that the Spirit of God has filled your heart when you have embellished His house with such great beauty and variety of workmanship. . . . Whatever in the arts you can learn, understand, or devise, is bestowed on you by the grace of the seven-fold Spirit.

THEOPHILUS, *ON DIVERS ARTS* (TWELFTH CENTURY)

These two excerpts are from texts that were created hundreds of miles and two hundred years apart. The anonymous members of a fraternity around Basra and Baghdad in the eastern Mediterranean called the Brethren of Purity (Ikhwān al-Ṣafā)—cited in the introduction of this book—wrote the first sometime in the tenth century. It is part of a famous, philosophically oriented compendium called the *Epistles of the Brethern of Purity*, which contains a total of fifty-two letters. The secret group espoused a complex syncretic philosophy that incorporated the Qur'an, Christianity, and Judaism, as well as a variety of ancient Greek philosophers, including Neoplatonists such as Plotinus (204/205–270 CE). Two centuries later, the anonymous Theophilus (thought by many scholars to have been Roger of Helmarshausen, a metalworker and Benedictine monk who lived around 1100 in central Germany) wrote the second citation within a three-part treatise on the art of painting, glasswork, and metalwork.[1]

These texts show in two different contexts the ways in which crafts could be connected to larger philosophical and cultural values. The eighth epistle of the

Ikhwān highly values the crafts (while also assigning particular crafts different positions in a hierarchy). It connects the crafts (each as a microcosm) to a universal soul and intellect (the macrocosm).[2] Theophilus, in the context of twelfth-century Christianity, exhorts artisans that whatever they achieve in art is bestowed by the grace of the seven-fold Spirit—the spirit of wisdom to know that created things proceed from God, the spirit of understanding through which they have received the capacity for practical knowledge, the spirit of council whereby they do not hide or work in secret but reveal and teach their crafts openly, the spirit of fortitude through which they shake off apathy and work hard, the spirit of knowledge that has provided practical knowledge, the spirit of piety through which they set limits on the work and refrain from coveting high payment, and the spirit of the fear of the Lord.[3] The writings of the Ikhwān and those of Theophilus exemplify the ways in which the skilled crafts could be deeply entwined with the life of the intellect and culture, in this case religious culture in the premodern centuries.

Crafts were also integral to social and economic life. Elites everywhere possessed fine clothing, jewelry, luxurious goods of all kinds from tableware to books, as well as richly ornamented houses and palaces as a way of displaying (and imposing) their political and social power and authority. And non-elites, who also used a panoply of material goods on a lesser scale, were often duly impressed. The social standing of the makers of these goods varied, depending on the time and place, and on the kinds of goods that they made.[4] In all cultures, there were also ascetics, including members of ascetic monastic groups who turned their back on worldly consumption altogether. Yet even ascetics, who rejected the accumulation and display of things, needed—in order to survive—a certain number of possessions, crafted in one way or another. (Even the rags that some people wore had gone through processes of textile and clothing fabrication.)

Artisans

Skilled artisans usually learned their craft in an apprenticeship—either formal, sometimes with a contract between the young person's parents and a skilled master, or informal, usually within the household and family. Although skilled work was often gendered, the gender roles of modern times (for example, those of North America in the 1950s) should not be imposed on premodern times. As we know from pictorial and documentary sources, women could be blacksmiths, miners, ore sorters, and were skilled in many crafts. In some industries—as Sharon Farmer has shown for the Parisian silk industry of the late thirteenth and early fourteenth centuries—women became not only modestly paid workers but prominent heads of workshops and prosperous entrepreneurs. Often, such artisans, male and female, trained and worked in workshops that were part of a family home, with products sold from the front. Sometimes wives and husbands both practiced the craft and, in any case, wives constituted an essential part of the household labor force. They might run the shop and assist in the fabrication of whatever was being made in the workshop. They might also keep accounts, acquire supplies, and bring the goods to the town market. Widows often carried on their deceased husband's trade, and in this capacity could supervise apprentices and journeymen, purchase materials, make contracts, and sell their wares. Urban women also worked as retailors selling their products (often food and drink) door-to-door, or on the street, or they could rent a stall and sell at the town market. Women were highly active in local markets. In northern European cities, for example, local retail markets functioned in which buyers and sellers exchanged money and diverse goods from ale to stockings, from cloth to pots and pans. Women largely controlled these markets.[5]

In the early Byzantine Empire, the centralized government attempted to regulate both manufacture and commerce. Villages traded their products in market towns, which in turn were linked to Constantinople. Emperors sought to consolidate control in order to maintain the flow of money from taxes, essential for maintaining their large bureaucracies and armies. Merchants and artisans were not always sharply distinguished, since artisans often sold their own goods in local markets. Those goods included jewelry, glassware, pottery, textiles including silk, perfumes, and leatherwork, as well as foodstuffs such as bread, olive oil, and vegetable oil.[6]

Byzantine artisan trades included carpentry, masonry, and building construction, along with ivory carving, painting, book illustration, and the fabrication of mosaics. In Constantinople, some trades, such as silk manufacture, were organized into professional guilds that were supervised by the state. An ordinance regulating these guilds appears in a surviving document called *The Book of the Eparch* (or *Prefect*) dating from the tenth century. The regulations protected guild members from competition, both from landowners involved in trade and from artisans and merchants outside the guild. They also governed the quality and quantity of production, and set profit margins for the sale of various staples. Centrally important was the legislation regarding silk manufacture, which aimed to prevent poor craftsmanship, attain uniformity in products, promote ethical conduct among workers, and ensure tax payments. In addition to supervising the local guilds, the eparch controlled the activities of foreign traders and, beginning in the tenth century, began to grant trading privileges to outsiders, including Italians from Amalfi and Venice. Scholars debate the degree of influence and significance of the state-controlled guilds.[7]

Historians have learned much about the organization of the medieval workplace in Cairo from the Geniza documents discussed in chapter 4. The typical workplace was a shop headed by a single artisan, a family, a clan, or several partners, usually not more than five. Crafts in Islamic lands were organized into guilds only in the fourteenth century. Practitioners of particular crafts in urban settings were usually concentrated together in the same area. Streets or quarters often were named after the trades that were carried out there. Often the same worker who made a product also sold it.[8]

The Geniza documents reveal a high degree of specialization. The most important industries consisted of textiles, dyeing, and clothing; metals, glass, and pottery; building construction; and food processing. Specialists worked in numerous small trades: makers of kohl sticks used to apply eyelid powder; makers of writing cases, mirrors, mats, fans, spindles, sieves, combs for hair and combs for flax; makers of beads; perforators of pearls; people who processed corals. In every household, at least one woman worked as a spinner.[9]

In her study of occupations in the medieval Islamic world, Maya Shatzmiller listed hundreds of occupations including extractive work such as agriculture, mining, and fishing; trades such as dye makers, soap makers, brickmakers, and textile workers; and food processors such as bakers and sausage makers. As in Byzantine and European lands, there was some division of labor by gender. Within silk manufacture, women took care of silkworms, a labor-intensive task. They spun flax, wool, cotton, and silk, and worked as embroiderers, weavers of brocades, and carpet makers, often with their children working alongside. Although commercial weaving was a male occupation, both men and women worked as dyers. Female craft skills and trades were transmitted within informal apprenticeship systems in which women trained young girls including their own daughters.[10]

In western European lands, many material objects were made on site in early medieval villages, manors, and monasteries. Villagers made shoes, bags, and harnesses from pigskin. They fabricated wool or flax cloth and clothing, and made wooden farm tools and cooking implements. Smiths who worked iron for tools first constructed their forges and anvils in forests to be near the necessary fuel (charcoal made from wood). The iron used for tools was mined elsewhere. Eventually the smith became a valued artisan in the village and manor, making horseshoes, plowshares, pots, pans, armor, and weapons.[11]

Monks and nuns and their assistants made many everyday items in monastic workshops, including wooden objects such as furniture and barrels, cloth and clothing, pottery, leather goods, and metal objects. Some monasteries also supported skilled artisans who produced finely worked liturgical objects, from candlesticks to embroidered garments. This specialized work is vividly described by the anonymous Theophilus, who, if he was Roger of Helmarshausen, is known for three highly decorated objects—a jewel-studded book cover and two portable altars.[12]

As the towns of Western Europe grew, specialized crafts proliferated. Tanners made leather out of skins, which they then provided to glovers, saddle makers, and parchment and vellum makers. Textile workers produced cloth—wool, linen, cotton, and silk. Metalworkers included goldsmiths, silversmiths, ironworkers, armorers, and smiths who worked with copper, tin, and pewter. Crafts included glassmaking for vessels and windows, pottery and tile making, painting, sculpting, and brickmaking. Carpenters and joiners made a variety of objects, from barrels to marriage chests to altarpieces.[13]

Craft guilds in Western Europe emerge in the record around 1100. The number and kinds of guilds varied widely from town to town as did their relative power and their relationships to political authority. Guild regulations governed apprenticeship as young boys (and sometimes girls) learned a craft through the stages of apprenticeship, journeyman, and then master, only the latter able to own their own shop. The guilds restricted the practice of the craft to its own members, ensured quality control of their products, and (often through associated confraternities) provided social benefits such as dinners, support for funerals of deceased members, and assistance to widows. Scholars disagree about whether or not guilds were beneficial. Some, such as Sheilagh Ogilvie, argue that they were largely harmful, preventing innovation and protecting the interests of members over consumers and would-be makers who were nonmembers (especially women). Others, such as S. R. Epstein and Maarten Prak, argue the opposite, suggesting that guilds were beneficial for the premodern economies of Europe especially in the area of skills training, and that the economic environment of guilds was conducive to technological innovation. For example, "tramping," the frequent requirement that journeymen practice the craft elsewhere for certain period of time, helped to disseminate craft techniques and generate innovation.[14]

The craft guilds also play a role in the broader issue of the dissemination of techniques and technologies from one generation to the next (usually through apprenticeship), but also across borders and from one geographic area to another. As Liliane Hilaire-Pérez and Catherine Verna noted, communities adopted particular crafts and techniques according to their own specific habits, needs, constraints, and geographies. In addition, craft cultures were influenced by the degree to which techniques were kept secret or shared. As Koen Vermeir has explicated, openness and secrecy should be understood as overlapping categories and not necessarily in opposition to each other. The dissemination of techniques was also influenced by the emergence in the fifteenth century of patents or privileges—limited monopolies awarded by rulers or city-states and governments for inventions or innovations. These were not equivalent to modern patents. For one thing, the recipient had only to possess knowledge of the innovation. There was no requirement of originality or personal invention.[15]

Mining and Metallurgy

The Byzantine Empire inherited a highly developed mining and metallurgical tradition from the Romans, including both large-scale operations and small production, such as peasants mining for metals as a side activity. Metals obtained included gold, silver, copper, and iron. The mining of gold and silver, particularly needed for the minting of coins, was extensive, especially in the Balkans. The Taurus Mountains in the Anatolian Peninsula provided abundant iron ores. Descriptions of metallurgical techniques exist in Byzantine "chemical" codices dating from the tenth to the fifteenth century. Metallurgists used kilns fired with charcoal and wood, and used bellows to increase the heat. The codices describe tempering (hardening) iron to make steel, fabricating bronze molds and reliefs for coins, refining gold and silver, and producing lead.[16]

Goldsmiths and silversmiths in the Byzantine Empire produced numerous objects such as dishes, lampstands, chalices, and jewelry. Objects made of precious metals were frequently melted down in subsequent decades or centuries to make new objects. Smiths used a panoply of traditional techniques—hammering, annealing (a process of heating to make the material more workable), casting, repoussé and chasing (hammering from the reverse side and the front side, respectively, to make a relief), and coldworking (working a metal at a temperature below its recrystallization temperature to strengthen it). They made thousands of cast objects, especially crosses, by the lost wax process: first they made a wax model, built a mold around it, hardened the mold, and then melted the wax, forming a cavity into which they poured the molten metal. After the metal hardened, they broke off the mold and finished the piece. The poured metals consisted of either bronze (an alloy of copper and tin), brass (copper and zinc), or lead.[17]

Archaeological findings on the Arabian Peninsula as well as textual evidence have revealed that the expansion of Islam from the seventh to ninth centuries was supported by extensive mining of gold, silver, copper, and iron, Evidence includes trenches, pits, tunnels, residual ores, tailings (waste material after the metal has been extracted), and piles of slag (the residue of ore, flux, furnace, and trace elements of the extracted metal), as well as traces of charcoal, which miners burned in kilns for smelting the ore. In addition, archaeologists have found medieval mining tools such as hammers, picks, and shovels, as well as the remains of smelters and of mining villages. The latter include houses and barracks for miners, suggesting in some cases as many as a thousand or more manual laborers (who were enslaved persons, prisoners, and servants). The proliferation of precious metals led to the flourishing crafts of jewelry-making and the fabrication of metallic ornaments. Most mines seem to have been held privately but were taxed by the nascent Islamic state.[18]

In later centuries in the Islamic world, mining was carried out either as an underground or an open cast operation. In the first, workers dug a vertical shaft down to a vein of ore and then excavated a horizontal passage—a procedure similar to *qanat* construction. In open cast mining, they dug an opening into the slope of a hill or mountain and then followed the vein along the side. In both kinds of mining, workers used pickaxes and chisels, and they used capstans to raise ores from deep shafts. They used oil lamps for light, regarding it as a danger signal when one went out. They drained deep mines by raising the water with waterwheel technologies and then used the water to irrigate fields.[19]

Metals mined and processed included gold, silver, copper, lead, and zinc (alloyed with copper to become brass). Workers used lead for the foundations of structures, canal linings, and roofing. They also used it to extract gold and silver from copper. Tin came from Malaysia or England. Metalworkers alloyed it with copper to make bronze for pots and pans and mirrors. For luxury objects, especially from the twelfth century, they created decorative surfaces, including techniques in which they inlayed brass vessels with gold and silver. Metallurgists working gold, silver, and copper often operated small, independent workshops. Metalworking skills were often handed down within families. Others worked in workshops clustered in one area, or in large commercial workshops, or state-owned factories. An example of the latter were state-run mints, which produced coins from recycled gold and silver bars, copper vessels, and scrap metal. Indeed, all over the Mediterranean and European world, increasingly rulers and cities created mints to produce their own coinage—a potent sign of their power and authority.[20]

Iron mines could be found in the Maghreb, in al-Andalus, in Egypt and Syria and throughout European lands. Before the fifteenth century, in wrought-iron production, metalworkers roasted the iron ore and quenched it in water. They mixed the ores with charcoal in what was called a bloomery furnace, and directed bellows to the center of the hearth. The resulting product is called the bloom. They purified and consolidated the bloom by a process of hammering, further heating, and hammering again to remove the slag. Archaeological sites such as al-Basra in Morocco have yielded high concentrations of hammer scale (plate-like spherical by-products of the hammering). In Islamic lands, metallurgists also produced cast iron (that melted and flowed into containers) in small blast furnaces (furnaces made hot enough to melt the iron).[21]

In early medieval Denmark, iron was produced from iron ore from bogs (wetland ecosystems characterized by peat) and—as in many other early medieval locales—was a household industry, especially in areas where the ore existed along with enough forest to provide fuel. Families smelted iron in late summer before harvests. These iron-smelting farmers could also be smiths, producing tools and other objects for local use. They made weapons such as swords, implements such as knives (of which thousands have been found in archaeological sites), and other tools. Alternatively, they hammered the bloom into iron bars for exchange in more distant places, where specialized, full-time blacksmiths flourished.[22]

Iron production has been documented in many other parts of Europe as well. For example, iron mills were established at Mount Amiata within the province of Siena in Italy in the second half of the thirteenth century. Both monastic communities and towns produced iron. Documents show that specific areas of forest were designated as wood to be used for fuel. In Viterbo in central Italy, iron ore was imported from the island of Elba and added to poorer quality local ores. Archaeologists have found remains of furnaces, slags, iron ore, and smelting operations.[23]

In Europe, the blast furnace was a medieval development. The furnaces were constructed by raising the height of the traditional bloomery furnace and increasing the force of the blast with power bellows, thereby achieving temperatures sufficiently high so that liquid iron could flow out to be cast as molten "pigs." A prototype seems to have been in use in central Swabia (present-day southeastern Germany) between the eleventh and the thirteenth century, making possible the regular production of cast iron. In northern Italy, there is evidence for blast furnaces from the thirteenth century. Blast furnaces were capital intensive and had to operate continuously for effective production. They proliferated in the fifteenth century especially because of an increased demand for cast-iron cannonballs. In addition, cast iron was used for armor, pots and pans and, after the 1540s, guns. From the fifteenth century, the skilled ironworkers who constructed and operated the furnaces were much in demand and sometimes immigrated to new regions for better pay.[24]

The rising population of fifteenth century European lands created a demand for metals that could not be met by traditional mining. One result was a central European mine boom that extended from about 1450 to 1550 with the output of silver, copper, and other metals increasing as much as fivefold. This greater productivity was the result of deeper mines that required capital investments, paid workers, and new, large-scale excavation and ore-processing equipment. The new capitalist mining operations were financed by the princes and patrician bankers of central Europe, along with smaller investors who bought shares. These mines and ore-processing facilities were characterized by deep mine shafts, many kinds of pumps and other water-removal equipment often powered by huge waterwheels, furnaces, and ventilating machines made with revolving fans and bellows, and the use of gunpowder to excavate mines.[25]

In the fifteenth century, two new ways to extract silver from its ores were developed. The first was liquation, the separation of silver from copper/silver ore using lead and relying on the different melting points of the metals and the affinity of silver for lead over copper. The second process was amalgamation, a process in which mercury was used to separate silver from other minerals.[26]

Goldsmiths worked gold and silver with forge and hammer or by casting. They usually forged vessels, dishes, and cast jewelry. To forge, they hammered an ingot into a sheet, cut it to size, and shaped it with a hammer. Annealing (heating and then cooling the metal) prevented hardening. They further treated the surface by planishing (smoothing) with special hammers and then polishing. In casting, they used crucibles for melting. They poured liquid metal into piece molds or cast it by the lost wax process. They attached precious stones to gold and silver with a solder, an alloy often containing lead. Goldsmiths enjoyed high status partly because they worked with the most precious of materials and partly because they were closely aligned with princes and rulers who consumed their products. Also the products of their workmanship inspired awe. To give just one example, as Pamela Smith has discussed, the Nuremburg goldsmith Wenzel Jamnitzer (1508–1585) made dazzling live castings of plants and animals that adorned his silver and gold vessels.[27]

Casting copper alloys required melting the metal and pouring it into a mold made of stone, clay, or metal. In Islamic lands, characteristic metal objects proliferated—fountain spouts in the form of stags, pouring vessels in the shape of birds, and pitchers with gazelle-headed spouts. Metalsmiths everywhere cast copper using the lost wax process. A later method was sand casting in which the mold was made

Axis superior A. Rota cuius pinnas riui impetus percutit B. Tympanum
dentatum C. Alter axis D. Tympanum quod ex fusis constat E.
Ferrum teres & curuatum F. Siphonum ordines G.

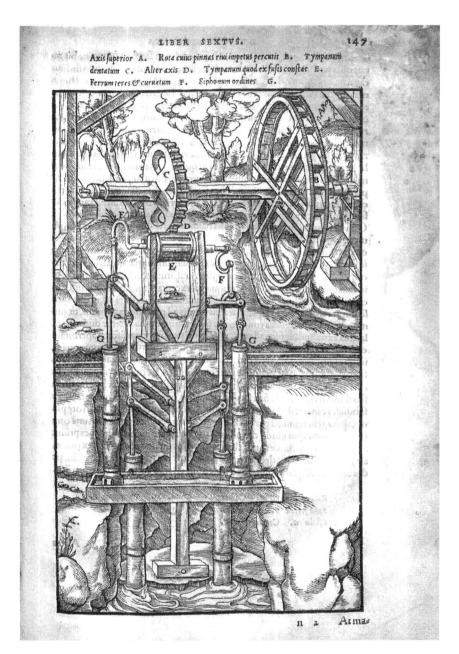

n 2 Arma

Figure 5.1. Piston pumps powered by undershot waterwheel to drain a mine. Georg
Agricola, *De re metallica libri XII* (Basel: Froben, 1561), 147. Courtesy Linda Hall
Library.

Figure 5.2. Oil painting of a goldsmith making a ring for a betrothed couple—*A Goldsmith in His Shop*, Petrus Christus, 1449. Metropolitan Museum of Art, online collection. Wikimedia Commons.

of compacted sand in molding boxes around a form of wood. Once cast, objects could be pieced together, if necessary, lathe-turned, filed, and polished. Bells, guns, and statues were the largest objects to be cast. Vannoccio Biringuccio, a Sienese mine overseer, provided a detailed description of casting bells and guns in his treatise, the *Pirotechnia*.[28]

The central European mine boom ended because of Spanish colonial mining in

Figure 5.3. Silver writing box with cast silver insects, amphibians, and plants—Wenzel Jamnitzer, c. 1560–1570 (detail). Kunsthistorisches Museum, Vienna. Photo by James Steakley, January 23, 2014. Wikimedia Commons.

the Americas. Huge quantities of silver were mined by highly exploited indigenous workers (whose prior mining knowledge was also exploited). Most notable were the silver mines at Potosí in present-day Bolivia. Shiploads of silver from the Americas flooded European and world markets, lowering the value of the metal and undermining the profitability of central European mines.[29]

Nevertheless, mining in central Europe continued as an industry fundamentally important to rulers of these lands and their courts, as well as to the population as a whole. As Thomas Morel has astutely demonstrated, central European silver mines became the focus of mathematical practices involving surveying underground to locate veins of ore and to map existing and potential galleries, tunnels, and shafts. This complex practice of "underground mathematics" developed separately from university-based, Euclidean geometry. Yet it formed the basis of growing skill on the part of mine surveyors and fueled a growing appreciation for precision, measurement, and practical mathematics that extended beyond the mines to the culture more generally.[30]

Mining and metallurgy everywhere had huge environmental consequences. As Richard Hoffman notes, for Europe, in many areas mining and metal processes, such as smelting, impacted the surrounding woodlands. Massive amounts of wood fuel were required for smelting, and timber was also needed for supports and props in mine shafts and in other construction. In the sixteenth century, continuously running blast furnaces were huge consumers of fuel. In addition, new mines required the clearance of surface vegetation, and extracting and washing the ores polluted surrounding waters. Where large-scale mining of silver, lead, and copper occurred in areas of the Harz mountains in central Germany, vegetal anomalies exist today that are a result of smeltering a thousand years ago. Soils hold so much toxic copper that some plants will not grow. In the late nineteenth century, large-scale cattle deaths were discovered to be a result of zinc poisoning due to flooding that had eroded medieval zinc waste dumps 60 kilometers (about 37 miles) away. Sites in Germany and England near medieval lead workings today contain lead concentrations ten times beyond modern safety limits.[31]

Such consequences were far from the minds of most premodern people. They instead were focused on acquiring and using the many objects that mining and metallurgy produced: anchors, anvils, aquamaniles, armillary spheres, armlets, astrolabes, basins, bells, belts, bowls, boxes, bracelets, braziers, buckets, canal linings, candlesticks, cannons, cannon balls, caskets, chains, chainmail, chalices, clamps, clasps, clocks, coins, compasses, cranks, crowbars, crucifixes, cups, daggers, dishes, door fittings, earrings, ewers, flasks, forceps, forges, forks, furnaces, gears, glasses, gold thread, grates, guns, hammers, hinges, hoes, horseshoes, incense burners, inkwells, keys, lamps, lampstands, lances, lanterns, latches, linings, locks, medallions, mirrors, mortars, nails, necklaces, needles, oil lamps, pans, pen cases, pins, pipes, pitchers, plate armor, plates, plowshares, pots, railings, relief sculptures, reliquaries, rings, roofing, saws, scabbards, scissors, screws, spears, spoons, statues, stirrups, stoves, trays, vessels, window grills, and wire, to name a few.

Textiles

Textiles were used not only for clothing but also for tents, rugs, wall hangings, and woven tapestries. Rugs were portable and could be moved around to change part of a dwelling from a sitting area to a dining area to sleeping quarters. Nomads and travelers used textiles for bags to carry goods and also for tents. Everywhere people used textiles for cushions and seat covers, beds and bedclothes, tablecloths, napkins, and towels. Clothing was made according to the customs of particular locations and norms of gender, class, and status. The four primary fibers used for textiles were wool and linen, native to the Mediterranean and Europe, and silk and cotton, both originating in Asia. Fibers such as hemp were also used, as well as mixed fibers such as linen and silk blends.[32]

In pre-Islamic times, linen was made in Egypt and was valued by the Arabs as well as by the Byzantine rulers of Egypt. It remained the stable cloth of Islamic lands for centuries. Linen is made from flax, which was grown in upper Egypt. To derive usable fibers from flax, workers dried the stalks, rotted them in warm water (retting), dried them again, and then placed them on stone or wooden blocks and pounded them with mallets (scrutching). They then pulled the fibers through a comb known as a hackle or heckle to remove any remaining woody material and separate the bundles of fibers. They spun the fibers into long threads and wove them into cloth, usually on vertical looms. When the Arabs conquered Egypt, they inherited a flourishing linen industry located in the towns and villages of lower Egypt. Egyptian weavers produced both ordinary linens and fine linen cloth, including rich brocades. Linen is a versatile textile, fine enough for veils and strong enough for ropes. Its gray-brown fibers must be bleached, and it is difficult to dye. When decorative colors were desired, workers often mixed the linen with another fiber such as cotton or silk. Egypt was the main center of production in the medieval centuries, although linen was also produced in Syria, Sicily, North Africa, and on the Iberian Peninsula.[33]

Like linen, cotton is a plant product that needs to be processed. During Islamic rule, Iraq, Syria, and Yemen in southwest Arabia became major centers of cotton production. Cotton was used for making cloth but also for stuffing quilts, mattresses, and pillows. Textile workers began by beating the "bolls" to remove the seeds, after which they prepared and spun the fibers, wove them into cloth, and then dyed and finished it. Cotton is versatile and could be made into a variety of fabrics from thin gauze to thick fabrics similar to wool. Both linen and cotton rags constituted the essential raw materials for the fabrication of paper. Cotton came to European lands from the eastern Mediterranean and was established as an industry in numerous northern Italian cites, especially in the north, from the early twelfth century. From there it spread north to French and German cities.[34]

Silk was a luxury fabric. For the Byzantines, one of the most important industries was silk manufacture. Silk was a highly valued commodity that could also serve as a symbol of authority. The highest grades of silk clothed the imperial family and its court, as well as church officials. Scholars debate the precise origins of Byzantine sericulture—the name given to raising silkworms. In any case, by the seventh century silk production was a principal feature of the Byzantine economy. Women carried out a significant part of the labor, including moriculture (the cultivation of the mulberry trees) necessary for the raising of silkworms. Silk production took place in various parts of the empire, such as Syria, Asia Minor, southern Greece, Italy, Phoenicia, and Egypt, but the heart of the industry was Constantinople, where silk cloth was made in both imperial factories and private workshops.[35]

Silk thread comes from the cocoons of silkworms. Moriculture, sericulture, and the production of high-quality silk yarn were complex procedures requiring great skill. Once silk workers obtained the thread from the cocoons, they twisted it, using a silk reel. After twisting, the silk could be dyed before putting it on the loom. Then it was ready for the imperial workshops or private silk weavers or for wealthy individuals who had weaving done for them in their own homes. A weaving workshop required carpenters to make and repair the looms, pattern makers to create the motifs for weaving, draw boys to operate the device that produced figures on the woven cloth, embroiderers, printers, and finally, tailors of silk clothing. There were probably different types of draw looms for different patterns and kinds of weaving; their precise structure and development remains a topic of ongoing study and discussion.[36]

When the Arabs conquered Syria, they took over the Byzantine silk industry there and expanded it. Sericulture gradually spread throughout Islamic lands. By the tenth century Islamic silk-weaving centers were distributed across the Mediterranean. Much of it was carried out in family-based cottage industries, although there also existed state workshops. Women were involved in all stages of silk production and monopolized the raising of silkworms. Between the ninth and twelfth centuries, Italian merchants brought Byzantine silks to Italy and Western Europe. In 1147, the Norman king of Sicily, King Roger (ruled 1130–1154) established silk weaving in Palermo, using both Byzantine and Islamic workers. In the twelfth and thirteenth centuries, silk weaving (based on the purchase of raw silk) was established in Lucca and Venice, and then spread to other centers. In Lucca, silk entrepreneurs carried the product to winders, spinners, dyers, and weavers, each who worked in his or her own separate workshop. In Venice by 1278 a guild of silk weavers was working to strict specifications. The Venetians purchased cocoons, women unraveled them, and men reeled the silk. Specialized workers degummed the silk, others prepared the warps for weaving. In Paris, as Sharon Farmer has shown, silk manufacturing arrived in the late thirteenth century and flourished for more than a century. It was founded on the skills of immigrants from northern Italy and, for complex reasons, it provided some women with higher status and pay than most female workers in Europe.[37]

The manufacture of wool cloth became the most important industry of medieval Europe. Most stages of production were carried out by specialists: shearing the sheep and cleaning the wool, combing or carding the wool, spinning, putting the yarn on the loom, weaving, dyeing, fulling (a process of strengthening the cloth by soaking it in certain substances and trampling or beating it), and finishing.[38]

Over time, the number of production methods increased. First combing was used when preparing wool and later carding was introduced. For most of the ancient and medieval centuries, workers prepared wool by combing it. Workers used two combs. They fixed one on a post or on their knee and placed the wool on its teeth, then they repeatedly pulled the second comb through the wool, disentangling it and separating long from short fibers. Carding was introduced as a second technique, probably borrowed during the late thirteenth century from the Islamic cotton industries of Spain or Sicily, and it was used for short-fibered wool. The card was a rectangular wooden board fitted with hundreds of short wire hooks. The worker used two cards, working the hooks in opposite directions on the wool, disentangling it but without separating the longer from the shorter fibers. It was very effective in combining different kinds of wools and imparted a felting quality to the wool, but it was initially controversial and often banned, probably because it allowed shorter fibers to be mixed with longer ones, thereby degrading the quality of luxury broadcloths.[39]

Spinning, which was accomplished by women, involved drawing out the fibers, twisting them together to form a continuous yarn, and winding them onto a spindle. In hand-spinning, spinners used a distaff, a tool that held the raw wool on a forked end and included a lower end for winding the spun yarn, and a spindle whorl, a wooden disc attached to a rod that served as a flywheel, allowing spinners to turn it rapidly as it moved toward the ground while they twisted the wool. The spinning wheel was introduced in the later twelfth or early thirteenth century. It increased productivity threefold but initially produced inferior yarns. The new machine was gradually improved, allowing it to be used more efficiently to produce more evenly wound yarns. Improvements included a mechanism to control the tension and a foot treadle, freeing the spinner's hands and allowing her to produce high-quality yarns.[40]

Once the yarn was spun, workers put it on a loom for weaving. They placed the warp yarns first, forming the foundation. Then, they passed the weft yarn back and forth through the shed (the opening separating the yarns of the warp). Throughout ancient and early medieval Mediterranean and European lands, weavers producing both linen and wool cloth used an upright, warp-weighted loom. They stretched the vertically hanging warp yarns (the foundation yarns) with weights of stone or other materials. The warps were divided by a shed rod, so that weavers, using a heddle, could pass the weft yarn through the shed. An important innovation was the introduction of the horizontal treadle loom, which probably entered Europe from the Byzantine Empire or Muslim Spain or Sicily (where it was used for producing

Figure 5.4. A. Byzantine image of woman weaving with horizontal loom (left) and woman spinning with a hand spindle (right). Bibliothèque Nationale de France, Paris, Grec 134, eleventh century. Photo by Maxim91. Wikimedia Commons. B. Woman spinning with a spinning wheel (left) and woman carding with two hand cards (right). Luttrell Psalter, fol. 193, lower margin, 1325–1335. British Library, London / Art Resource.

cottons in the early eleventh century). The horizontal loom produced a better-quality, more densely woven cloth; could produce longer cloths; and increased productivity. The limitation of the horizontal loom—its inability to produce cloth wider than the arm stretch of a weaver—was overcome with the development of the horizontal broadloom, which was operated by two weavers sitting side by side.[41]

After weavers had made the cloth, further specialists finished it. Finishing in-

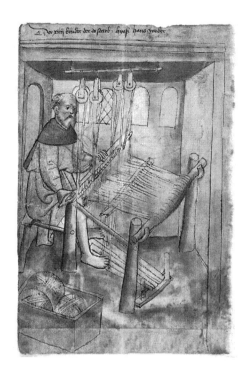

Figure 5.5. Monk weaving on a horizontal loom. Germanisches National Museum, Nuremberg, Hausbücher der Nürmberger Zwölf-brüderstiftungen [Housebook of the Nuremberg Twelve-Brothers Monastery], Nuremburg City Library, Nuremberg, Amb. 317.2, fol. 4v (Mendel I), c. 1425. Photographer unknown. Wikimedia Commons.

cluded fulling to clean and thicken the fabric. Fullers soaked the cloth in a barrel with fuller's earth (a clay-like material containing hydrous aluminum silicates among other substances) and usually urine. They trampled it, hung it and brushed it with teasels (the prickly seedhead of the teasel plant), then put it back in the barrel and trampled it again. This process shrunk and felted the wool, making it far stronger. Gradually this labor-intensive process was replaced by the fulling mill. With the mill, fulling became one of the first mechanized industrial processes in Europe. The mill included a large, two-story building and was powered by a water-wheel, either overshot or undershot, which rotated a drum that lifted two heavy, oak trip-hammers one after the other and then released them to drop with great force on the woolen cloth in a trough below. Fulling mills spread throughout Europe in the twelfth and thirteenth centuries.[42]

Part of the finishing process involved raising the loose fibers of the cloth (the nap) with teasels to enable shearing. In fifteenth-century England, this process was mechanized by the introduction of a water-powered gig mill. In this machine, the teasels were set into a rotating cylinder that was dragged across the surface of the cloth, raising the nap as it moved on a revolving belt. Cloth makers used the gig

Figure 5.6. Fulling mill. Note the cams on the shaft turned by the waterwheel that lift the fulling hammers, which then fall on the cloth. Vittorio Zonca, *Novo Teatro di machine et edificii per varie et sicure operationi* (Padua: P. Bertelli, 1607), fol. 42v (detail). Courtesy Linda Hall Library.

mill for cheap cloths, but they continued to hand finish fine woolens, fearing that the gig mill would damage them.[43]

Dyeing was a specialty unto itself, which could be accomplished on the raw wool, on the yarn, or on the cloth (usually after the final shearing). Most dyes, which included those made of various plants, mollusks, or insects, required a mordant to fix it. The most important mordant was alum, which was found almost exclusively in Phocaea on the western coast of the Anatolian Peninsula, and after 1275 was a Genoese monopoly until a new supply was found in the 1460s in Tolfa in central Italy.[44]

As textile manufacture changed from a local craft industry to one that produced commodities for long-distance trade, labor and power relationships changed to the disadvantage of women. Women remained involved in spinning and other tasks within a "putting out" system in their own homes, paid (very poorly) by the piece. Men, especially drapers and wool merchants, came to control most of the other parts of the process. The industry came to be organized into four major crafts—weaving, fulling, dyeing, and finishing. Drapers, who were usually men but could also be women, were key entrepreneurial actors. They bought the wool, had it sorted, beat, washed, and greased by their own employees, or the wool merchant's. They then "put it out" to combers or carders, collected it, and distributed it to spinners, both those who spun by hand and those who had spinning wheels, almost all of whom were women working in their own homes with their own equipment. They collected the yarns for the warpers and winders, also mostly women who worked in the draper's workshop. Warpers set up the warps on the loom; winders inserted the weft in the shuttles. The draper was a master weaver but also employed other weavers, usually apprentices and journeymen.[45]

Tailors and other skilled workers made cloths into clothing, as well as other products such as bed linens. Making clothes and accessories could be highly specialized. For example, as Carole Collier Frick has shown for Florence, many specialists contributed: male and female tailors; craftswomen who made headcloths; hosiers who made body hose, stockings, and socks; undershirt/blouse makers; hat makers, including specialists making straw hats and specialists making felt hats; seamstresses who sewed up cloth cut by tailors; cap makers; doublet makers (a doublet was a tightly fitting, upper body garment worn by men); hat foundation makers; glove makers; embroiderers; menders of torn clothing; buttonhole makers; and veil makers to name just a few. This range of specialties was not unusual. Also typical were those who fixed and sold used clothing and rags.[46]

While this example is from Florence, cloth and clothing did not remain local products. Clothing was bought, sold, and exchanged over long distances, creating shared cultures of materials and clothing. Recent investigations of marriage and dowry documents from between the tenth and twelfth centuries, for example, shows a shared cloth culture between southern Italy, Egypt, and the eastern Mediterranean.[47]

Ceramics

Throughout the Mediterranean and Europe, artisans produced ceramic objects—including drain pipes, roof tiles, tableware, cooking pots, and storage vessels. The study of ceramic fragments (which often constitute archaeological markers for both time and place) has been key to understanding commerce and interchange over vast distances.[48]

Ceramic is clay (of which there are many varieties) that has been transformed by firing into a rocklike, dense material. Potters historically have made three main kinds of ceramic objects. The most common is earthenware, or common clay, familiar to us through planters, pipes, and tiles. It is any clay that becomes ceramic at about 750 to 900 degrees Celsius. It often has a reddish tint, caused by traces of iron, and in premodern times was used extensively for cooking, water storage, and aspects of building construction such as roofing. The second is stoneware, which is fired at about 1,000 degrees Celsius and is harder and more durable. It consists of a stonelike, fully vitrified substance that is impervious to liquids, and has long been used for food storage and service. Stoneware spread from China as kiln technology improved. The third kind of ceramic is porcelain. It consists of china clay or kaolin and petuntse (a type of limestone called porcelain stone or china). Porcelain originated in China and was exported in great quantities from the ninth century. In the Mediterranean and Europe, it was greatly admired and imitated—but not successfully fabricated until after 1600.[49]

Potters glazed much of their wares. Glaze is a kind of glass made to fuse with ceramic in the firing. It consists of mostly silica, some clay, colorants such as iron or copper, and a fluxing agent to lower the melting temperature of the silica and other ingredients.[50]

Byzantine potters continued ancient practices, including the use of lead-based glazes. In general, these potters created their wares for everyday use not as a luxury product. Pottery was often distributed by commercial ships, making shipwrecks an important source of knowledge. Other sources include the archaeological investigation of Byzantine kilns and the dumps surrounding them.[51]

Potters used clays that they had collected from riverbanks. In Islamic lands, by the twelfth century, they were also making fritware from a stone paste consisting of sand, quartz, and other substances mixed with water. Potting techniques included modeling slabs and coils of clay by hand, pressing clay into molds, and throwing, that is shaping clay on a potter's wheel. Shaped objects were left to dry in the sun and then fired in a kiln. Particularly important for the dry climate of the eastern Mediterranean were earthenware vessels for the storage of water, called *habbs*, some of which were highly decorated. Another type of vessel was called a *gadus*—the pots used for the wheel of the *saqiya*. Potters could apply a glaze to their vessels or tiles, which functioned as a seal. They made glazes by using red lead, which lowers the temperature at which the glaze mixture turns to glass. In the twelfth century, an alkaline glaze similar to glass was developed made from a mixture of soda or potash. Glazes could be colored and decorated, techniques leading to the development of fine ceramic wares.[52]

Islamic potters developed distinctive methods for decorating pots and other ceramic objects. For example, in the Abbasid dynasty (750–1258), in one method of decorating glazed ceramic objects, artisans painted designs onto opaque white glazed vessels, using a cobalt-pigmented blue glaze and then firing it. This opaque white glazed ware was widely copied in China and then in Europe.[53]

Potters in northern Europe and England were also influenced by ceramics from the East. Chinese porcelains arrived in Europe in the fourteenth century and proliferated thereafter, inspiring many attempts at imitation. As everywhere, pottery making required digging the clay and making objects by hand, in molds, or on a pottery wheel. Potters turned their wheels using one hand or they used a kick wheel, which left both hands free. To make tiles, potters put clay into molds or forms on a flat surface, which they then stacked to dry. They often glazed unfired tiles with lead combined with other materials for color. For example, the addition of copper produced a green glaze. The fabrication of fine tiles for floors and walls developed particularly in Spain under Moslem influence. In England in the fourteenth and fifteenth centuries, there is evidence for an increased use of floor tiles for manor and town houses and for royal buildings. Kilns were fired with wood, coal, or peat.[54]

The importation to Italy in the eleventh and twelfth centuries of tin-glazed pottery from the Islamic world brought the development of maiolica—highly decorated, tin-glazed painted and incised plates, pitchers, and other wares. Originating in al-Andalus, the art of maiolica migrated to the Italian Peninsula, where numerous centers of production emerged primarily in central Italy. New glazes and colors were added to enhance the historical and mythical scenes painted on the pottery,

Figure 5.7. A. Playing card depicting female potter making a pot on a wheel. Wood-cut. Kunsthistorisches Museum, Vienna, Hofämterspiel, Kunstkammer 5105. © KHM-Museumsverband. B. Potter making clay pots on a wheel, and man digging clay and a furnace for firing pots (background). Woodcut by J. Amman. Single Sheet, Welcome Collection 34967i. Courtesy of the Welcome Institute, London.

making these ceramics a popular luxury item both locally and as a commodity in long-distant markets.[55]

Glass

Glass manufacture was a major industry of the Greco-Roman and Byzantine worlds, as it continued to be in Islamic and European lands. Glass is made by firing a mixture of sand and soda or potash with the addition of metal oxides for coloring. Glassworkers first made a glass-paste and then made objects with simple tools, including a blowpipe made of cane, iron, or bone, pincers to hold the material, and shears for cutting it. They also used a solid iron rod called a pontil for holding the object while they shaped it. Glass kilns had various shapes depending on their specific purpose. In Byzantium, for example, objects made from glass included many kinds of receptacles, such as bowls and cups, flat glass for windows and tiles, and glass jewelry. Glass tesserae (tiny pieces) for mosaics were used extensively all over the empire from the sixth to the fourteenth century. As we know from many

Figure 5.8. A. Albarello (ceramic medicinal jar) with decoration, painted blue and white with glaze, Syrian, 1400–1450. Louvre, Département des Arts de L'Islam, UCAD 4288. Photo by Marie-Lan Nguyen, 2006. Wikimedia Commons. B. Blown glass vase, gilded and enameled, Egypt or Syria, 1310–1330. Corning Museum of Glass, Corning, New York, No. 55.1.36 (detail). Photo by Andre Carrotflower, February 12, 2022. Wikimedia Commons.

sources, including the cargo of the "glass ship," at Serçe Limani, mentioned in chapter 4, glassmakers sometimes started with broken glass (cullet) rather than the raw materials of silica, lime, and soda.[56]

Around these basic methods, glassmakers developed further techniques. In an early Islamic procedure, indebted to prior Greco-Roman traditions, glassmakers produced thinly blown, clear glass vessels decorated with finely spiraled or pinched threads. They also produced lustre-painted glass bowls. In another technique, they used molds with counter-sunk patterns, usually made of clay or wood, into which they poured the molten glass. In a third technique, glassmakers initially created a clear blank, around which they blew a gather (a viscous melted batch of ingredients).

They created a design in relief by cutting away the surrounding surface. In twelfth-century central Anatolia, a tradition of enameled and gilded glasswork developed, which was greatly admired and copied by Crusaders, who carried the objects and the technique itself back to Western Europe. This Anatolian glasswork influenced Venetian glassmaking.[57]

Venice became a center of European glassmaking. From the thirteenth century, the industry was located on the island of Murano. Murano glassmakers created crystal (*cristallo*) in the fifteenth century, as well as mirrors, and a white-colored or "milk" glass (*lattimo*). It became the center of European luxury glass. As Francesca Trivellato has noted, although women were excluded from the guilds of skilled glassmakers on the lagoon island of Murano, they were the mainstay of glass bead manufacture in Venice itself. Across Europe glassmakers fabricated vessels of all kinds, distillation equipment for alchemical and medicinal purposes, and windows. Scientific analysis of glass remains from the archaeological investigation of a glass furnace in Comacchio in northern Italy (an important early medieval port town on the Adriatic Sea south of Venice) shows that glass producers used recycled glass, much of it from the eastern Mediterranean.[58]

Stained glass windows flourished as a major art form in Europe from about the twelfth to the sixteenth century. They became notable features of Gothic cathedrals but were also used in lay buildings such as town halls and palaces. As Lech Kalinowski has pointed out, stained glass windows are a part of architecture. Their existence is dependent on external, natural light shining through them. To make the windows, glaziers blew the molten glass into the form of a cylinder, cut it along its length, and flattened it into a sheet. Or they transferred the blown glass on the blowpipe onto a pontil iron and spun the iron rapidly to produce a circular sheet. Sometimes the color of the glass was imparted by adding oxides such as manganese, cobalt, and copper to the molten glass. In the earlier centuries, glassworkers drew a working drawing on a whitewashed table, placed a sheet of glass on it, and followed the drawing as they painted the glass and cut the lead. Gradually, cartoons (designs drawn on paper) or patterns were used instead. The glaziers cut the glass into the desired shapes, painted pigments onto the pieces, and then fired them in an annealing furnace. They then joined the colored pieces together with lead. They completed the entire window before inserting it into the window opening.[59]

Leatherwork

Leather is the preserved hide of animals that can be divided into two categories—hides, taken from the skins of large animals such as cattle and horses, and skins, taken from smaller animals such as pigs and goats. In both cases, the material consists of three layers: an outer layer of hair, the main skin structure called the corium, and the subcutaneous or flesh layer. Leatherworkers made the corium layer into leather by a process of tanning. Leatherworkers usually were divided into those that turn skins and hides into leather, and those that turn the leather into articles for sale. (However, in Scandinavian lands, leatherworkers often prepared their own leather before making it into objects.) Tanners tanned leathers from hides, and curriers worked the hides into various kinds of leather suitable for different objects. Specialists making one or more objects included shoemakers, who made shoes and boots; cobblers, who repaired footwear, sometimes for resale; belt makers; purse makers; harness makers; bottle makers; glovers (many of whom were women); parchment makers; sheathers, who made sheaths for knives; scabbard makers, who made scabbards for swords; and saddlers. Saddles, which were designed to take the weight off the horse's spine, consisted of a wooden structure (called a tree) covered with leather. Leatherworkers and carpenters worked together to make saddles.[60]

Tanners obtained their hides from butchers and then washed them, often in streams or rivers. They removed the hair either by letting the skins putrefy and then easily pulling it out, or by folding the hides and soaking them in pits filled with lime or urine. They then scraped the hide. They tanned it in another pit, using a vegetable tanning agent, usually oak bark. The hide remained in the pit with the tanning agent for six months to two years. The tanner then dried the leather and usually sold it to a currier who converted it into the kinds of leather needed for various objects. Artisans distinct from tanners and curriers worked lighter leathers for gloves and other soft objects, using different processes. In England, they were called whittawyers or tawyers. These artisans treated the skins with alum and oil rather than oak bark.[61]

Undoubtedly, the most ubiquitous leather product was shoes. Historic leather is preserved only in anaerobic conditions, but where such conditions prevail, archaeologists have discovered thousands of shoes and shoe fragments. For example, in Turku (a city located in eastern Sweden in the premodern centuries, now southern Finland) a 1990s excavation produced ten thousand leather artifacts of various kinds, including waste scraps from workshops, many dating from the late fourteenth and early fifteenth century. Most were related to shoes. Other objects included the leather remains of sheaths, scabbards, and grip coverings (leather coverings placed on the hilt or handle of swords and daggers to improve the user's grip). In his detailed studies of these leather remains, Janne Harjula notes that most

Figure 5.9. Adult left foot ankle shoe with a one-piece upper and a one-piece sole, Saxo-Norman, tenth–early eleventh century. Museum of London, No. NN15476. With permission of the Museum of London.

were made with calfskin and sewn with hemp, a plant grown in the same locale. Shoemakers made many types of shoes, such as one-piece shoes (made from a single piece of leather), thong shoes (closed by thongs running externally around the ankle or leg part of the shoe), and side-laced shoes, to name a few.[62]

Most shoemakers made shoes on lasts, wooden blocks in the shape of a foot. They used awls (a tool with a handle and sharp, pointed blade) for making holes in the leather. They also used knives and shears for cutting leather, spindles for twisting the waxed thread (usually hemp) that they sewed with, and sewing needles. Most shoemakers made ordinary shoes, but shoes of elite individuals could be highly ornamented with stamped leather designs, and velvet and braided fabrics, and could be gilded in gold leaf. Gilding leather was the work of specialists.[63]

Decorative Arts and Crafts

Aesthetic pleasure, splendor of display, and religious piety were often evident in the production of the many arts of Mediterranean and European lands. These include portable objects such icons, ivories, and panel paintings, as well as sculptures, both three-dimensional and relief, and wall decorations, including mosaics and paintings.

In the Byzantine Empire, wall mosaics were the most important way of decorating Christian monuments. Mosaics are made of thousands of tesserae, which were usually made of glass, with as many as forty different colors. They could also be made of stone, terra-cotta, or silver. Before workers set the tiny pieces into the wall, they covered its surface with thick layers of lime plaster. They painted the design into the wet plaster, often using as many as four colors, which would be visible through the glass tesserae. Often mosaic workers scrunched precious metals such as gold between the tesserae.[64]

Mosaic painters also worked as fresco painters. As with mosaics, workers prepared the wall with lime plaster, laying it in horizontal courses and including chopped straw in the final layer to increase the adhesion of the paint. They usually put the underdrawing on the plaster with a red earth pigment known as *sinopia*. They applied the paint to the wet surface, but faces and flesh parts of the figures were done last, often *al secco* (dry) because the plaster had dried. The painter used seven or eight basic pigments, which an assistant on the scaffold often combined to make hue variations before the application to the wall. Painters often worked in family teams, the craft being passed down from father to son.[65]

Byzantine artisans who frescoed walls in the summer months, often painted icons (images of sacred figures, painted on wood) in the winter. Icon painters used boards made most often from poplar, elder, birch, lime, or cypress. For larger panels, they joined several sections of wood. The painting required several steps: sizing the panel with a glue made from boiled animal skin and bones, and sometimes then covering it with linen; covering the panel with thin layers of gesso (chalk or gypsum made with glue and water); smoothing and polishing the final layer; making a drawing on the panel with charcoal or incising it into the gesso. Finally, they painted the icons with pigments mixed with an egg-yolk-based medium, or alternatively, in a waxed-based medium called encaustic. Icons could also be decorated with mosaics.[66]

One of the most highly developed Byzantine crafts—also practiced across North Africa and in Europe—was ivory carving. Ivory carvers obtained their raw material from the tusks of African elephants. They created exquisite plaques, boxes, and small images and objects. In Constantinople, they probably worked in their own homes. Many ivories were used as icons of venerated figures, such as Jesus, Mary, and saints. They carved the tusks with a pick-like or chisel-edged tool. Artisans

Figure 5.10. Byzantine ivory depicting the crucifixion, flanked by the Virgin Mary and St. John, tenth–eleventh century, Constantinople. Photo by Walters Art Museum, No. 71.65.

Figure 5.11. Dome of the Rock, interior view, Jerusalem. Photo by Virtutepetens, January 25, 2018. Wikimedia Commons.

carved plaques from the large part of the tusk, applying their tools at different angles and smoothing their work with successively finer files. Detailing required a fine blade or pick for delicate lines such as facial features and folding drapery. Ivory carvers finished their pieces by polishing them to remove tool marks and to enhance contrast between details and background. Sometimes they added coloring. Tenth-century, finely carved ivory pixides (small, cylindrical boxes with knobbed tops used for toiletries), found in al-Andalus, survive in substantial quantities.[67]

In the Islamic world, artisans developed distinctive decorative motifs, and they used such patterns across mediums such as wood, stone, and fabrics. Early examples include the wooden tie beams in the Dome of the Rock (completed in 691) that are covered with sheet metal made of a copper alloy and ornamented with repoussé (low relief hammered in from the reverse side) and displaying vegetal and geometric designs that are gilded and painted. In the thirteenth century a formation known as arabesque developed. It consisted of designs made up of intertwining geometrical and/or vegetal motifs. Artisans used arabesque patterns widely in wood, stone, and stucco.[68]

Wood carving was a highly developed craft. In eleventh-century Egypt, wood was especially valued because of its scarcity, and a lively tradition of wood relief carving flourished. Carved wooden relief panels were often used as decorative elements in buildings. Marquetry, in which artisans attached pieces of dark and light wood and other materials such as bone to a surface to make specific designs, was a highly skilled craft. A major center of wood carving and furniture making was twelfth-century Aleppo in present-day northern Syria.[69]

In Western Europe, sculptors carved images from wood or stone. Much medieval sculpture was architectural in that it was joined with the carved moldings of buildings or designed for specific spaces within them. In addition, tomb sculpture became an important area of sculptural activity. Medieval stone and wooden sculpture was often painted, usually by specialists rather than by the sculptor. Life-size figure sculpture emerged in the mid-twelfth century. Usually, the sculptor carved the figure in his workshop and then placed it in its designated position. The sculptor (and/or the patron) chose the stone based on availability and preference—including various types of limestone, alabaster, and marble (each of which has specific characteristics), and the very hard stone, porphyry. For tomb sculpture, alabaster became a preferred stone. For a freestanding sculpture, sculptors first rough-cut the block with an ax or adze. They then further shaped the stone using points, punches, and picks, which they struck with a hammer or mallet. They used a variety of chisels for differing effects: claw chisels for rapid but controlled removal of material, a flat chisel for carving rigid channels. They also used drills to excavate the stone and to create special effects. They smoothed with rasps or rifflers which were metal tools with rough surfaces, or they used sand or emery. They polished marble and alabaster with pumice. As they did with stone, sculptors of wood chose their material for its properties—softwoods such as pine are easier to carve, whereas hardwoods such as oak and limewood are more difficult to carve but more durable.[70]

In Western Europe, painting—whether frescos on walls, or on wooden altarpieces, or eventually in the sixteenth century, on canvas panels—was based on workshop practice in which a master instructed and was assisted by apprentices and journeymen. Many projects were complex and involved diverse artisans. For example, in making a wooden altarpiece, which could have folding wings (especially in the Netherlands), a carpenter first built the structure. The altarpiece could be carved or painted (or both). Pigments (substances imparting color) for painting required extensive preparation, although some pigments could be acquired at the local apothecary or from a spice merchant. Otherwise, pigments required mixing with a substrate (such as water and treated glue), grinding, and grading. Artisans also made drawing materials by hand—charcoal, by slowly roasting twigs on the embers of a fire; pens, from the quills of goose feathers. They made paintbrushes with the hairs of animals such as squirrels and stoats, or they used hog bristles. To make tempera paint (pigment bound with egg), they either kept chickens or bought eggs from a poulterer. They made glue for size by boiling goat feet or parts of other animals.[71]

After receiving the altarpiece from the carpenter, painters prepared it further by planing it down to make it flat and remove any grease. They filled flaws and cracks with sawdust and glue, beat down nail ends, and covered rough areas with tin foil. They covered the whole surface with animal glue, laid on strips of linen cloth, and smoothed and dried them. They applied layers of gesso. They created designs, transferred them to the surface, and then painted it, either with tempera or with oil paint.[72]

Fresco painting was used as a technique for wall-painting. Painters cleaned and dampened the wall and applied a layer of coarse plaster—a procedure that could be completed several years before actual painting began. The design was usually first drawn on paper (the cartoon). It was cut up into pieces and attached to the wall and used to create an outline of the image on the wall. Painting the wall involved laying down lime and then applying pigment. Lime that dried before being painted was removed, and the process was begun again the next day. As they dried, the pigments became an integral part of the wall.[73]

Until the fourteenth century, paint usually consisted of tempera. It was the main paint medium for icons in the Greek and Russian Orthodox churches. Painters also used it on panel paintings, especially from the thirteenth to fifteenth centuries in Italy, and throughout other regions of Europe. Cennino Cennini (c. 1370–c. 1440), from a town near Florence, described the technique in detail his late fourteenth-century treatise, *Il libro dell'arte*. Artisans needed to apply the tempera thinly to avoid surface cracking. The paint hardened into a waterproof surface.[74]

After the mid-thirteenth century, painters in northern Europe began to use oil paint, mixing linseed or sometimes walnut oil with their pigments. Oil paint could be applied thickly and, because it dries slowly, it allowed blending and soft modeling. Gradually oil painting spread south to the Mediterranean. It was a method well suited to canvas panels (rather than wooden surfaces).[75]

In early fifteenth-century Florence, Filippo Brunelleschi (1377–1446) invented the artist's mathematical perspective, which was developed throughout the next two centuries, along with realism, naturalism, and attention to observation and to classical forms, which comprised what came to be known as Renaissance art and architecture. This development in the arts produced painters, sculptors, and architects who became admired for their ingenuity, some of who became famous in their own lifetimes—Leonardo da Vinci (1452–1519); Michelangelo Buonarotti (1475–1564); Albrecht Dürer (1471–1528); Andrea di Pietro della Gondola, known as Palladio (1508–1580); and Mimar Sinan (1489 or 1490–1588), architect of Suleiman the Magnificent.[76] Some of these men have been made heroes in our own times. Yet they should be understood in the context of their times, not ours. There were also, it should be emphasized, thousands of painters who were not famous, many of whom are unknown to us, at least by name. Famous or unknown, all were trained in workshops. Most sought and found patrons. Many of the best-known, including those mentioned above, not only practiced their craft but also practiced, in various ways, the craft of writing. They lived in a time when the status of certain kinds of practices and handwork was rising, and they were both the protagonists of this elevation and its beneficiaries.

* * *

Crafts and industries produced hundreds of kinds of products and objects. One particular genre consisted of instruments and machines—for the most part constructed out of metal, wood, and glass. It is to these that we now turn.

CHAPTER SIX

Instruments and Machines Including Weapons

I have studied the books of the earlier [scholars] and the works of the later [craftsmen]—masters of ingenious devices with movements like pneumatic [movements] and water machines for the constant and solar hours, and the transfer by bodies of bodies from their natural positions. I have contemplated in isolation and in company the implications of proofs. I considered the treatment of this craft for a period of time and I progressed, by practising it, from the stage of book learning to that of witnessing, and I have taken the view on this matter of some of the ancients and those more recent [scholars]. I was fervently attached to the pursuit of this subtle science and persisted in the endeavour to arrive at the truth. The eyes of opinion looked to me [to] distinguish myself in this beloved science. Types of [machines] of great importance came to my notice, offering possibilities for types of marvelous control.

IBN AL RAZZĀZ AL-JAZARĪ, *THE BOOK OF KNOWLEDGE OF INGENIOUS MECHANICAL DEVICES*

Al-Jazarī was in the service of Nāsir al Dīn, king of the Diyār Bakr—a region above the Tigris River, now a part of southeastern Turkey. Around 1206, he wrote and illustrated a book about mechanical devices. Its fifty chapters, each treating a single device, included water and candle clocks, vessels with complex spouts, fountains, water-raising machines, and automata—machines that appear to be self-operating or with control mechanisms designed to follow a sequence of operations. He explains the construction of these devices in detail and illustrates each one. In the above quote, al-Jazarī is discussing his own progress in the study of such devices and machines, and the passage makes plain his fervent enthusiasm for the topic. It shows that while machines and mechanical devices were useful for practical tasks, they could also have cultural roles, serving, for example, as objects of passion and delight.

Instruments (devices for taking measurements) and machines (externally powered devices for doing work) are integral to most technologies. Many instruments and machines first emerged in the ancient world and were subsequently absorbed and developed in later centuries in Byzantine, Islamic, and Western European cultures. Arab scholars during the eighth and ninth centuries undertook to master the learning of the cultures with which they had come into contact, including the Byzantine and Sasanian (Persian) empires and India. They sought out classical Greek and Hellenistic writings as well as texts in Syriac and other languages, and translated them into Arabic. Often, they added extensive commentaries. Eventually they incorporated the Greek corpus of learning and wrote subsequent works that extended that learning in highly original ways. Translations included Greek and Hellenistic writings on mechanics and mechanical devices, including the pseudo-Aristotelian *Mechanical Problems*, written by a follower of Aristotle, probably in the fourth century BCE. The author defined mechanics as a discipline combining mathematics with physics, or to put it another way, theory (the mathematics of mechanical motion) and practice (actual machines and devices). Writings on mechanical devices included those by Philo of Alexandria (third century BCE) and Hero of Alexandria (second century CE). All such extant writings had been translated into Arabic by the ninth century.[1]

Another ancient work treating machines was the Latin *De architectura* by the Roman architect/engineer Vitruvius. The only complete architectural treatise extant from the ancient world, Vitruvius's treatise included Book 10 devoted to machines.[2]

Beyond these traditions of writing and images, actual working machines were everywhere in use in the medieval and early modern centuries. These are known to modern scholars through documents, tracts and treatises, images, and archaeological investigations of physical remains at particular sites, as well as machines and machine parts now located in museums.

Machinery

Machinery can be discussed in terms of what it was used for and how it was powered. We have already encountered machines powered by humans and animals, by water, and by fire. They include plows, water-lifting machines, and the most common medieval machine—the mill—driven by animals or by moving water. Textile machines include the spinning wheel, the vertical and, later, the horizontal loom, the fulling mill (increasingly common in the fifteenth century), and the gig mill (a machine that raised the nap on cloth as part of the finishing process). Machines for building construction included weight-lifting machines (cranes) that could be powered by humans on treadmills, or by animals such as horses turning capstans. The printing press is a machine, as are kilns and smelting ovens, the latter two powered by combustion.

Another source of power was the wind, already discussed as it pertains to sailing ships, but used as well for windmills. From 1300, windmills (called Persian windmills) in what is present-day Iran were constructed as vertical axis wind turbines, with the sails or blades enclosed in walls so that the wind entered only one side, turning the wheel as in a turbine. The millstones for grinding the grain could be held above the sails or below. In this area of the world between June and October, the wind blows steadily from the north, making such windmills highly efficient during these months.[3]

In Europe, windmills appeared around the twelfth century. There, the winds blow in variable directions, making it necessary to turn the machine toward the wind. Builders constructed two kinds of windmills. In post-mills, the entire structure was mounted on a single vertical post, which the miller turned on a central pivot to face the sails into the wind. In tower mills, the machinery was built inside a fixed structure, with the windshaft and sails mounted on a cap on top, which the miller rotated. The design and material of the sails of windmills, whatever their configuration, were often influenced by the sails of ships. Windmills were not only used for grinding grain but also for such tasks as sawing lumber and draining land (famously in the Low Countries).[4]

In the fifteenth and sixteenth centuries, machines of all kinds increasingly appeared in handwritten notebooks with hand-drawn images, as well as in printed books—an indication of the cultural interest and appeal of machines beyond their usefulness in the practical world. For example, in the 1430s, the Sienese notary Mariano Taccola (1382–c. 1453) created notebooks that illustrated hundreds of machines, including many kinds of mills and pumps. A younger Sienese, the architect/engineer Francesco di Giorgio (1439–1501) wrote treatises on architecture and engineering that included illustrations and descriptions of machines, such as various types of mills. The artist/engineer Leonardo da Vinci (1452–1519) was an acquaintance of Francesco di Giorgio and owned one of his treatises. Leonardo's notebooks contain beautiful images of machines. One of these notebooks, the Madrid Codex I is filled with drawings of machine elements such as gears and springs, accompanied by discussions of how they worked, as well as more theoretical discussions of the principles of mechanics.[5]

In the sixteenth century, a variety of printed books illustrated with images of machines proliferated. The *De re metallica*, by Georg Agricola (1494–1555), included

Figure 6.1. A. Two vertical axis windmills, Sistan, Iran. Photo by MorvaridMeraj, March 29, 2017. Wikimedia Commons. B. Reconstructed post windmill from 1638, originally from Essen, Germany. Open Air Museum, Cloppenburg, Germany. Photo by Heinz-Josef Lücking, September 21, 2008. Wikimedia Commons. C. Tower windmill, Kinderdijk, Netherlands. Photo by Willard84, April 17, 2017. Wikimedia Commons.

full-page woodcuts of mining operations and machinery (see figure 5.1). The engineer Domenico Fontana (1543–1607) made clear the methods used for his successful transfer of the Vatican obelisk from the side of St. Peter's to the piazza in front—by describing them in writing but also by showing in dramatic illustrations the machinery and how it was used. In the second half of the sixteenth century, "theaters of machines" appeared. Authors such as Jacques Besson (c. 1540?–1573), Agostino Ramelli (1531–c. 1610) (see figure 4.6), and Vittorio Zonca (1568–1603) (see figure 5.6) created books that contained hundreds of detailed, engraved images of mills, pumps, waterwheels, cannons, and other kinds of machines, along with descriptive texts. Such books, often dedicated to elite rulers, point to a culture in which machines had come to represent power and authority, as well as the knowledge of the men who created them (at least some of whom were skilled engineers).[6]

Instruments Including Automata and Clocks

The Arabic word *hiyal* describes the whole range of topics entailing mechanics and machinery. Based on Greek sources, an Islamic tradition of mechanical writings emerged during the ninth century. The three Mūsā brothers—Muhammed, Ahmad, and al-Hasan, known collectively as the Banū Mūsā—were famous for their engineering works in Baghdad and Samarra. They were sons of a noted astronomer and traveled to the Byzantine Empire to bring back books for translation from Greek and Syriac into Arabic. They wrote books on mathematics and astronomy, as well as on mechanics. Their *Book of Ingenious Devices* (written around 850) describes six fountains and eighty-three trick vessels including pitchers that intermittently dispense measured quantities, vessels that replenish themselves after a small amount of liquid is removed, and vessels from which a mixture of liquids would pour from separate spouts. Such vessels worked through different combinations of siphons, pipes, valves, floats, pulleys, lever arms balanced on axles, rack-and-pinion gears, cranks, miniature waterwheels, and balances. Each machine was a kind of automata.[7]

Islamic interest in precision instruments extended to astronomical instruments used in observatories built throughout Islamic lands. Skilled astronomers utilized quadrants (an instrument, usually in the shape of a quarter circle, with a sighting mechanism for taking angular measurements of altitude) and other instruments for determining the altitudes of celestial bodies. They also used armillary instruments for measuring the ascensions and declensions of those bodies, and instruments that measured the angular distance between them. Another significant instrument, the planispheric astrolabe functioned as an observational tool that enabled astronomers to determine the time of sunrise and sunset, and the positions of the

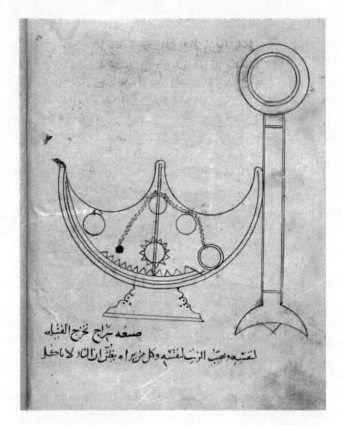

Figure 6.2. Lamp with self-moving wick. The end of the wick (upper left) is moved along a channel at the bottom of the vessel. The float (right) sinks as the oil decreases, causing the rod to turn, causing the wheel at the bottom to move along the teeth, moving the wick forward. From Banū Mūsā Ibn-Šākir: *Kitāb al-Hiyal*. Staatsbibliothek zu Berlin. Preussischer Kulturbesitz, Orientabteilung, Ms. Or. quart 739, fol. 68v.

celestial bodies, including the rising time of fixed stars or constellations. The astrolabe consists of a two-dimensional model of the celestial sphere (assuming that the Earth was the center of the universe). Instrument makers constructed it using stereographic projection—a geometric method in which the points on the celestial sphere were placed on the flat surface of the instrument. Most astrolabes were made for a fixed latitude and could be used only in a single location.[8]

The astrolabe was just one of a group of instruments, in the service of the mosque and the Islamic religion, used to calculate the correct time of prayers, as

Figure 6.3. Back of planispheric astrolabe, ninth century, North Africa. Nasser D. Khalili Collection of Islamic Art, Accession No. SCI 430. Wikimedia Commons.

well as the *qibla*—the direction (towards Mecca) to be faced during prayer. The aspects of practical astronomy relevant to the Islamic religion expanded to a wider domain in the thirteenth century. Astronomers were often employed by mosques and madrasas (schools for Islamic instruction), but they also were independent teachers and scholars. These and other Islamic scholars wrote many books on such topics, often for the purpose of teaching. One example is a book from Mamluk Egypt, a treatise in 120 chapters on instruments by a fourteenth-century astronomer, *Najm al-Dīn al-Mīṣrī* (fl. c. 1300–1350). The book describes various kinds of astrolabes, quadrants, and sundials, each of which is illustrated. It may have been used as a reference manual for astronomy students.[9]

Practical mechanics continued to hold much interest. Three centuries after the writings of the Banū Mūsā, al-Jazarī created his illustrated treatise, *The Book of Knowledge of Ingenious Mechanical Devices*, quoted at the beginning of this chapter. Indebted in part to the Banū Mūsā, most of the devices he describes and illustrates are automata (moving devices that mimic natural forms), water clocks, and water-raising machines. Al-Jazarī's clocks embody numerous mechanical concepts and technologies, including feedback control methods and the use of paper models to establish intricate designs. He completed his work with the explicit intent of enabling later artisans to reconstruct the machines. The work was copied and translated, exerting influence both East and West.[10]

Al-Jazarī's interest in automata was shared by many in both Byzantine and Islamic lands. To impress visitors to imperial palaces, artisans constructed devices such as roaring lions, singing birds with flapping wings, moving beasts, and animated human figures, as well as gushing fountains and clocks. Artisans made moving animals with beaten copper, brass, or wood, which were operated by mechanical means such as gears and pullies, or by hydraulic means using channels, syphons, and valves. Relatively unknown to the West, European travelers to the eastern Mediterranean encountered such automata with surprise and sometimes amazement. Automata reached the Latin Christian West in the early ninth century when the caliph of Baghdad, Harun al-Rashid (ruled 786–809), sent to the Holy Roman Emperor, Charlemagne (ruled 768–814), an elaborate water clock with moving figures. As Elly Truitt has shown in rich detail, thereafter, automata were increasingly discussed in texts and, she notes, constructed in European, especially French-speaking, lands. Automata became part of medieval literature and of courtly life. As inanimate objects that moved by themselves, they challenged the boundaries between life and death, and between art and nature. As seemingly inert objects that played sudden tricks (such as squirting or emitting noises), they existed on the borderline

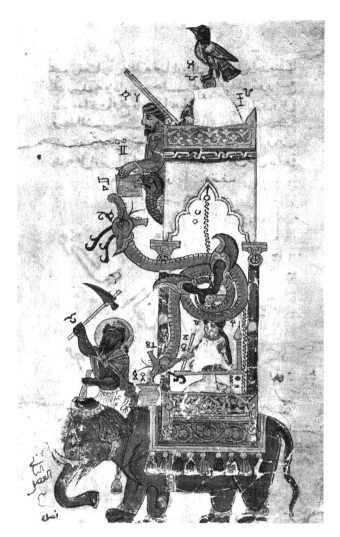

Figure 6.4. Water clock in the shape of an elephant. There is a bucket of water inside the elephant, and inside this bucket, floating on the water, is an empty bowl with a small hole at the bottom; this is the timing mechanism. As the bowl slowly fills with water and sinks (which takes a specific amount of time), the bowl pulls a string that is attached to a see-saw mechanism in the tower above, which releases a ball that drops into the mouth of the serpent, causing the serpent to tip forward. This forward motion pulls the sunken bowl out of the water (to start the process again) and causes the driver to raise his hand and hit a drum, marking the passage of time. From al-Jazarī, *The Book of Knowledge of Ingenious Mechanical Devices*, 1315. Wikimedia Commons.

between the natural and human-made world and between something true and something fraudulent. Therein lay their fascination.[11]

Influenced by Islamic gardens, some elite figures in the West also created extensive gardens featuring automata. The first was constructed for the chateau of Hesdin in Artois near the modern border of France and the Netherlands. Burgundian dukes had the garden designed in the late thirteenth century, and it was expanded in the fourteenth and fifteenth centuries. Automata in the garden included artificial birds and monkeys, artificial people (androids), trick fountains, and a timekeeping device. The mechanisms of the automata, which required extensive maintenance, included cranks and gear trains as well as traditional mill technologies.[12]

A subsequent example of a garden featuring automata was the famous garden of Pratolino near Florence, created in the 1570s by Francesco I de' Medici, duke of Florence (ruled 1574–1587) and his chief architect, Bernardo Buontalenti (1531–1608). The garden was on two levels and was filled with water-powered automata, fountains, and squirting water tricks. One level was above ground where visitors could stroll and become startled and amazed by dramatic automata and water events. The second level was underground where technicians secretly followed the visitors, working the keys, valves, and cranks to activate the mechanisms.[13] Such gardens, depending on their time and location, possessed symbolic value. Their effects seemed magical to visitors and enhanced the power and authority of the rulers and other elite individuals who designed them.

Clocks

Clocks included sundials, water clocks, and mechanical clocks. Sundials, which have been in use for more than five thousand years, could be made in various forms. They were constructed with a flat plate with hour lines (the dial) and a gnomon that casts a shadow onto the dial. The shadow moves through the hour lines as the Sun appears to move through the sky. Sundials were ubiquitous in the Mediterranean world. Their disadvantage was that they could not be used at night or on cloudy days.[14]

The water clock (clepsydra) measured the passing of time by the gradual, evenly regulated flow of water from one container to another over a set period of time. The water clock took many forms, some of which were described in detail by al-Jazarī and in other Arabic books. (Precise timekeeping was important in the Islamic religion because of the requirement for prayer at five specific times of day.) Water clocks could be small or monumentally large. Many water clocks in Islamic lands were accompanied by automata, such as chirping birds. In Western European lands, water clocks are recorded by the tenth century. Monastic establishments, needing to establish the hours for the eight daily prayers required by the Benedictine Rule, often maintained water clocks, sometimes with attached bell-striking mechanisms. By the twelfth century, there was a substantial market in Europe for water clocks (which probably arrived via al-Andalus). They had their drawbacks—the water could freeze in the winter and evaporate in the summer, rendering them inaccurate or inoperable.[15]

A time-measuring instrument that operated by the same principle as the water clock, but used sand, was the hourglass, also called the sandglass. A seemingly simple device, the sandglass measured specific units of time, whether a fraction of an hour, a full hour, or periods longer than eight hours. The instrument was used widely in Europe—city councils employed them to determine the start of council meetings, the amount of time that could elapse before late council members would lose the fees they received for attendance, the time that elapsed before they were fined, and the time spent on each item of the meeting agenda. In other contexts, sandglasses regulated rounds of tournament games and the length of sermons. They measured the hours that miners worked underground and limited the time in which an individual could be subjected to torture. In addition to its many practical uses, the sandglass became a powerful iconographic symbol for the passing of time.[16]

The mechanical clock, indebted to older traditions of water clocks and automata, appeared in the Latin West in the late thirteenth century. It was weight driven and worked by means of a verge and foliot regulating mechanism, which allowed the gradual descent of the weight and thus the slow motion of the clock. Early mechanical clocks were large iron mechanisms that were installed in towers. In the fourteenth century, a clock that struck the hours amazed the citizens of Milan. In Padua, the father and son Jacopo (1290–1359) and Giovanni (1330–1388) da Dondi, both physicians, created an astronomical clock for the palace of the ruling Carrara family. It displayed a calendar showing the holy days and it told time. By the late fourteenth century many cities possessed public clocks. They were a source of civic pride and often featured moving figures. By the mid-fifteenth century, a spring mechanism was adopted to the clock, replacing the weight-driven clock. Springs allowed greater accuracy and permitted the development of smaller clocks, usually beautifully ornate objects that ran for weeks and allowed the development of timepieces that could be worn on the body (i.e., watches).[17]

Figure 6.5. Mechanical clock that works a bell in the tower, Salisbury Cathedral, c. 1386, restored 1956. Photo by Rwendland, September 22, 2012. Wikimedia Commons.

The famous Strasbourg astronomical clock, in the (present-day) French city, was built between 1571 and 1574 and can still be seen today. It demonstrates how complicated these structures could be. The sixteenth-century clock replaced an older fourteenth-century version. The construction of the new clock was overseen by Conrad Dasypodius (1529/1531–1601), a professor of mathematics and scholar of ancient architectural and mechanical writings, including those of Vitruvius and Hero of Alexandria. Dasypodius wrote a detailed description of the mechanism and its construction. The clock comprised an elaborate set of images, automata, and instruments. It included a table of eclipses for the years 1573–1605, an astrolabe, a clock that struck the hours and quarter hours, a carillon that played every four hours (after which a rooster crowed twice), a celestial globe, an ecliptic ring (showing the Sun's apparent path around the Earth during the year), a hand for the seven planets and the moon along with zodiacal signs, and a calendar disk showing the months and days with letters for Sundays and the saints' days.[18]

The vogue for mechanical clocks signaled changing cultural attitudes towards time. Mechanical clock time was marked in equal segments through night and day, regulating human life in new ways. This new concept of quantified hours did not replace the more traditional human experience of daily time regulated by sunrise and sunset and the seasons. Nor did it replace the traditional view of cosmic and divine time in the Christian universe. Rather, traditional and new concepts of time intermingled, influencing culture, literature, and daily life in complex ways.[19]

Navigation and Navigational Instruments

Navigation in the broader sense can mean traveling in a vessel on water, but in the narrower sense, used here, it refers to finding the position of the ship and guiding it toward its intended destination. This task required specialist skills, including knowledge of land and sea breezes, of wind patterns, of sea currents, of coastlines, and it required understanding the meanings of qualities of the water (such as its color) and knowledge of birds and fish. It often required a sounding line to measure the depth of the water. Sailors also used the rising and setting of the stars to determine their latitude. Since antiquity, mariners had relied on sightings of the North Star to determine the four cardinal points—north, south, east, west.[20]

The magnetic sea compass originated in China, was acquired by the Arabs, and arrived in Western Europe by the late thirteenth century, although the precise chronology is by no means clear. Also, by the thirteenth century, portolan charts came into use; they provided characteristics of the coast, shoreline, and shoals and other obstructions on particular sailing routes. These charts included wind roses

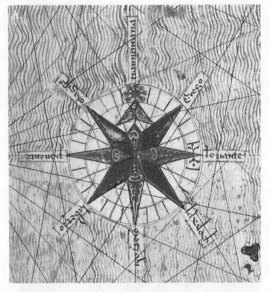

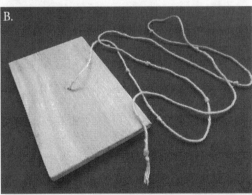

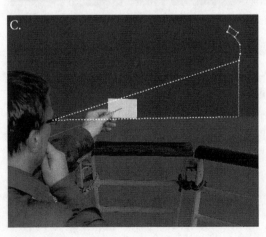

Figure 6.6. A. Compass rose. From *Mapamundi: The Catalan Atlas of the Year 1375* (detail). Newberry Library, case oversize G1026.M37 1978. B. Wooden *kamāl*, replica of instrument used in medieval Arabic navigation. Instrument and photo by Bordwall, May 27, 2015. Wikimedia Commons. C. Demonstration of how a *kamāl* was used when aiming at the North Star to determine latitude. Photo by Markus Nielbock, June 3, 2016. Wikimedia Commons. D. Using the cross staff (Jacob staff) to find the altitude of the sun. From John Seller, *Practical Navigation, Or an Introduction to the Whole Art*, 7th ed. (London: J.D. and Richard Mount, 1694), 162.

showing the cardinal directions (north, south, east, west) and intermediate points. The Arabs used an instrument called the *kamāl* to measure the altitude of celestial bodies from the horizon or sea level, and thereby establish latitude. It consisted of a plate of wood, through which a knotted string was threaded. The navigator held it at fixed distances from the eye, using the knots on the string, and used the height of the plate to measure angular altitudes of celestial bodies. In addition, in the Mediterranean and Atlantic, two instruments came into use in the mid fifteenth-century to establish latitude at sea. Mariners used the cross-staff, also called the Jacob's staff, and the mariner's astrolabe to determine the elevations of the North Star above the horizon and the altitude of the Sun at noon. Both required charts that made numerical adjustments to the numbers read off the instruments.[21]

Oceanic voyages gave particular impetus to the development of navigational instrumentation and also encouraged intense communication between practiced navigators and university-educated mathematicians and scholars. The voyages posed challenges quite different from more traditional Mediterranean and North Sea navigation. Early oceanic voyages were fraught with danger. As ships left familiar coastal waters for unknown seas, the traditional charts and known landmarks that guided coastal navigation could no longer be used. In the early fifteenth century, the Portuguese began efforts to circumnavigate the African continent. Soon Portuguese astronomers were teaching navigators position-finding by the identification of celestial bodies such as the Sun and the North Star. (South of the equator, where the North Star can no longer be seen, the star constellation called the Southern Cross was used). Latitude could then be determined with the aid of tables, such as one showing the Sun's declination throughout the year.[22]

The discovery of magnetic variations in the mid-fifteenth century and the development of instruments and techniques to measure that variation at sea in the sixteenth aided in locating latitude. Yet sea charts remained inaccurate because they were made by transferring geometric projections onto a flat surface and failed to take account of the curvature of the Earth. The problem was solved in the mathematical work of Pedro Nuñez (1502–1578) of Portugal, who demonstrated that on a sphere, a rhumb line (a line of constant compass heading) is not straight as it is on a plane, but is a spiral ending at the pole. Rhumb lines cut all meridians (imaginary north-south lines on the Earth's surface that connect the geographic poles) at a constant angle. It was a Flemish instrument maker and cartographer, Gehard Mercator (1512–1594) who drew such lines for the first time on a terrestrial globe. In 1569, Mercator published the first sea chart in which the directions were true.[23]

As navigation gradually became a mathematical art, it also increasingly came under the purview of institutions, such as the Casa de la Contratación in Seville and the Armazéns da Guiné e Índia in Lisbon. It was taught as a discipline in classrooms and became a topic of books. Conflict arose concerning the value of practiced navigators versus those that learned navigation and its relevant mathematics in the classroom. One side saw hands-on navigators, those who had learned at sea in apprenticeship kinds of positions, as the highly skilled and preferable officers at sea. Others argued that an education in navigation, cartography, and mathematics in the classroom was the preferred background for navigators. It was an argument that continued into the eighteenth century. Despite which side the officers on any particular ship were on, it should not be thought that Mercator's chart was immediately put to use. In the centuries that are the focus of this book, navigators used charts and instruments, but they just as often used the tacit knowledge that they had gained by long experience at sea.[24]

Practical Mathematics and Artisanal Practitioners

Measurement, precision, and practical mathematics became increasingly significant in the sixteenth century in navigation but also in other disciplines such as architecture and military arts. Around mid-century, the practical mathematician Niccolò Tartaglia (1499/1500–1557) invented both the gunner's quadrant (*squadra*) to assist in aiming cannons, and a surveying instrument to measure distance and height (of towers for example), by means of sighting and triangulation. He was one of many who created a variety of new sighting instruments and surveying compasses (and wrote about them). Near the end of the century, Galileo Galilei (1554–1642) invented a military compass known in modern terminology as the sector. The value of precision measurement was increasingly appreciated, as was the artisanal skill required to make such instrumentation. Cities such as Nuremberg in south Germany became well known for their instrument and clock makers.[25]

In other cities, such as London, instrument makers' shops became locales for substantive conversations between the skilled instrument makers and natural philosophers as the philosophers made their purchases. As Jim Bennett especially has pointed out, mathematical practitioners and mathematical instruments cannot be easily separated into categories related to modern disciplines, such as astronomy, navigation, and cartography. Rather the instruments were the tools of mathematical practitioners who could use them for a variety of tasks, such as surveying, architecture and building construction, latitude finding, cartography, astronomy, and navigation. Mathematics was used in a variety of disciplines and practices. Claiming to be a mathematical practitioner could also raise social status, leading some to claim to be such, even when they did not actually use mathematics much in their practices.[26]

Mathematical practitioners are at the center of a thesis about the influence of artisans on the "Scientific Revolution" of the sixteenth and especially seventeenth centuries. The "Scientific Revolution" (an idea being scrutinized and questioned by historians of science, thus the quotes) refers to the changing approaches and views of the natural world as developed by Copernicus, Vesalius, Galileo, Descartes, and Newton, to name just a few of its heroes. The influence of artisans on this "revolution" was importantly argued by Edgar Zilsel (1891–1944) in a series of articles written in the 1930s and early 1940s. Zilsel argued that the values and practices of what he called "superior artisans"—observations, experiment, precision measurement, and hands-on practice—influenced the development of new approaches to the study of the natural world that led to modern science.[27]

Zilsel's thesis was ignored for decades (in part because of its Marxist orientation) but then became highly influential while also being modified and, at the same time, broadened. As Leslie Cormack has noted, changes in the ways sixteenth- and seventeenth-century people investigated the natural world were related to social and cultural changes. Pamela Smith has analyzed, in depth, both artisanal objects and artisanal writings. She and her students have replicated in the laboratory craft procedures derived from the instructions in an anonymous, sixteenth-century artisanal and technical manuscript. This work has led to far better understanding of what such texts were saying and what their authors understood about what they did. She has coined the term "artisanal epistemologies" to point to the knowledge that skilled artisans possessed. In my own work, I have pointed to "trading zones" between artisanally trained and learned (usually university-trained) people in which each had knowledge and skill that could benefit the other. The scholarship of Henrique Leitão and Antonio Sánchez [Martínez], among others, have greatly increased our knowledge of practitioner/learned interactions within Portuguese and Spanish oceanic navigation.[28] These developments were predicated on the rising status of objects, a rising appreciation of the skill involved in making them, and the rising social status of certain kinds of makers such as architect/engineers, painters, sculptors, instrument makers, and skilled navigators.

Machines of War

Warfare was endemic during the centuries that are the focus of this book. Military technologies and organization differed between Byzantium, Islam, and the Latin West, but capabilities were often more or less evenly matched. As combatants confronted one another, they sometimes shifted technologies and strategies, adopting the ways of their enemies as their own. Warfare could function as a conduit for the transmission of technical knowledge, albeit of a destructive sort.[29]

Offensive weaponry and defensive fortifications lie at the heart of any discussion of military technology, yet hardware and fortifications, tactics and strategy are never separate and distinct from political organization, logistics, and leadership. In addition, war was divided by gender. During the medieval and early modern centuries (and for centuries thereafter), war was fought primarily by men. But in the Western European states, women could become lords of fiefs, or act as such in the absence of their husbands, and they sometimes acquired military responsibilities as well. In all regions, women were involved in provisioning armies with food, clothing, and shelter. Apart from this, however, warfare profoundly affected noncombatants, whether they were men, women, or children. Onerous taxes for warfare were borne by peasants, and a field that a group of peasants had just planted could be ravaged by battle. Crops and vineyards might be destroyed, animals taken away, innocent victims could be killed, maimed, sexually assaulted, or forced into a lifetime of slavery.[30]

In eastern Mediterranean lands during the seventh and eighth centuries, the Byzantines and then the Arabs learned much about mounted archery from their nomadic enemies in central Asia—the Huns and the Avars. These skilled archers used a small, powerful weapon called a short bow. It had an inner core of wood, a layer of horn on the side facing the archer, and an outer layer of sinew. Bowstrings were often made of silk. Byzantine archers did not shoot from moving horses as the nomads did, but they did adopt some of their bow-drawing techniques. The Arabs adopted some of these same techniques.[31]

Bows had a variety of shapes and characteristics across the Mediterranean, in northern Europe and in England. In Spain, the Christian *reconquistador* armies used mounted archers, who were able to shoot their arrows backward over the rear of their horses, a technique probably taken from their Muslim enemies. From the early fourteenth century, English archers used longbows in their wars with France. The basic form of the longbow had an ancient history (contrary to the many assertions that it was a new weapon in the fourteenth century). Longbow archers drew the string to their ear rather than their chest, allowing the use of a longer arrow that had a greater range and twice the impact of an arrow shot from an ordinary bow. However, the actual effect of the longbow (as opposed to tactics or other factors) is debated among military historians. What is clear is that the English use of the longbow declined in the fifteenth century. In part this was because its skilled

A.

B.

C.

D.

Figure 6.7. A. Depiction of a battle with horsemen using short bows. Chromolithograph from a tenth-century Slavonian manuscript (detail) in the Vatican Library. HIP / Art Resource, New York. B. Archers practicing with longbows. Luttrell Psalter, 1325 (detail). Wikimedia Commons. C. Crossbow with windlass and foot stirrup. From Ralph Payne-Gallwey, *The Crossbow* (London: Longmans, Green, 1903), fig. 25. D. Archer with crossbow with foot stirrup. From Ralph Payne-Gallwey, *The Crossbow* (London: Longmans, Green, 1903), fig. 76.

use required training from childhood, and for complex reasons the pool of skilled longbow archers increasingly shrank.[32]

A very different kind of bow, one with a mechanical release mechanism, was the crossbow—a powerful weapon, which was gradually improved in the later medieval centuries. Crossbows may have survived from ancient Roman times or they may have been introduced from China in the tenth century. In any case, they eventually became the primary weapon of archers. The crossbow was not widely used in Islamic lands until the twelfth century, but in al-Andalus it came to be preferred. Some crossbows had a stirrup on the end of the stock, into which the archer placed his foot before drawing the string by means of a hook on the end of a rope attached to his belt. The first crossbows were wooden. Later ones could be composite bows made of sinew, horn, and wood. In the fifteenth century steel crossbows appeared, increasing the range and impact of the weapon tremendously. The bolts of these weapons had to be loaded by mechanical means such as a windlass.[33]

Throughout the Mediterranean and Europe, handheld weapons—such as swords, daggers, spears or lances, javelins, war axes, a weapon called the staff-weapon (a combination of spear or lance with an ax, blade, or hammer), maces, and war hammers—each possessed a complex history and a variety of forms. For example, the Byzantines adopted spears, both thrusting and throwing weapons, from the Avars, with wrist-straps at the center of the shaft. Arab spears had long reed shafts, the best of which came from India and the Persian Gulf. For close combat, Arabs from the early Abbasid period used maces, axes, and most importantly, swords. The Vikings used mainly long-range archery, spears, axes, and swords. (The development of the Viking war axe may well have been influenced by the Arab weapon.) In Scandinavian lands, Viking burials for males sometimes included weapons such as swords, axes, and spears.[34]

Perhaps the most ubiquitous handheld weapon was the sword. The Byzantines preferred a long-bladed cavalry sword, while Muslim fighters carried a shorter infantry sword. Archaeologists have grouped Viking swords into twenty-six general types and twenty special types. The two main parts of the weapon consisted of the blade, made of a single piece of iron (sometimes edged with steel), and the grip for the hand. The blade's form and length varied from one army to another and depended in part on the enemy to be faced. Smiths fabricated high-quality swords in a method known as pattern welding. First they made the patterned blade by twisting rods of steeled iron, hammering and forging them to create the core, and then adding a sharp cutting edge. They polished it to reveal the pattern. Swords made in this way were highly prized.[35]

By the early seventh century, horseback riders used stirrups throughout the Eurasian steppes, and they gradually were adopted in Western Europe. The use of the stirrup in eighth century Francia was the focus of a heated debate in the history of technology. The debate concerned Lynn White Jr.'s thesis, elaborated in *Medieval Technology and Social Change*, that the arrival of the stirrup and its promotion by the ruler of the Franks, Charles Martel (ruled 718–741), was a technological event of transcendent significance. It allowed, White argued, the emergence of the heavily armed knight who could engage in mounted shock combat with a lance couched under his arm. Because he could brace himself on his stirrups, he could charge into an opponent with the full force of his horse, piercing the armor of his enemy with his lance without being thrown from his horse.[36]

White further suggested that the high cost of both the equipment and the horses needed for this type of fighting led to feudalism—the social arrangement whereby vassals swore fealty to their lords, pledging service in battle. In exchange the lord provided the vassal (now a knight) with a fief, often comprising land, or a manor (discussed in chapter 1). Income from this fief provided the knight with the resources he needed to maintain a warhorse and equip it and himself for battle. As evidence for this, White pointed to the confiscation by Charles Martel of church lands and his gift of some of those lands to his followers.[37]

Historians debated White's thesis extensively and ultimately rejected it. Bernard Bachrach, one of White's most thorough critics, questioned whether stirrups were widely used in the eighth and ninth centuries, or even highly valued, and he questioned whether Charles Martel's land policies, which actually resembled those of his predecessors quite closely, should be called feudalism at all. He added that the development of mounted shock combat, which he placed in the eleventh and twelfth centuries, depended not just on stirrups but rather on a combination of elements, such as a particular kind of saddle with a rigid back plate (called a cantle), a high pommel (the protuberance on the top front of the saddle), and double girthing or breast collars.[38]

White developed his thesis in response to a particular idea of feudalism—that it was created for military purposes and survived as long as it was the only way the king could acquire a military force. Aside from its technological determinism and other issues, White's thesis becomes problematic because the concept of feudalism itself came to be a debated. Historians disagreed on what exactly it was, when it developed, and whether it was a useful concept at all.[39]

The heavily armed knight became, nevertheless, a significant feature of warfare in Western Europe by the eleventh century. His extensive equipment required the skilled craftwork of the armorer and the metalworker. The hauberk (mail shirt) was made of some 25,000 rings. The sword, lance, and other equipment were equally elaborate and costly.[40]

The proliferation of castles and their transition from wood to stone, along with the augmentation both in the East and the West of urban fortifications, led to an increase in the importance of siege warfare. New siege weapons were developed, including mobile siege towers and battering rams, and especially artillery machines.[41]

By the eleventh century, an artillery machine, developed originally in China, based on the rotation of a lever became a regular feature of medieval siege warfare. The generic term for this machine was the trebuchet. Workers constructed it from wood, including a long tapering, rotating beam that pivoted around an axle that was attached to a tower and base. The axle divided the beam into two unequal sections. The longer arm ended in a sling into which stone boulders would be placed. In the traction trebuchet, between 40 and 125 ropes would be attached to the opposite end of the beam. The machine was then powered by a team of people (occasionally documents indicate women as well as men)—one to a rope—who pulled in unison to discharge the stone up to about 150 meters (about 492 feet). The weakness of the traction trebuchet was that the pull exerted by the team was not always consistent, either because of their unequal strengths, or because of a lack of unity, or because the group had become decimated by attacking enemies. The counterweight trebuchet, developed later in the eastern Mediterranean, addressed this weakness, having a fixed counterweight, usually a box filled with stones or other heavy materials, which was released to discharge the weapon. Between the development of the traction and the counterweight trebuchets, hybrid forms were utilized.[42]

The trebuchet was a powerful siege weapon. In addition, by changing the size of the counterweight and the pivotal length, the range could be varied, thereby making it far more accurate. As the machine spread westward, it helped to shift the balance of power from besieged to besieger. In response, castle design was modified with higher walls and towers. The trebuchet was superseded only by the gradual development of gunpowder artillery beginning in the fourteenth century.[43]

Another weapon, the pike, became highly effective when combined with the tactics of a highly disciplined army. Pikes were elongated spears held by tightly massed soldiers who moved in formation, and who were trained to face the charges of cavalry without breaking. The Swiss were the most renowned practitioners of pike warfare in the fifteenth century. Late medieval warfare was in a state of change in which traditional means, such as those involving heavily armed knights, were

Figure 6.8. Counterweight trebuchet. From Konrad Kyeser, *Bellifortis*, sixteenth-century ms, Biblioteca Nazionale Centrale, Florence, Fondo Nazionale, II.III. 317, fol. 14r.

deployed against new military forms such as pikemen. Military leaders often combined new tactics with new combinations of weaponry. Yet, as Bert Hall has pointed out, rulers did not have a full range of choices in deploying an army. The English could deploy longbowmen (at least for a time) because their social and cultural conditions could produce a sufficient number of men with the necessary extensive training from childhood. The Swiss and the Flemish, small countries with cohesive populations, were able to organize and train armies of pikemen, but the French with its larger land area and more diverse population, could not.[44]

Gunpowder and gunpowder artillery entered slowly into this complex military arena. Gunpowder itself, a product of Chinese alchemical experimentation, consists of a mixture of saltpeter, sulfur, and charcoal. In Byzantine and Islamic lands, combatants used various incendiary mixtures, some gunpowder, some not. These included the so-called Greek fire, the exact composition of which has been long a topic of debate (and undoubtedly varied from one place to another). It probably was based on naphtha (a flammable liquid hydrocarbon or petroleum mixture), along with other substances such as quicklime and saltpeter. Usually employed by ships, this liquid mixture was preheated and pressurized below deck, and then shot from tubes or siphons. The Arabs developed their own incendiary substance, called *naft*, which variously consisted of a petroleum mixture or gunpowder. In Syria and Egypt, armies were using gunpowder artillery by the thirteenth century.[45]

Gunners first deployed gunpowder weapons in European warfare between 1325 and 1425. During this period, these machines were heavy, inaccurate, and had a slow rate of fire, making them inferior to the longbow and the crossbow. A major advance in gunpowder manufacture occurred during the fifteenth century with the development of corning or granulating the powder, making it less unpredictable and less volatile during transportation—thus safer. After some years of experimentation in which various kinds of artillery, including large-caliber bombards, were tried, gradually, the single-piece, muzzle-loading cannon came to dominate. Small guns also developed, including the arquebus and the musket for infantry, and later the handheld pistol.[46]

Along with the technological development of various kinds of guns came the skilled profession of the gunner. Gunners at times were trained in special artillery schools organized by states, as has been shown for the school in Seville established by the Spanish monarchy in the late sixteenth / early seventeenth centuries, and for England. Gunners could promote their expertise to patrons by emphasizing their mathematical skills (relevant to aiming cannons for example), even though, as Steven Walton has argued, such skills could be more useful in gaining employment than in actual practice. The rise of the gunner was accompanied by literally hundreds of pamphlets, manuals, and books, many illustrated, about topics that ranged from gunpowder recipes to kinds of guns to the full range of skills needed for the successful practice of shooting on both land and sea.[47]

As gunpowder artillery increasingly came into use, traditional medieval fortifications and city walls were found to be inadequate; their tall, flat surfaces turned out to be perfect targets for artillery, a shortcoming that became shockingly evident during the French invasions of Italy in the 1490s. Soon, modifications to fortifications emerged in response. First, defenders of castles and cities cut gunports into the defensive walls, so that they also could use gunpowder artillery effectively as a defensive weapon. Then they added structures such as artillery towers and earthworks, which were rather inexpensive and doubled the effectiveness of medieval curtain walls. In the late fifteenth and sixteenth centuries, the polygonal bastion fort was developed. It featured triangulated surfaces and raised gun platforms, ditches, detached forts called ravelins, and artificial slopes. These structures were far more effective in defending against gunpowder artillery than had been the tall, flat, stone walls of medieval fortifications and towns.[48]

The development of gunpowder artillery and bastion fortification comprise two of the elements involved in a debate concerning a "military revolution" in the sixteenth century and beyond. A historian of early modern Sweden, Michael Roberts, first used the term in 1955, referring to developments in Swedish armies during the Thirty Years War (1618–1648). Geoffrey Parker considerably broadened the meaning of the term to refer to radical changes in warfare in the sixteenth century that marked the difference between the medieval and modern worlds. For Parker, the changes involved gunpowder artillery, bastion fortification, and huge increases in the size of armies to besiege the new fortifications effectively, as well as changes in the arming of ships at sea. Parker's thesis of a military revolution has been vigorously debated, with some scholars suggesting that it presents an oversimplified and unduly accelerated picture of military change.[49]

Warfare, like other complex processes, involved far more than issues concerning technology. Military forces, whether small, marauding mercenary bands or increasingly large armies, could exert devastating effects on villages and fields in the countryside, even without the occasional burning of villages and crops to destroy supply sources of enemy forces. Armies in these centuries might be thought of as a traveling city, made up of men, women, and children. Women sometimes fought and occasionally led armies (most famously, Joan of Arc). Most women however did not fight but ran the huge supply train that was an essential part of all premodern armies. They foraged for, purchased, or traded, and then prepared food. They fabricated, laundered, and repaired clothing and shoes and cared for the sick. They were wives and sometimes prostitutes, and cared for children. They were as much a part of late medieval armies as were men.[50]

* * *

The instruments and machines discussed in this chapter have often been usefully studied as isolated objects, possibly because many specimens are still extant, often as exhibits in museums. However, in their time of actual use, as historians recognize, they constituted only one aspect of complex political, cultural, and social processes that included material production, time-telling, and navigation, and the organized violence of warfare. They can only be understood in this wider context in which they were constructed and used—something that can be said for all material and technological artifacts.

Conclusion

Why Premodern Technology Matters

> It's a matter of life. Without this water, the farmers can't grow anything,
> the village can't survive.
>
> ANTONIO JESÚS RODRIGUEZ GARCIA,
> TWENTY-FIRST-CENTURY FARMER
> IN THE VILLAGE OF PITRES IN THE
> ALPUJARRA MOUNTAINS IN SOUTHERN SPAIN

Southern Spain is particularly vulnerable to the effects of global warming, including extreme heat, a scarcity of water, and desertification. These problems were exacerbated by the trend from the 1960s, promoted by the agrifood industry, of abandoning traditional (medieval, Islamic) systems of irrigation and water management in favor of reservoirs and other systems incompatible with traditional methods. The medieval Arabic hydraulic system, including hundreds of miles of channels, called *acequias*, which had been dug, kept clear of debris, and managed village by village, were abandoned. Although the new hydraulic systems brought about enormous agricultural productivity, they are now found to be unsustainable. They have failed to deal with (and, indeed, have partially caused) extreme water scarcity. As a solution, villagers, university scholars, and volunteers are undertaking the arduous task of locating the traditional canals, clearing them, and putting them back into use. During the mountain snowmelt of the spring, this intricate, traditional system with its many canals slows the water down. Rather than the water rushing into rivers and lakes, which then dry up in the summer, the water is diverted into hundreds of channels. It is absorbed slowly into aquifers and turns up months later in springs downslope that can irrigate crops during the dry season. The knowledge, especially of elders in village communities, of where the canals are and how they were managed has been crucial to this effort.[1]

Premodern technology matters. In this case, knowledge of it may be crucial for the agricultural viability of southern Spain, and other regions as well. This is not to say that all of the technologies discussed in this book have practical relevance today, but knowledge of past technologies allows for a more sophisticated understanding of the lives of ordinary people in the past. It fosters an appreciation of the difficulties and accomplishments of those people as they interacted with and mastered their physical world. As David Nye remarked, "Technology matters because it is inseparable from being human."[2] The study of past technologies matters because it helps to connect us to our own past, and at the same time provides information that might be crucially relevant to the world we live in today.

This book is a synthesis of technologies in Mediterranean and European lands between 600 and 1600—technologies in the broadest sense of the term. I have tried to avoid grand linear narratives, teleological approaches, and chronicles of "progress," "invention," "revolution," and "modernization"—all concepts and approaches that have infused much traditional history of technology. Such approaches simplify, but they also distort. They tend to neglect, to ignore, and to not look for evidence that does not fit into the paradigm being favored. Teleological explanations interpret activities and processes in terms of a presumed ultimate goal. One grand linear narrative has been "the rise of the West." Another, for example, has been the development of banking in fourteenth-century northern Italy leading directly to the rise of modern capitalism. "Progress" might be seen in terms of economic surpluses, but the question must always be asked, "Progress for whom?" The search for inventions, characteristic of the early discipline of the history of technology, tended to isolate these inventions from surrounding societies, cultures, and technologies. The idea of revolution, also pervasive in the history of technology, tends to isolate the presumed revolutionary technology from its surrounding context.

In this book I have tried to avoid jargon and obfuscation. But I have also tried to indicate some of the complexities of technologies in different times and locations. These complexities come out of differing physical environments but also out of the ways in which different communities and cultures adopt and use technologies. Technology is not isolated from other aspects of human society and culture— such as the development of manorialism, or the influence of religion, or assumptions about gender and social status on the use and development of particular technologies.

This is a book without heroes and without "men of genius" (and they were almost always men). It is without monumental inventions, although innovations abound. This is a book about the technological activities of ordinary people and the ways in which they manipulated the physical and natural world to maintain and advance themselves and their families and communities. The names of such people are mostly unknown. A common lament of historians is that we have much less access to ordinary people in the past than to elites. It is true that most extant writing comes from the more or less elite strata of society. (This is not the case with archaeological findings, the study of material objects, or the use of genetics, dendrochronology, and other methods of understanding the past.)

Nevertheless, I suggest that knowing in some detail what a person did in their craft or technological life, and how they did it, is one way of understanding a person's life, whether or not we know that person's name. A spinner who has learned how to spin wool from early childhood with a spindle whorl; a weaver of silk, linen, or wool; a shoemaker who has learned to make shoes as an adolescent; families—men, women, and children—who carry out every phase of the harvest every year; a navigator who guides ships with sails powered by the wind—all possess bodily skills that are part and parcel of themselves and undoubtedly of their identities as well.

In the postindustrial and increasingly virtual worlds in which we live, and especially in the world of scholarship, we tend to underestimate, overlook, and minimize the complex tasks of mastery that it took (and takes, or would take, today) to acquire and practice such skills. This was one of the themes that the great economic historian Larry Epstein had planned to develop had his untimely death not prevented it.[3] He was concerned with guilds and the importance of the guilds in imparting skill—but the point is more generally true as well. Whether guilds exist or not in a particular time and place, whether training involves formal apprenticeship or informal practice, the skills acquired are not only difficult to acquire but become intrinsic to the lives of people who do acquire and use them.

The material world and the technologies that humans use to manipulate and shape that world are profoundly embedded in historical cultures. This reality is often ignored by scholars, teachers, and students. The embeddedness of material culture and technology in daily life is true for all people, whether men, women, or children, whether wealthy and powerful, poor and powerless, or something in between, whether living in a castle or a palace or a hut, whether in a city or a village. Just as technology is a part of human life today, it is also part of human history.

Crafts and technologies were created by human skill and ingenuity and in addition were in themselves part of a larger culture of knowledge, including more formalized knowledge such as reading and writing, and knowledge as represented by books and by educational institutions such as madrasas in the Islamic world and universities in the medieval West. Craftwork is also mental work. This opens our thinking to the world of "artisanal epistemologies," as discussed by Pamela Smith, or to the "mindful hand," as in the title of an importance collection of studies. I have suggested "trading zones" to describe places such as printshops, Roman ruins, arsenals, mines, and instrument shops where apprenticeship-trained artisans and university-trained scholars communicated, each possessing knowledge and skills of benefit to the other. Whether or not a particular space at a specific time can be considered a "trading zone" can be determined by historical investigation.[4]

From another point of view, Marcus Popplow has suggested an alternative to the traditional dichotomies of "science and technology," or "theory and practice." He suggests instead an approach that emphasizes first the "formalization" of technical knowledge, such as acquiring skill through apprenticeship, guild rules, technical drawings, privileges for invention, treatises, teaching in schools, and scientific academies; and second, a stage of "interaction" in which this formalized knowledge is spread into broader contexts such as cities, courts, or academies.[5]

In Western European historiography, the relationship between artisanal cultures and knowledge cultures has been influenced by the "Zilsel thesis" in which Edgar Zilsel proposed that certain kinds of "superior artisans" had an important influence on the "Scientific Revolution."[6] Yet the concept of the "Scientific Revolution" itself, as traditionally conceived, has been rejected by many (by no means all) historians of science. These historians have studied many practices not previously considered "scientific" (such as alchemy and astrology). They have embraced globalism accompanied by a view of numerous centers of investigation and diverse ideas about the nature of the world, and they have studied various practices, such as observing and cataloging plants, based on those ideas. It has been argued that as a traditional category, the "Scientific Revolution" narrative has made other traditions invisible, and has had a hegemonic effect, reinforcing the dominance of Western Europe.[7]

This is not the place to review the ongoing discussions among scholars concerning the "Scientific Revolution." Rather it is to suggest that the relationships of artisans (and their knowledge and practices) to learned cultures might be usefully separated from concerns about "revolutions" of any kind, including the "scientific"; that they be considered within more local cultural and regional contexts. Such a focus involves an interest in discovering the ways in which the worlds of making were related to more "learned" worlds and to broader cultural issues having to do with religion, or with philosophy, or with learning in particular places and within particular institutions. One advantage of unhooking artisanal practices from broadly themed narratives of revolution (or of progress) is that those practices in Mediterranean and European lands would become far more comparable to seemingly similar practices in other parts of the world.[8] Another is that it would encourage deeper investigation and focus on local understandings of technological practices and the ways in which they formed a part both of the material and nonmaterial lives of people in specific times and places. It is a history of artisanal crafts and technologies in local context, historicized, understood as part of the lives of people everywhere, a history useful even today, a history that matters.

Notes

Introduction

Epigraph: Ikhwān al-Ṣafā, "Epistle 8 on the Practical Crafts," trans. Nader El-Bizri in *Epistles of the Brethren of Purity: On Composition and the Arts: An Arabic Critical Edition and English Translation of Epistles 6–8*, by Ikhwān al-Ṣafā, ed. and trans. Nader El-Bizri and Godefroid de Callataÿ (Oxford: Oxford University Press, 2018), 137–165, esp. 149–151, quote on 149–150.

1. Ikhwān, "Epistle 8 on the Practical Crafts," 137–165. For an illuminating discussion of the fifty-two epistles of the Ikhwān (especially "Epistle 8") and their engagement with craftsmanship, see Margaret S. Graves, *Arts of Allusion: Object, Ornament, and Architecture in Medieval Islam* (New York: Oxford University Press, 2018), 26–58.

2. See David Edgerton, *The Shock of the Old: Technology and Global History since 1900* (Oxford: Oxford University Press, 2007); Edgerton, "What Is the Historiography of Technology About?," and Pamela O. Long, "The Craft of Premodern European History of Technology," both in *Technology and Culture* 51 (July 2010): 680–697 and 698–714, respectively. For the origins of such views in the nineteenth century, see Eric Schatzberg, *Technology: Critical History of a Concept* (Chicago: University of Chicago Press, 2018), esp. 55–71.

3. Schatzberg, *Technology*; and Francesca Bray and Barbara Hahn, "'The Goddess Technology Is a Polyglot': A Critical Review of Eric Schatzberg, *Technology: Critical History of a Concept*," *History and Technology* 38, no. 4 (2022): 275–316.

4. For "distributed cognition," studies with particular relevance to the history of technology include Edwin Hutchins, *Cognition in the Wild* (Cambridge, MA: MIT Press, 1995), who develops the concept within the framework of a detailed analysis of both contemporary and traditional navigation; and Chandra Mukerji, *Impossible Engineering: Technology and Territoriality on the Canal du Midi* (Princeton: Princeton University Press, 2009), who uses the framework within a history of the construction of the Canal du Midi in seventeenth-century France.

5. See esp. Richard Sennett, *The Craftsman* (New Haven: Yale University Press, 2018), 37, 53–118; Pamela H. Smith, *The Body of the Artisan: Art and Experience in the Scientific Revolution* (Chicago: University of Chicago Press, 2004); and Lissa Roberts, Simon Schaffer, and Peter Dear, eds., *The Mindful Hand: Inquiry and Invention in the Late Renaissance to Early Industrialisation* (Amsterdam: Koninklijke Nederlandse Akademie van Wetenschappen, 2007).

6. Lawrence Shapiro and Shannon Spaulding, "Embodied Cognition," in *The Stanford Encyclopedia of Philosophy*, ed. Edward N. Zalta, June 25, 2021, https://plato.stanford.edu/archives/win2021/entries/embodied-cognition/.

7. See Graves, *Arts of Allusion*, esp. 26–68; and Pamela O. Long, *Openness, Secrecy,*

Authorship: Technical Arts and the Culture of Knowledge from Antiquity to the Renaissance (Baltimore: Johns Hopkins University Press, 2001), esp. 30–35.

8. Elspeth Whitney, *Paradise Restored: The Mechanical Arts from Antiquity through the Thirteenth Century*, Transactions of the American Philosophical Society, vol. 80, pt. 1 (Philadelphia: American Philosophical Society, 1990), esp. 75–127; and Jerome Taylor, ed. and trans., *The Didascalicon of Hugh of St. Victor: A Medieval Guide to the Arts* (New York: Columbia University Press, 1961), esp. 74–79. And see the astute discussion of Elly R. Truitt, *Medieval Robots: Mechanism, Magic, Nature, and Art* (Philadelphia: University of Pennsylvania Press, 2015).

9. Smith, *Body of the Artisan*; and Pamela H. Smith, *From Lived Experience to the Written Word: Reconstructing Practical Knowledge in the Early Modern World* (Chicago: University of Chicago Press, 2022).

10. Pamela O. Long, *Artisan/Practitioners and the Rise of the New Sciences* (Corvallis, OR: Oregon State University Press, 2011), 94–126; and Long, "Trading Zones in Early Modern Europe," *Isis* 106 (December 2015): 840–847.

11. Such relationships are proving to be productive foci of investigation in other areas of the world as well. See, for example, Hyeok Hweon Kang, "Crafting Knowledge: Artisan, Officer, and the Culture of Making in Chosŏn Korea, 1392–1910" (PhD diss.: Harvard University, 2020).

12. See esp. Ricardo Córdoba de la Llave, ed., *Craft Treatises and Handbooks: The Dissemination of Technical Knowledge in the Middle Ages* (Turnhout: Brepols, 2013); Pascal Dubourg Glatigny and Hélène Vérin, *Réduire en art: La technologie de la Renaissance aux Lumières* ([Paris]: Éditions de la Maison sciences de l'homme, 2008); William Eamon, *Science and the Secrets of Nature: Books of Secrets in Medieval and Early Modern Culture* (Princeton: Princeton University Press, 1994); Guido Frison and Giulia Brun, "*Compositiones Lucenses* and *Mappae Clavicula*: Two Traditions or One?: New Evidence from Empirical Analysis and Assessment of the Literature," *Heritage Science* 6 (2018), article 24; Robert Halleux, *Le savoir de la main: Savants et artisans dans l'Europe pré-industrielle* (Paris: Armand Colin, 2009); Elaine Yen Tien Leong, *Recipes and Everyday Knowledge: Medicine, Science, and the Household in Early Modern England* (Chicago: University of Chicago Press, 2018); Long, *Openness, Secrecy, Authorship*; Long, "Manuals," in *Information: A Historical Companion*, ed. Ann Blair et al. (Princeton: Princeton University Press, 2021), 589–593; and Smith, *From Lived Experience to the Written Word*.

13. See James Belich, John Darwin, and Chris Wickham, "Introduction: The Prospect of Global History," in *The Prospect of Global History*, ed. James Belich et al. (Oxford: Oxford University Press, 2016), 3–22, esp. 4–5.

14. Catherine Holmes and Naomi Standen, "Introduction: Towards a Global Middle Ages," in "The Global Middle Ages," ed. Holmes and Standen, special issue, *Past & Present*, supplement 13 (Oxford: Oxford University Press, 2018), 12–44, esp. 3 and 23–25.

15. For an introduction to the Crusades, see Jonathan P. Phillips, *Holy Warriors: A Modern History of the Crusades* (New York: Random House, 2010).

16. See esp. Michael Decker, *Tilling the Hateful Earth: Agricultural Production and Trade in the Late Antique East* (Oxford: Oxford University Press, 2009), 7–27; Corisande Fenwick, *Early Islamic North Africa: A New Perspective* (London: Bloomsbury, 2020); Mark Whittow, "Geographic Survey," in *The Oxford Handbook of Byzantine Studies*, ed. Elizabeth Jeffreys with John Haldon and Robin Cormack (Oxford: Oxford University Press, 2008), 219–231; and Chris Wickham, *Framing the Early Middle Ages: Europe and the Mediterranean, 400–800* (Oxford: Oxford University Press, 2005), 29–32. For Atlantic Ocean bordering lands, see particularly Barry Cunliffe, *Facing the Ocean: The Atlantic and Its Peoples* (Oxford: Oxford University Press, 2001), esp. 465–553.

17. Henri Pirenne, *Mohammed and Charlemagne*, trans. Bernard Miall (London: Allen & Unwin, 1968). Critiques include Richard Hodges, "Henri Pirenne and the Question of Demand

in the Sixth Century," and Paolo Delogu, "Reading Pirenne Again," both in *The Sixth Century: Production, Distribution and Demand*, ed. Richard Hodges and William Bowden (Leiden: Brill, 1998), 2–14 and 15–40, respectively. See also Peter Burke, *The French Historical Revolution: The Annales School, 1929-2014*, 2nd ed. (Stanford: Stanford University Press, 2015), esp. 23–24.

18. See Michael McCormick, *Origins of the European Economy: Communications and Commerce, AD 300-900* (Cambridge: Cambridge University Press, 2001); and Wickham, *Framing the Early Middle Ages*, 821–823.

19. Lucien Febvre, "Réflexions sur l'histoire des techniques," and Marc Bloch, "Avènement et conquête du moulin à eau," both in *Annales d'histoire économique et histoire* 7 (1935): 531–535 and 538–563, respectively. Bloch's article has been translated as "The Advent and Triumph of the Watermill," in *Land and Work in Mediaeval Europe: Selected Papers by Marc Bloch*, trans. J. E. Anderson (New York: Harper Torchbooks, 1969), 136–168. For the Annales school, see esp. Burke, *French Historical Revolution*; and for the Annales issue on the history of technology, see Pamela O. Long, "The *Annales* and the History of Technology," *Technology and Culture* 46 (January 2005): 177–186.

20. Fernand Braudel, *The Mediterranean and the Mediterranean World in the Age of Philip II*, trans. Siân Reynolds, 2 vols. (New York: Harper & Row, 1972); Peregrine Horden and Nicholas Purcell, *The Corrupting Sea: A Study of Mediterranean History* (Oxford: Blackwell, 2000), 36–43 for a discussion and critique of Braudel; and Jessica L. Goldberg, *Trade and Institutions in the Medieval Mediterranean: The Geniza Merchants and their Business World* (Cambridge: Cambridge University Press, 2012), 22–29. See also David Abulafia, *The Great Sea: A Human History of the Mediterranean* (Oxford: Oxford University Press, 2011), esp. 241–452; Burke, *French Historical Revolution*, 36–72; W. V. Harris, ed., *Rethinking the Mediterranean* (Oxford: Oxford University Press, 2005); Horden and Purcell, *The Boundless Sea: Writing Mediterranean History* (London: Routledge, 2020); and Horden and Sharon Kinoshita, eds., *A Companion to Mediterranean History* (Chichester, West Sussex: John Wiley & Sons, 2014).

21. Lynn White Jr., *Medieval Technology and Social Change* (Oxford: Oxford University Press, 1962); and White, *Medieval Religion and Technology* (Berkeley: University of California Press, 1978). For assessments of White's work, see esp. Bert S. Hall, "Lynn White's Medieval Technology and Social Change after Thirty Years," in *Technical Change: Methods and Themes in the History of Technology*, ed. Robert Fox (London: Harwood, 1996), 85–101; Alex Roland, "Once More into the Stirrups: Lynn White jr. [*sic*], Medieval Technology and Social Change," *Technology and Culture* 44 (July 2003): 574–585; and Shana Worthen, "The Influence of Lynn White jr.'s [*sic*] *Medieval Technology and Social Change*," *History Compass* 7, no. 4 (2009): 1201–1217.

22. Lynn T. White Jr., "The Historical Roots of our Ecologic Crisis," *Science*, March 10, 1967, reprinted in White, Machina ex Deo: *Essays in the Dynamism of Western Culture* (Cambridge, MA: MIT Press, 1968). For a detailed assessment, see Richard C. Hoffmann, *An Environmental History of Medieval Europe* (Cambridge: Cambridge University Press, 2014), esp. 87–94.

23. Cunliffe, *Facing the Ocean*, 465–553; and Erik Thoen et al., "Series Introduction: A New Rural History of North-West Europe," in *Making a Living: Family, Income and Labour*, ed. Eric Vanhaute, Isabelle Devos, and Thijs Lambrecht, vol. 3 in *Rural Economy and Society in North-Western Europe, 500-2000*, ed. Erik Thoen (Turnhout: Brepols, 2011), quote on xiii.

24. Holmes and Standen, "Introduction: Toward a Global Middle Ages"; and see Jonathan Shepard, "Networks," and Glen Dudbridge, "Reworking the World System Paradigm," both in Holmes and Standen, *Global Middle Ages*, 116–157 and 297–316, respectively.

25. Horden and Purcell, *Corrupting Sea*, 176–178, for landscape archaeology; and see for southern French territories, Aline Durand and Philippe Leveau, "Farming in Mediterranean France and Rural Settlement in the Late Roman and Early Medieval Periods: The Contribution

from Archaeology and Environmental Sciences in the Last Twenty Years (1980–2000)," in *The Making of Feudal Agricultures?*, ed. Miquel Barceló and François Sigaut (Leiden: Brill, 2004), 177–253; and for central Europe, Péter Szabó, *Woodland and Forests in Medieval Hungary*, BAR S1348 (Oxford: Archaeopress, 2005), 33–35. For materiality, see esp. Tehmina Goskar, "Material Culture," in Horden and Kinoshita, *Companion to Mediterranean History*, 281–295; Timothy J. LeCain, *The Matter of History: How Things Create the Past* (Cambridge: Cambridge University Press, 2017); Pamela H. Smith, Amy R. W. Meyers, and Harold J. Cook, eds., *Ways of Making and Knowing: The Material Culture of Empirical Knowledge* (New York: Bard Graduate Center, 2014); Christy Anderson, Anne Dunlop, and Pamela H. Smith, *The Matter of Art: Materials, Practices, Cultural Logics, c. 1250–1750* (Manchester: Manchester University Press, 2014); Pamela H. Smith, director, "The Making and Knowing Project," https://www.makingandknowing.org/.

26. See for example, Perrine Mane, *Le travail à la campagne au Moyen Age: Étude iconographique* (Paris: A. and J. Picard, 2006).

27. See esp. Adam Robert Lucas, "Narratives of Technological Revolution in the Middle Ages," in *Handbook of Medieval Studies: Terms, Methods, Trends*, ed. Albrecht Classen (Berlin: De Gruyter, 2010), 967–990.

28. For a cogent introduction and many local studies of the relationships of environmental and human factors in the fourteenth century, see esp. Martin Bauch and Gerrit Jasper Schenk, "Teleconnections, Correlations, Causalities between Nature and Society? An Introductory Comment on the 'Crisis of the Fourteenth Century,'" in *The Crisis of the 14th Century: Teleconnections between Environmental and Societal Change?*, ed. Martin Bauch and Gerrit Jasper Schenk (Berlin: De Gruyter, 2020), 1–16. See also John Aberth, *An Environmental History of the Middle Ages: The Crucible of Nature* (London: Routledge, 2013); A. T. Grove and Oliver Rackham, *The Nature of Mediterranean Europe: An Ecological History* (New Haven: Yale University Press, 2001); Hoffmann, *Environmental History of Medieval Europe*; Martin Knoll and Reinhold Reith, eds., *An Environmental History of the Early Modern Period: Experiments and Perspectives* (Zurich: LIT Verlag, 2014); and Alan Mikhail, *Under Osman's Tree: The Ottoman Empire, Egypt and Environmental History* (Chicago: University of Chicago Press, 2017). See also for a cogent overview, Stephen Mosley, *The Environment in World History* (London: Routledge, 2010); Tim Soens, ed., *A Cultural History of the Environment*, vol. 3, *The Oceanic Age, 1200–1550* (London: Bloomsbury, forthcoming); and Erik Thoen and Tim Soens, eds., *Struggling with the Environment: Land Use and Productivity*, vol. 1 in *Rural Economy and Society in North-Western Europe, 500–2000*, ed. Erik Thoen (Turnhout: Brepols, 2015). For the Great Famine, see esp. William Chester Jordan, *The Great Famine: Northern Europe in the Early Fourteenth Century* (Princeton: Princeton University Press, 1996); and Philip Slavin, *Experiencing Famine in Fourteenth-Century Britain* (Turnhout: Brepols, 2019).

29. See esp. Aberth, *Environmental History*, 141–232; Hoffmann, *Environmental History*, 174–181 and passim; Pamela O. Long, "Work, Power, Energy," in Soens, *Cultural History of the Environment*, forthcoming; Mikhail, *Under Osman's Tree*, 111–150; and Vaclav Smil, *Energy and Civilization: A History* (Cambridge, MA: MIT Press, 2017), 49–222.

30. Hoffmann, *Environmental History*, 196–227; and Smil, *Energy and Civilization*, esp. 127–224.

31. Smil, *Energy and Civilization*, 157–163.

32. Richard W. Bulliet, *Cotton, Climate, and Camels in Early Islamic Iran: A Moment in World History* (New York: Columbia University Press, 2009); Ronnie Ellenblum, *The Collapse of the Eastern Mediterranean: Climate Change and the Decline of the East, 950–1072* (Cambridge: Cambridge University Press, 2012); and for an extended critique, Johannes Preiser-Kapeller, "A Collapse of the Eastern Mediterranean?," *Jahrbuch Österreichischen Byzantinistik* 65 (2015): 195–242. See also Aberth, *Environmental History*, 26–28 and 49–51; Bruce M. S. Campbell, *The*

Great Transition: Climate, Disease and Society in the Late-Medieval World (Cambridge: Cambridge University Press, 2016); Grove and Rackham, *Nature of Mediterranean Europe*, 130–140; Brian Fagan, *The Ice Age: How Climate Made History, 1300–1850* (New York: Basic Books, 2000); Hoffmann, *Environmental History*, esp. 320–341; and Johannes Preiser-Kapeller and Ekaterini Mitsiou, "The Little Ice Age and Byzantium within the Eastern Mediterranean, ca. 1200–1350: An Essay on Old Debates and New Scenarios," in Bauch and Schenk, *Crisis of the 14th Century*, 190–220.

33. For the Justinianic plague, see esp. Lester K. Little, ed., *Plague and the End of Antiquity: The Pandemic of 541–750* (Cambridge: Cambridge University Press, 2007); and Little, "Plague Historians in Lab Coats," *Past & Present*, 213 (November 2011): 267–290. For the later Black Death, see esp. James Belich, "The Black Death and the Spread of Europe," in Belich et al., *Prospect of Global History*, 93–105; Monica H. Green, "The Four Black Deaths," *American Historical Review* 125 (December 2020): 1601–1631; and Green, ed., *Pandemic Disease in the Medieval World: Rethinking the Black Death* (Kalamazoo, MI: Arc Medieval Press, [2015]).

34. Recent work includes Maria Ågren, ed., *Making a Living, Making a Difference: Gender and Work in Early Modern European Society* (Oxford: Oxford University Press, 2017); Anna Bellavitis, *Il lavoro delle donne nelle città dell'Europa moderna* (Rome: Viella, 2016); Madeleine Pelner Cosman, *Women at Work in Medieval Europe* (New York: Checkmark Books, 2000); Anna Esposito, *Donne del Rinascimento a Roma e dintorni* (Rome: Roma nel Rinascimento, 2013); Catriona Macleod, Alexandra Shepard, and Maria Ågren, *The Whole Economy: Work and Gender in Early Modern Europe* (Cambridge: Cambridge University Press, 2023); Maya Shatzmiller, *Labour in the Medieval Islamic World* (Leiden: Brill, 1994), esp. 347–368; Merry E. Wiesner-Hanks, *Women and Gender in Early Modern Europe*, 4th ed. (Cambridge: Cambridge University Press, 2019); and Maria Paola Zanoboni, *Donne al lavoro nell'Italia e nell'Europa medievali (secoli XIII–XV)* (Milan: Editoriale Jouvence, 2016).

35. See for example, Wickham, *Framing the Early Middle Ages*, 153–379, for elites and 383–588, for peasants.

36. See esp. Geraldine Heng, *The Invention of Race in the European Middle Ages* (New York: Cambridge University Press, 2018); and Lynn T. Ramey, *Black Legacies: Race and the European Middle Ages* (Gainesville: University Press of Florida, 2014).

37. McCormick, *Origins of the European Economy*, esp. 733–777. And see also Hannah Barker, *That Most Precious Merchandise: The Mediterranean Trade in Black Sea Slaves, 1260–1500* (Philadelphia: University of Pennsylvania Press, 2019); Stephen P. Bensch, "From Prizes of War to Domestic Merchandise: The Changing Face of Slavery in Catalonia and Aragon, 1000–1400," *Viator* 25 (1994): 63–93; Robert C. Davis, *Christian Slaves, Muslim Masters: White Slavery in the Mediterranean, the Barbary Coast, and Italy, 1500–1800* (Houndmills, Basingstoke: Palgrave Macmillan, 2003); Sally McKee, "Slavery," in *The Oxford Handbook of Women and Gender in Medieval Europe*, ed. Judith Bennett and Ruth Karras (Oxford: Oxford University Press, 2013), 281–294; Craig Perry et al., *World History of Slavery*, vol. 2, *AD 500–AD 1420* (Cambridge: Cambridge University Press, 2021); Youval Rotman, *Byzantine Slavery and the Mediterranean World*, trans. Jane Marie Todd (Cambridge, MA: Harvard University Press, 2009); and Rotman, "Migration and Enslavement: A Medieval Model," in *Migration Histories of the Medieval Afroeurasian Transition Zone: Aspects of Mobility between Africa, Asia and Europe, 300–1500 C.E.*, ed. Johannes Preiser-Kapeller, Lucian Reinfandt, and Yannis Stouraitis (Leiden: Brill, 2020), 387–412.

CHAPTER 1: Food Production

Epigraphs: Andrew Dalby, trans., *Geoponika: Farm Work: A Modern Translation of the Roman and Byzantine Farming Handbook* (London: Prospect Books, 2011), 325 (bk. 18.1).

Dorothy Oschinsky, *Walter of Henley and Other Treatises on Estate Management and Accounting* (Oxford: Clarendon Press, 1971), 307–385, quote on 337 (ch. 97–98).

1. See esp. Michael Decker, *Tilling the Hateful Earth: Agricultural Production and Trade in the Late Antique East* (Oxford: Oxford University Press, 2009), 263–271, who provides an in-depth discussion of the text, the relevant context, and its usefulness as a source for Byzantine agriculture.

2. Oschinsky, *Walter of Henley*, 144–148.

3. See esp. John Aberth, *An Environmental History of the Middle Ages: The Crucible of Nature* (London: Routledge, 2013), 11–76; Richard C. Hoffmann, *An Environmental History of Medieval Europe* (Cambridge: Cambridge University Press, 2014), 113–154; Alan Mikhail, *Under Osman's Tree: The Ottoman Empire, Egypt and Environmental History* (Chicago: University of Chicago Press, 2017), 1–15 for an introduction; and the essays for the earlier centuries (500–1750) in Erik Thoen and Tim Soens, eds., *Struggling with the Environment: Land Use and Productivity*, vol. 1 in *Rural Economy and Society in North-Western Europe, 500–2000* (Turnhout: Brepols, 2015). For soil in the environment, see David R. Montgomery, *Dirt: The Erosion of Civilizations* (Berkeley: University of California Press, 2007); and Steven Mosley, *The Environment in World History* (London: Routledge, 2010), 56–81.

4. Chris Wickham, *Framing the Early Middle Ages: Europe and the Mediterranean, 400–800* (Oxford: Oxford University Press, 2005), 258–302, 383–441 (for case studies in specific locales), and 519–558.

5. Decker, *Tilling the Hateful Earth*, 80–83. The apt phrase, "razor's edge of hunger" is Decker's.

6. Peregrine Horden and Nicholas Purcell, *The Corrupting Sea: A Study of Mediterranean History* (Oxford: Blackwell, 2000), esp. 175–230; and see also Wickham, *Framing the Early Middle Ages*, 535–550. For storage, see esp. Alfonso Vigil-Escalera Guirado, Giovanna Bianchi, and Juan Antonio Quirós, eds., *Horrea, Barns and Silos: Storage and Incomes in Early Medieval Europe* ([Bilbao]: Universidad del País Vasco, 2013).

7. See esp. Decker, *Tilling the Hateful Earth*, 28–79; Michel Kaplan, *Les hommes et la terre à Byzance du VIᵉ au XIᵉ siècle: Propriété et exploitation du sol* (Paris: Publication de la Sorbonne, 1992), 256–272; Alexander P. Kazhdan, "The Peasantry," in *The Byzantines*, ed. Guglielmo Cavallo, trans. Thomas Dunlap, Teresa Lavender Fagan, and Charles Lambert (Chicago: University of Chicago Press, 1997), 43–73, esp. 43–46; Angeliki E. Laiou and Cécile Morrisson, *The Byzantine Economy* (Cambridge: Cambridge University Press, 2007), 12–13; Jacques Lefort, "The Rural Economy, Seventh–Twelfth Centuries," and Angeliki E. Laiou, "The Agrarian Economy, Thirteenth–Fifteenth Centuries," in *The Economic History of Byzantium from the Seventh through the Fifteenth Century*, ed. Angeliki E. Laiou, 3 vols. (Washington, DC: Dumbarton Oaks, 2002), 1: 231–310 and 311–375, respectively.

8. Michael Decker, "Agriculture and Agricultural Technology," in *The Oxford Handbook of Byzantine Studies*, ed. Elizabeth Jeffreys with John Haldon and Robin Cormack (Oxford: Oxford University Press, 2008), 397–405, esp. 400–401; Decker, *Tilling the Hateful Earth*, 97–112; Kaplan, *Les hommes et la terre*, 25–46; and Kazhdan, "The Peasantry," 47.

9. Decker, "Agriculture and Agricultural Technology," 397–405; Kaplan, *Les hommes et la terre*, 25–46; Kazhdan, "The Peasantry," 47; Laiou, "Agrarian Economy," esp. 319–352; and Lefort, "Rural Economy," 248–252 and 253–254.

10. Decker, "Agriculture and Agricultural Technology," 400; Decker, *Tilling the Hateful Earth*, 262–271; and Dalby, *Geoponika*, 118–120 (4.12–14, vines), 193 (9.16, olive trees), 211 (10.23–24, pear), 214–215 (10.37, pomegranate, myrtle, citron), 215 (10.39, plums), 216 (10.41, cherries), 219 (10.52–53, figs), 221–222 (10.57, 10.62, almonds), 222–223 (10.63, chestnuts), 223–224 (10.65, wal-

nuts, pistachios), 224–225 (10.69, mulberries), 226–230 (10.75–77, kinds of grafting for various trees), 236 (11.3, baytree), and 240–241 (11.18, roses).

11. Anthony Bryer, "The Means of Agricultural Production: Muscles and Tools," and Lefort, "The Rural Economy," in Laiou, ed., *Economic History of Byzantium*, 1: 101–113 and 234–235, respectively; Decker, "Agriculture and Agricultural Technology," 398–399; Decker, *Tilling the Hateful Earth*, 88–92; Kaplan, *Les hommes et la terre*, 46–55; and Kazhdan, "The Peasantry," 48.

12. Decker, "Agriculture and Agricultural Technology," 398–399; Decker, *Tilling the Hateful Earth*, 88–92; and Kaplan, *Les hommes et la terre*, 46–55.

13. Bryer, "Means of Agricultural Production," 104–106; Decker, "Agriculture and Agricultural Technology," 399; Decker, *Tilling the Hateful Earth*, 194–195 and (for crop rotation and fallowing) 216–227; and Horden and Purcell, *Corrupting Sea*, 234–237. For trenching, see Dalby, *Geoponika*, 134 and 137–138 (5.20 and 5.26, vines), and 188–190 (9.9 and 9.10, olive trees).

14. Bryer, "Means of Agricultural Production," 109–110; Decker, "Agriculture and Agricultural Technology," 399; Decker, *Tilling the Hateful Earth*, 92–97; and Kazhdan, "The Peasantry," 48–49.

15. Bryer, "Means of Agricultural Production," 110–112; Decker, "Agriculture and Agricultural Technology," 402–403; and Kazhdan, "The Peasantry," 49.

16. The extensive scholarship on ancient and medieval olive oil production includes Etan Ayalon, Rafael Frankel, and Amoss Kloner, eds., *Oil and Wine Presses in Israel from the Hellenistic, Roman and Byzantine Periods*, BAR International Series, 1972 (Oxford: Archaeopress, 2009); Ahmet, K. "A Middle Byzantine Olive Press Room at Aphrodisias," *Anatolian Studies* 51 (2001): 150–167; and Decker, *Tilling the Hateful Earth*, 149–173.

17. Decker, *Tilling the Hateful Earth*, 121–148; Ayalon, Frankel, and Kloner, eds., *Oil and Wine Presses*; Rafael Frankel, "Introduction," in Ayalon, Frankel, and Kloner, *Oil and Wine Presses*, 1–18, esp. 2–3; Kaplan, *Les hommes et la terre*, 69–74; and Lefort, "Rural Economy," 254–256.

18. See Walter Ashburner, "The Farmer's Law," *Journal of Hellenic Studies* 32 (1912): 68–95, quote on 90 (clause 30); and see Wickham, *Framing the Early Middle Ages*, 462–464.

19. Bryer, "Means of Agricultural Production," 102–104; Kaplan, *Les hommes et la terre*, 74–79; Kazhdan, "The Peasantry," 52–55; Lefort, "Rural Economy," 263–267; and for pigeons, Decker, *Tilling the Hateful Earth*, 60.

20. Angeliki E. Laiou, "Women in Byzantine Society," in *Women in Medieval Western European Culture*, ed. Linda E. Mitchell (New York: Garland Publishing, 2016), 81–94, esp. 90–94; Alice-Mary Talbot, "Women," in Cavallo, ed., *The Byzantines*, 117–143, esp. 126–127, 129; and Wickham, *Framing the Early Middle Ages*, 556–558.

21. Alan K. Bowman and Eugene Rogan, "Agriculture in Egypt from Pharaonic to Modern Times," in *Agriculture in Egypt from Pharaonic to Modern Times*, ed. Bowman and Rogan (Oxford: Oxford University Press, 1999), 1–32; and Wickham, *Framing the Middle Ages*, 22–25.

22. Bowman and Rogan, "Agriculture in Egypt," 4–6; Mikhail, *Under Osman's Tree*, 111–130; Gladys Frantz-Murphy, "Land-Tenure in Egypt in the First Five Centuries of Islamic Rule (Seventh–Twelfth Centuries AD)," and James G. Keenan, "Fayyum Agriculture at the End of the Ayyubid Era: Nabulsi's *Survey*," both in Bowman and Rogan, eds., *Agriculture in Egypt*, 237–266 and 287–299, respectively; and Wickham, *Framing the Early Middle Ages*, 22–25, 275–280.

23. Alan Mikhail, *Nature and Empire in Ottoman Egypt: An Environmental History* (Cambridge: Cambridge University Press, 2011).

24. David Mattingly with a contribution from Crispin Flower, "Romano-Libyan Settlement: Site Distribution and Trends," in *Farming the Desert: The UNESCO Libyan Valleys Archaeological Survey*, ed. Graeme Barker, vol. 1, *Synthesis* (London: UNESCO, the Department of Antiquities,

Tripoli, and the Society for Libyan Studies, 1996), 159–190, esp. 162–163 (which shows the complexity of settlement patterns and the difficulty of interpreting them); Corisande Fenwick, *Early Islamic North Africa: A New Perspective* (London: Bloomsbury, 2020), 81–82; and Wickham, *Framing the Early Middle Ages*, 18–19.

25. Mattingly with Flower, "Romano-Libyan Settlement," esp. 155–158 and 170–190; and Fenwick, *Early Islamic North Africa*, 81–82.

26. Fenwick, *Early Islamic North Africa*, 86–88, 94.

27. Fenwick, *Early Islamic North Africa*, 93; and Marijke van der Veen, Annie Grant, and Graeme Barker, "Romano-Libyan Agriculture: Crops and Animals," in Barker, ed., *Farming the Desert*, 228–263.

28. Fenwick, *Early Islamic North Africa*, 95–99; and van der Veen, Grant, and Barker, "Romano-Libyan Agriculture," 228–263.

29. See esp. Andrew M. Watson, *Agricultural Innovation in the Early Islamic World: The Diffusion of Crops and Farming Techniques, 700–1100* (Cambridge: Cambridge University Press, 1983).

30. Michael Decker, "Plants and Progress: Rethinking the Islamic Agricultural Revolution," *Journal of World History* 20 (June 2009): 187–206. For a cogent discussion of the influence of the Watson thesis, see Paolo Squatriti, "Of Seeds, Seasons, and Seas: Andrew Watson's Medieval Agrarian Revolution Forty Years Later," *Journal of Economic History* 74 (December 2014): 1205–1220.

31. See esp. the chapters for the earlier centuries (500–1750) in the four-volume series, which includes Erik Thoen and Tim Soens, eds., *Struggling with the Environment*, vol. 1 in *Rural Economy and Society in North-Western Europe, 500–2000*, ed. Erik Thoen (Turnhout: Brepols, 2015); Leen Van Molle and Yves Segers, eds., *The Agro-Food Market: Production, Distribution and Consumption*, vol. 2 in *Rural Economy and Society in North-Western Europe, 500–2000*, ed. Erik Thoen (Turnhout: Brepols, 2013); Eric Vanhaute, Isabelle Devos, and Thijs Lambrecht, eds., *Making a Living: Family, Income and Labour*, vol. 3 in *Rural Economy and Society in North-Western Europe, 500–2000*, ed. Erik Thoen (Turnhout: Brepols, 2011); and Bas J. P. van Bavel and Richard W. Hoyle, eds., *Social Relations: Property and Power*, vol. 4 in *Rural Economy and Society in North-Western Europe, 500–2000*, ed. Erik Thoen (Turnhout: Brepols, 2010). For northern French and northern German lands, see Bernard Delmaire and Fulgence Delleaux, "Northern France, 1000–1700," and Klaus-Joachim Lorenzen-Schmidt, "Northwest Germany, 1000–1750," both in Thoen and Soens, *Struggling with the Environment*, 149–184 and 309–338, respectively. More generally, see Wickham, *Framing the Early Middle Ages*, 465–481.

32. See esp. Jean-Pierre Devroey and Anne Nissen[-Jaubert], "Early Middle Ages, 500–1000," in Thoen and Soens, *Struggling with the Environment*, 10–68, esp. 34–41; Christopher Dyer, *Making a Living in the Middle Ages: The People of Britain, 850–1520* (New Haven: Yale University Press, 2002), 23–25; and Adriaan Verhulst, *The Carolingian Economy* (Cambridge: Cambridge University Press, 2002), 31–37.

33. Lynn White Jr., *Medieval Technology and Social Change* (Oxford: Oxford University Press, 1962), 39–78. Balanced critiques include Horden and Purcell, *Corrupting Sea*, 232–234; and Michael Toch, "Agricultural Progress and Agricultural Technology in Medieval Germany: An Alternative Model," in *Technology and Resource Use in Medieval Europe: Cathedrals, Mills, and Mines*, ed. Elizabeth Bradford Smith and Michael Wolfe (Aldershot: Ashgate, 1997), 158–169. See also Bert Hall, "Lynn White's *Medieval Technology and Social Change* after Thirty Years," in *Technological Change: Methods and Themes in the History of Technology*, ed. Robert Fox (Amsterdam: Overseas Publishers Association, 1996), 85–101.

34. See Devroey and Nissen, "Early Middle Ages, 500–1000," 41–46; Joachim Henning, "Did the 'Agricultural Revolution' Go East with Carolingian Conqest? Some Reflections on Early

Medieval Rural Economics of the Baiuvarii and Thuringi," in *The Baiuvarii and Thuringi: An Ethnographic Perspective*, ed. Janine Fries-Knoblach and Heiko Steuer with John Hines (San Marino: Boydell Press, 2014), 331–359; and Paolo Squatriti, *Weeds and the Carolingians: Empire, Culture, and Nature in Frankish Europe, AD 750–900* (Cambridge: Cambridge University Press, 2022), 53–55.

35. Karl Brunner, "Continuity and Discontinuity of Roman Agricultural Knowledge in the Early Middle Ages," in *Agriculture in the Middle Ages: Technology, Practice, and Representation*, ed. Del Sweeney (Philadelphia: University of Pennsylvania Press, 1995), 21–49; Georges Comet, *Le paysan et son outile: Essai d'histoire technique des cereals (France, VIIIe–XVe siècle)* (Rome: École Française de Rome, 1992), 47–82; Comet, "Technology and Agricultural Expansion in the Middle Ages: The Example of France North of the Loire," and Georges Raepsaet, "The Development of Farming Implements between the Seine and the Rhine from the Second to the Twelfth Centuries," both in *Medieval Farming and Technology: The Impact of Agricultural Change in Northwest Europe*, ed. Grenville Astill and John Langdon (Leiden: Brill, 1997), 11–39 and 40–68, respectively; François Sigaut, "L'Evolution des techniques," and Pascal Reigniez, "Histoire et techniques: L'outil Agricole dans la periode du Haut Moyen-Âge (V^e–Xe S.)," both in *The Making of Feudal Agricultures?*, ed. Miquel Barceló and François Sigaut (Leiden: Brill, 2004), 1–31, esp. 22–23 for the scythe, and 33–113, respectively; Toch, "Agricultural Progress," 158–169; and François Sigaut, "Crops and Agricultural Developments in Western Europe," in *Early Agricultural Remnants and Technical Heritage (EARTH): 8,000 Years of Resilience and Innovations*, ed. Patricia C. Anderson and Leonor Peña-Chocarro (Oxford: Oxbow Books, 2014), 107–112.

36. Horden and Purcell, *Corrupting Sea*, 263–269; Squatriti, *Weeds and the Carolingians*, 40; David Stone, *Decision-Making in Medieval Agriculture* (Oxford: Oxford University Press, 2005), esp. 3–21 and 231–276.

37. Adrian Verhulst, "Economic Organization," and Chris Wickham, "Rural Society in Carolingian Europe," both in *The New Cambridge Medieval History*, vol. 2, *c. 700–c. 900*, ed. Rosamond McKitterick (Cambridge: Cambridge University Press, 1995), 481–509 and 510–537, respectively. A detailed examination of crop rotation in one region underscores the complexity of the issue: Eric Thoen, "The Birth of 'The Flemish Husbandry': Agricultural Technology in Medieval Flanders," in Astill and Langdon, *Medieval Farming and Technology*, 69–88, esp. 74–77.

38. Paul J. Gans, "The Medieval Horse Harness: Revolution or Evolution? A Case Study in Technological Change," in *Villard's Legacy: Studies in Medieval Technology, Science and Art in Memory of Jean Gimpel*, ed. Marie-Thérèse Zenner (Aldershot: Ashgate Publishing, 2004), 175–187; John Langdon, *Horses, Oxen, and Technological Innovation: The Use of Draught Animals in English Farming from 1066 to 1550* (Cambridge: Cambridge University Press, 1986), 4–21; Georges Raepsaet, "Les prémices de la mécanisation agricole entre Seine et Rhin de l'antiquité au 13e siècle," *Annales. Histoire, Sciences Sociales* 50 (July–August 1995), 911–942; and Raepsaet, *Attelages et technique de transport dans le monde gréco-romain* (Brussels: Laboratoire d'Archéologie Classique de l'Université Libre de Bruxelles, 2002).

39. For the complicated and variable kinds of tenancy, see Bas van Bavel and Richard Hoyle, "Introduction: Social Relations, Property and Power in the North Sea Area, 500–2000," in Bavel and Hoyle, *Social Relations: Property and Power*, 1–47, as well as the other essays devoted to the earlier centuries in this volume; and see Thoen and Soens, *Struggling with the Environment*. See also Bas J. P. van Bavel, *Manors and Markets: Economy and Society in the Low Countries, 500–1600* (Oxford: Oxford University Press, 2010); Verhulst, "Economic Organisation," 488–499; and Wickham, *Framing the Early Middle Ages*, 259–302.

40. Jean-Pierre Devroey and Anne Nissen Jaubert, "Family, Income and Labour around the North Sea, 500–1000," in Vanhaute, Devos, and Lambrecht, *Making a Living*, 4–44, esp. 19–23; and Verhulst, "Economic Organisation," esp. 488–499. See also, Jean-Pierre Devroey, Alexis

Wilkin, and Alban Gautier, "Agricultural Prodution, Distribution and Consumption around the North Sea, 500–1000," in Molle and Segers, *Agro-Food Market*, 12–65, esp. 44–58.

41. Madonna J. Hettinger, "So Strategize: The Demands of the Day of the Peasant Woman in Medieval Europe," in Mitchell, *Women in Medieval Western European Culture*, 47–63; Phillipp Schofield and Jane Whittle, "Britain, 1000–1750," Gérhard Béaur and Larent Feller, "Northern France, 1000–1750," and Isabelle Devos, Thijs Lambrecht, and Richard Paping, "The Low Countries, 1000–1750," all in Vanhaute, Devos, and Lambrecht, *Making a Living*, 47–69, esp. 53–56, 98–125, esp. 19–112, and 157–183, respectively; Jane Whittle, "Rural Economies," in *The Oxford Handbook of Women and Gender in Medieval Europe*, ed. Judith M. Bennett and Ruth Mazo Karras (Oxford: Oxford University Press, 2013), 311–326; and Mary E. Wiesner-Hanks, *Women and Gender in Early Modern Europe*, 4th ed. (Cambridge: Cambridge University Press, 2019), 118–125. For the Netherlands, see Bavel, *Manors and Markets*, esp. 124–161.

42. Verhulst, "Economic Organisation," 481–483; and for a discussion from an environmental point of view, Aberth, *Environmental History*, 92–97. For drainage of peat bogs in the Netherlands, see Peter Hoppenbrouwers, "Agricultural Production and Technology in the Netherlands, c. 1000–1500," in Astill and Langdon, *Medieval Farming and Technology*, 89–113; and Hoffman, *Environmental History of Europe*, 167–168. For the draining of the English fens in the late sixteenth century, see Eric H. Ash, *The Draining of the Fens: Projectors, Popular Politics, and State Building in Early Modern England* (Baltimore: Johns Hopkins University Press, 2017).

43. See esp. Mark Overton, *Agricultural Revolution in England: The Transformation of the Agrarian Economy, 1500–1800* (Cambridge: Cambridge University Press, 1996); and Alexandra Sapoznik, "Britain, 1000–1750," in Thoen and Soens, *Struggling with the Environment*, 71–107, esp. 90–98.

44. Overton, *Agricultural Revolution*, 147–168; and Sapoznik, "Britain, 1000–1750," 91–94.

45. See esp. Hoffmann, *Environmental History*, 113–154.

46. For an overview, see Mosley, *Environment in World History*, 31–55; and see Aberth, *Environmental History*, 77–97; Bruno Andreolli and Massimo Montanari, eds., *Il bosco nel medioevo* (Bologna: Cooperativa Libraria Universitaria Editrice Bologna, 1988); Simonetta Cavaciocchi, ed., *L'uomo e la foresta secc. XIII–XVIII* (Prato: Istituto Internazionale di Storia Economica "F. Datini," 1995); Hoffmann, *Environmental History*, 119–133; Oliver Rackham, *Trees and Woodland in the British Landscape: The Complete History of Britain's Trees, Woods, and Hedgerows*, rev. ed. (London: J. M. Dent, 1990), 164–173; and Péter Szabó, *Woodland and Forests in Medieval Hungary*, BAR S1348 (Oxford: Archaeopress, 2005), esp. 19–25, 87–97, and 119–127.

47. Aberth, *Environmental History*, 87–92; and Hoffmann, *Environmental History*, 188–195.

48. Hoffmann, *Environmental History*, 184–188; Rackham, *Trees and Woodland*, esp. 4–10; and Szabó, *Woodland and Forests*, esp. 20–21 and 57–84.

49. Mateusz Falkowski, "Fear and Abundance: Reshaping of Royal Forests in Sixteenth-Century Poland and Lithuania," *Environmental History* 22 (October 2017): 618–642. And see also Karl Appuhn, *A Forest on the Sea: Environmental Expertise in Renaissance Venice* (Baltimore: Johns Hopkins University Press, 2009); and John T. Wing, *Roots of Empire: Forests and State Power in Early Modern Spain, c. 1500–1750* (Leiden: Brill, 2015), esp. 1–84.

50. For a global overview, see Brian Fagan, *Fishing: How the Sea Fed Civilization* (New Haven: Yale University Press, 2017). For medieval Europe, see esp. Richard C. Hoffmann, *The Catch: An Environmental History of Medieval European Fisheries* (Cambridge: Cambridge University Press, 2023), 55–58 for religious and other reasons for catching fish; and Hoffmann, "Medieval Fishing," in *Working with Water in Medieval Europe: Technology and Resource Use*, ed. Paolo Squatriti (Leiden: Brill, 2000), 331–393.

51. Hoffmann, *The Catch*, 268–315.

52. Ruthy Gertwagen, "Towards a Maritime Eco-history of the Byzantine and Medieval

Eastern Mediterranean," in *The Inland Seas: Towards an Ecohistory of the Mediterranean and the Black Sea*, ed. Tønnes Bekker-Nielsen and Ruthy Gertwagen (Stuttgart: Franz Steiner, [2016]), 341–368.

53. Richard C. Hoffmann, *Fishers' Craft and Lettered Art: Tracts on Fishing from the End of the Middle Ages* (Toronto: University of Toronto Press, 1997); Hoffmann, "Medieval Fishing," 343–352; and Hoffmann, *The Catch*, esp. 111–121.

54. Hoffmann, *Fishers' Craft*, esp. 43–45; Hoffmann, "Medieval Fishing," 352–364; and Hoffmann, *The Catch*, 121–128.

55. Hoffmann, "Medieval Fishing," 373–393; and Hoffmann, *The Catch*, 121–128.

56. Paolo Squatriti, *Water and Society in Early Medieval Italy, AD 400–1000* (Cambridge: Cambridge University Press, 1998), 97–123.

57. Sabine Florence Fabijanec, "Fishing and the Fish Trade on the Dalmatian Coast in the Late Middle Ages," in Bekker-Nielsen and Gertwagen, *Inland Seas*, 369–386.

58. See esp. Maryanne Kowaleski, "The Early Documentary Evidence for the Commercialisation of the Sea Fisheries in Medieval Britain," and Jennifer F. Harland et al., "Fishing and Fish Trade in Medieval York: The Zooarchaeological Evidence," in *Cod and Herring: The Archaeology and History of Medieval Sea Fishing*, ed. James H. Barrett and David R. Orton (Oxford: Oxbow Books, 2016), 23–35 and 172–204, respectively.

59. For the technologies of fishing in medieval England, see esp. Michael Aston, ed., *Medieval Fish, Fisheries and Fishponds in England*, 2 vols. (Oxford: BAR Publishing, 1988); Delle Hooke, "Use of Waterways in Anglo-Saxon England," in *Waterways and Canal Building in Medieval England*, ed. John Blair (Oxford: Oxford University Press, 2007), 37–54, esp. 45–54; and for a case study on one river, Terry Moore-Scott, "Medieval Fish Weirs on the Mid-Tidal Reaches of the Severn River (Ashleworth-Arlingham)," *Glevensis* 42 (2009): 31–44.

60. See esp. Fagan, *Fishing*, 243–256; Hoffmann, *The Catch*, 258–402; Poul Holm, "Commercial Sea Fisheries in the Baltic Region, c. AD 1000–1600," and James H. Barrett, "Medieval Sea Fishing, AD 500–1500: Chronology, Causes and Consequences," both in Barrett and Orton, *Cod and Herring*, 12–22 and 250–272, respectively; and Kowaleski, "Early Documentary Evidence."

61. Hoffmann, *The Catch*, esp. 378–383; Alf Ragnar Nielssen, "Early Commercial Fisheries and the Interplay among Farm, Fishing Station and Fishing Village in North Norway," and Arnved Nedkvitne, "The Development of the Norwegian Long-Distance Stockfish Trade," both in Barrett and Orton, *Cod and Herring*, 42–49 and 50–79, respectively.

62. Helge Sørheim, "The Birth of Commercial Fisheries and the Trade of Stockfish in the Borgundfjord, Norway," in Barrett and Orton, *Cod and Herring*, 60–70.

63. Fagan, *Fishing*, 254–255; and Hoffmann, *The Catch*, 326–329.

CHAPTER 2: Hydraulic Technologies

Epigraph: Yossef Rapoport and Ido Shahar, ed. and trans., *The Villages of the Fayyum: A Thirteenth-Century Register of Rural, Islamic Egypt* (Turnhout: Brepols, 2018), 41.

1. See esp. Brendan James Haug, "Watering the Desert: Environment, Irrigation, and Society in the Premodern Fayyūm, Egypt" (PhD diss.: University of California, Berkeley, 2012), 79–82; Yossef Rapoport and Ido Shahar, "Irrigation in the Medieval Islamic Fayyum: Local Control in a Large-Scale Hydraulic System," *Journal of Economic and Social History of the Orient* 55 (2012): 1–31; Rapoport, *Rural Economy and Tribal Society in Islamic Egypt: A Study of al-Nābulusī's Villages of the Fayyum* (Turnhout: Brepols, 2018); and for an edition of al-Nābulusī's text, Rapoport and Shahar, *Villages of the Fayyum*.

2. Karl Wittfogel, *Oriental Despotism: A Comparative Study of Total Power* (New Haven: Yale University Press, 1956); and Rapoport and Shahar, "Irrigation in the Medieval Islamic Fayyum." See also Haug, "Watering the Desert," 7–32, who cogently shows that Wittfogel's

views about the central state and irrigation in Egypt are connected to colonial engineering ideologies from the eighteenth to the twentieth century.

3. See esp. Gideon Avni, "Early Islamic Irrigated Farmsteads and the Spread of *Qanāts* in Eurasia," *Water History* 10 (2018): 313–338; Negar Sanaan Bensi, "The Qanat System: A Reflection on the Heritage of the Extraction of Hidden Waters," in *Adaptive Strategies for Water Heritage: Past, Present and Future*, ed. Carola Hein (Cham: Springer, 2020): 41–56; Julien Charbonnier and Kristen Hopper, "The *Qanāt*: A Multidisciplinary and Diachronic Approach to the Study of Groundwater Catchment Systems in Archaeology," *Water History* 10 (2018): 3–11; Thomas F. Glick, *Islamic and Christian Spain in the Early Middle Ages*, 2nd. rev. ed. (Leiden: Brill, 2005), 255–257; and Donald R. Hill, *Islamic Science and Engineering* (Edinburgh: Edinburgh University Press, 1993), 181–183. For Karajī's treatise, see Kaveh Niazi, "Karajī's Discourse on Hydrology," *Oriens* 44 (2016): 44–68.

4. This description is primarily based on Avni, "Early Islamic Irrigated Farmsteads," 314.

5. Antonio Rotolo, "Drainage Galleries in the Iberian Peninsula during the Islamic Period," *Water History* 6 (2014): 191–210; and see also Ramón Martinez-Medina, Encarnación Gil-Meseguer, and José Gómez-Espin, "Research on Qanats in Spain," *Water History* 10 (2018): 339–355.

6. Glick, *Islamic and Christian Spain*, 257–258; Hill, *Islamic Science and Engineering*, 92–93; Hill, "Mechanical Technology," in *Science and Technology in Islam*, pt. 2: *Technology and Applied Sciences*, ed. A. Y. al-Hassan, Maqbul Ahmed, and A. Z. Iskandar (Paris: UNESCO, 2001), 165–192, esp. 165; and D. Fairchild Ruggles, "Waterwheels and Garden Gizmos: Technology and Illusion in Islamic Gardens," in *Wind and Water in the Middle Ages: Fluid Technologies from Antiquity to the Renaissance*, ed. Steven A. Walton (Tempe: Arizona Center for Medieval and Renaissance Studies, 2006), 69–88, esp. 73–75.

7. Michael Decker, *Tilling the Hateful Earth: Agricultural Production and Trade in the Late Antique East* (Oxford: Oxford University Press, 2009), 199–203; Hill, *Islamic Science and Engineering*, 94–95; Hill, "Mechanical Technology," 166–167; and Ruggles, "Waterwheels and Garden Gizmos," 72–73.

8. Decker, *Tilling the Hateful Earth*, 189–203; Glick, *Islamic and Christian Spain*, 258–261; and Hill, "Mechanical Technology," 167–169. See also Ana Duarte Rodrigues and Magdalena Merlos Romero, "Noras, Norias and Technology-of-Use," in *The History of Water Management in the Iberian Peninsula between the 16th and 18th Centuries*, ed. Ana Duarte Rodrigues and Carmen Toribio Marin (Cham: Springer, 2020), 331–350. The concept of "technology of use" comes from David Edgerton, *The Shock of the Old: Technology and Global History since 1900* (Oxford: Oxford University Press, 2011).

9. Marc Bloch, "Avènement et conquête du Moulin à eau," *Annales d'histoire économique et histoire* 7 (1935): 538–563, translated as "The Advent and Triumph of the Watermill," in *Land and Work in Mediaeval Europe: Selected Papers by Marc Bloch*, trans. J. E. Anderson (New York: Harper & Row, 1969), 136–168; M. I. Finley, *The Ancient Economy*, 2nd ed., updated by Ian Morris (Berkeley: University of California Press, 1999); Jean Gimpel, *The Medieval Machine: The Industrial Revolution of the Middle Ages* (Hammondsworth, Middlesex: Penguin Books, 1976), esp. 1–28; and Lynn White Jr., *Medieval Technology and Social Change* (Oxford: Oxford University Press, 1962), esp.79–89. For a comprehensive discussion of this and related scholarship on the water mill, see Miquel Barceló, "The Missing Water-Mill: A Question of Technological Diffusion in the High Middle Ages," in *The Making of Feudal Agricultures?*, ed. Miquel Barceló and François Sigaut (Leiden: Brill, 2004), 255–314.

10. See esp. Kevin Greene, "Technological Innovation and Economic Progress in the Ancient World: M. I. Finley Reconsidered," *Economic History Review* 53 (2000): 29–59; Örjan Wikander, "Sources of Energy and Exploitation of Power," in *Engineering and Technology in*

the Classical World, ed. John Peter Oleson (Oxford: Oxford University Press, 2008), 136–157 (a good summary); and Wikander, "The Watermill," in *Handbook of Ancient Water Technology,* ed. Örjan Wikander (Leiden: Brill, 2000), 371–410. For a critique of Bloch's essay, see Paolo Squatriti, "'Advent and Conquests' of the Water Mill in Italy," in *Technology and Resource Use in Medieval Europe: Cathedrals, Mills, and Mines,* ed. Elizabeth Bradford Smith and Michael Wolfe (Aldershot: Ashgate, 1997), 125–138. See also George Brooks, "The 'Vitruvian Mill' in Roman and Medieval Europe," and Adam Lucas, "The Role of the Monasteries in the Development of Medieval Milling," both in Walton, *Wind and Water,* 1–38 and 89–127, respectively; Richard Holt, "Mechanization and the Medieval English Economy," in Smith and Wolfe, eds., *Technology and Resource Use,* 139–57; John Langdon, *Mills in the Medieval Economy: England 1300–1540* (Oxford: Oxford University Press, 2004); Adam Robert Lucas, *Wind, Water, Work: Ancient and Medieval Milling Technology* (Leiden: Brill, 2006); and Lucas, "Industrial Milling in the Ancient and Medieval Worlds: A Survey of the Evidence for an Industrial Revolution in Medieval Europe," *Technology and Culture* 46 (January 2005): 1–30.

11. See esp. Barceló, "Missing Water-Mill," esp. 276–296; Ricardo Córdoba de la Llave, "Some Reflections on the Use of Water Power in Al-Andalus," in *Economia e Energia Secc. XIII–XVIII,* Atti della Trentaquattresima Settimana di Studi, ed. Simonetta Cavaciocchi (Florence: Le Monnier, [2003]); Robert Cresswell, "Of Mills and Waterwheels: The Hidden Parameters of Technological Choice," in *Technological Choices: Transformation in Material Cultures since the Neolithic,* ed. Pierre Lemonnier (London: Routledge, 1993), 181–213; Thomas F. Glick and Helena Kirchner, "Hydraulic Systems and Technologies of Islamic Spain: History and Archaeology," in *Working with Water in Medieval Europe: Technology and Resource-Use,* ed. Paolo Squatriti (Leiden: Brill, 2000), 266–329; and Hill, *Islamic Science and Engineering,* 105–113.

12. In addition to the references in the previous note, see Glick, *Islamic and Christian Spain,* 261–267.

13. Brooks, "Vitruvian Mill," esp. 26–32; David Crossley, "The Archaeology of Water Power in Britain before the Industrial Revolution," in Smith and Wolfe, *Technology and Resource Use,* 109–124; and Paolo Squatriti, *Water and Society in Early Medieval Italy, AD 400–1000* (Cambridge: Cambridge University Press, 1998), 126–159.

14. Thomas F. Glick and Luis Pablo Martinez, "Mills and Millers in Medieval Valencia," Janet Loengard, "Lords' Rights and Neighbors' Nuisances: Mills and Medieval English Law," and Roberta Magnusson, "Public and Private Urban Hydrology: Water Management in Medieval London," all in Walton, *Wind and Water,* 189–211, 129–152, and 171–187, respectively; and Squatriti, *Water and Society,* 150–159.

15. Decker, *Tilling the Hateful Earth,* 178–184; Faisal Husain, "Sediment of the Tigris and Euphrates Rivers: An Early Modern Perspective," *Water History* 13 (2021): 13–32; and for a general overview, Hill, *Islamic Science and Technology,* 170–186.

16. Decker, *Tilling the Hateful Earth,* 184–189. See also Frank Braemer et al., "Conquest of New Lands and Water Systems in the Western Fertile Crescent (Central and Southern Syria)," *Water History* 2 (2010): 91–114, which comprises an archaeological and historical study of a long chronological period but points to the importance of cisterns that collected rainwater on the roof of every house as well as the significance of local village control and maintenance in the medieval period.

17. Decker, *Tilling the Hateful Earth,* 193–197.

18. Alan Mikhail, *Nature and Empire in Ottoman History: An Environmental History* (Cambridge: Cambridge University Press, 2011), esp. 1–123; and Mikhail, *Under Osman's Tree: The Ottoman Empire, Egypt and Environmental History* (Chicago: University of Chicago Press, 2017), 19–33.

19. Mikhail, *Nature and Empire,* 38–81.

20. Mikhail, *Nature and Empire*; Mikhail, *Under Osman's Tree*, esp.1–33 and 59–64; Rapoport and Shahar, "Irrigation in the Medieval Islamic Fayyum," 1–6; and see also Haug, "Watering the Desert," 76–106.

21. Haug, "Watering the Desert," 76–106; and Rapoport and Shahar, "Irrigation in the Medieval Islamic Fayyum," 14–28

22. Mikhail, *Nature and Empire*, 38–81; and Mikhail, *Under Osman's Tree*, 19–33.

23. Graeme Barker, "Castles in the Desert," in *Farming the Desert: The UNESCO Libyan Valleys Archaeological Survey*, vol. 1, *Synthesis*, ed. Graeme Barker (London: UNESCO, the Department of Antiquities, Tripoli, and the Society for Libyan Studies, 1996), 1–20.

24. David D. Gilbertson and Chris Hunt, "Romano-Libyan Agriculture: Walls and Flood-water Farming," in Barker, *Farming the Desert*, 191–225.

25. Graeme Barker and David D. Gilbertson with Chris O. Hunt and David Mattingly, "Romano-Libyan Agriculture: Integrated Models," David D. Gilbertson, "Explanations: Environment as Agency," David Mattingly, "Explanations: People as Agency," Graeme Barker with David D. Gilbertson, "Farming the Desert: Retrospect and Prospect," all in Barker, *Farming the Desert*, 265–290, 291–317, 319–342, and 343–363, respectively.

26. Glick and Kirchner, "Hydraulic Systems and Technologies of Islamic Spain."

27. Christopher Gerrard, "Contest and Cooperation: Strategies for Medieval and Later Irrigation along the Upper Huecha Valley, Aragón, North-East Spain," *Water History* 3 (2011): 2–28.

28. Squatriti, *Water and Society*, 79–95.

29. See esp. Hadrian Cook et al., "The Origins of Water Meadows in England," *Agricultural History Review* 51 (2003): 155–162; and Hans Renes et al., "Water Meadows as European Agricultural Heritage," in *Adaptive Strategies for Water Heritage: Past, Present and Future*, ed. Carola Hein (Cham: Springer, 2020), 107–130.

30. Squatriti, *Water and Society*, 76–79.

31. Petra J.E.M. van Dam, "Sinking Peat Bogs: Environmental Change in Holland, 1350–1550," *Environmental History* 6 (January 2001): 32–45; Richard C. Hoffmann, *An Environmental History of Medieval Europe* (Cambridge: Cambridge University Press, 2014), 167–168 and 202–203; Peter Hoppenbrouwers, "Agricultural Production and Technology in the Netherlands, c. 100–1500," in *Medieval Farming and Technology: The Impact of Agricultural Change in Northwest Europe*, ed. Grenville Astill and John Langdon (Leiden: Brill, 1997), 89–114, esp. 96–101; Iason Jongepier et al., "The Brown Gold: A Reappraisal of Medieval Peat Marshes in Northern Flanders (Belgium)," *Water History* 3 (2011): 73–93; William H. TeBrake, "Taming the Waterwolf: Hydraulic Engineering and Water Management in the Netherlands during the Middle Ages," *Technology and Culture* 43 (July 2002): 475–499; and Erik Thoen and Tim Soens, "The Low Countries, 1000–1750," in *Struggling with the Environment: Land Use and Productivity*, ed. Erik Thoen and Tim Soens, vol. 1 in *Rural Economy and Society in North-Western Europe, 500–2000*, ed. Erik Thoen (Turnhout: Brepols, 2015), 221–258, esp. 233–243.

32. Dam, "Sinking Peat Bogs"; Hoffmann, *Environmental History*, 202–203; Hoppenbrouwers, "Agricultural Production," 96–101; Jongepier et al., "Brown Gold," 73–93; Arne Kaijser, "System Building from Below: Institutional Change in Dutch Water Control Systems," *Technology and Culture* 43 (July 2002): 521–548; and TeBrake, "Taming the Waterwolf," 475–499.

33. TeBrake, "Taming the Waterwolf," 481–485.

34. Petra J.E.M. van Dam, "Ecological Challenges, Technical Innovations: The Modernization of Sluice Building in Holland, 1300–1600," *Technology and Culture* 43 (July 2002): 500–520; and Kaijser, "System Building from Below."

35. Eric H. Ash, *The Draining of the Fens: Projectors, Popular Politics, and State Building*

in Early Modern England (Baltimore: Johns Hopkins University Press, 2017), esp. 1–80 for the decades covered by the present book.

36. For numerous studies of water systems in French monasteries, see Léon Pressouyre et al., eds., *L'hydraulique monastique: Milieux, reseaux, usage* (Grâne: Créaphis, 1996); and see Sheila Bonde and Clark Maines, "Inside the Black Box: The Technology of Medieval Water Management at the Charterhouse of Bourgfontaine," *Technology and Culture* 53 (July 2012): 625–670; and Squatriti, *Water and Society*, 19–21.

37. Roberta J. Magnusson, *Water Technology in the Middle Ages: Cities, Monasteries, and Water Works after the Roman Empire* (Baltimore: Johns Hopkins University Press, 2001), 55–63.

38. Magnusson, *Water Technology*, 63–97 and 101–105.

39. Magnusson, *Water Technology*, 55–63; and Michael P. Kucher, *The Water Supply System of Siena, Italy: The Medieval Roots of the Modern Networked City* (New York: Routledge, 2005).

40. Squatriti, *Water and Society*, 66–76.

41. Brian Fagan, *The Little Ice Age: How Climate Made History, 1300–1850* (New York: Basic Books, 2020), ix–xviii and 47–97.

42. See esp. Pamela O. Long, "Responses to a Recurrent Disaster: Flood Writings in Rome, 1476–1606," in *Disaster in the Early Modern World: Examinations, Representations, Interventions*, ed. Ovanes Akopyan and David Rosenthal (Abingdon: Routledge, 2024), 225–245; Long, *Engineering the Eternal City: Infrastructure, Topography, and the Culture of Knowledge in Late Sixteenth-Century Rome* (Chicago: University of Chicago Press, 2018), 19–61; and Maria Margarita Segarra Lagunes, *Il tevere e Roma: Storia di una simbiosi* (Rome: Gangemi Editore, 2004), 249–303. The first treatise on Tiber River flooding was written after the terrible flood of 1530 by Luis Gómez, the Spanish prelate: Luis Gómez, *The Floods of the Tiber with Additional Documents on the Tiber Flood of 1530*, trans. Chiara Bariviera, Pamela O. Long, and William L. North (New York: Italica Press, 2023).

43. Christian Rohr, "Floods of the Upper Danube River and Its Tributaries and Their Impact on Urban Economies (c. 1350–1600): The Examples of the Towns of Krems/Stein and Wels (Austria)," *Environment and History* 19 (May 2013): 133–148; Rohr, "Ice Jams and Their Impact on Urban Communities from a Long-Term Perspective (Middle Ages to the 19th Century)," in *The Power of Urban Water: Studies in Premodern Urbanism*, ed. Nicola Chiarenza, Annette Haug, and Ulrich Müller (Berlin: De Gruyter, 2020), 197–212, esp. 202–204; Christoph Sonnlechner, Severin Hohensinner, and Gertrud Haidvogle, "Floods, Fights and a Fluid River: The Viennese Danube in the Sixteenth Century," *Water History* 5 (2013): 173–194.

44. See esp. James A. Galloway, "Coastal Flooding and Socioeconomic Change in Eastern England in the Later Middle Ages," *Environment and History* 19 (May 2013): 173–207; and Tim Soens, "Flood Security in the Medieval and Early Modern North Sea Area: A Question of Entitlement?," *Environment and History* 19 (May 2013): 209–232.

CHAPTER 3: Urbanism, Building Construction, and Urban Water Supplies

Epigraphs: Ibn Battuta, *The Travels of Ibn Battuta, A.D. 1325–1354*, trans. H.A.R. Gibb with revisions and notes from the Arabic text edited by C. Defrémery and B. R. Sanguinetti, vol. 1 (1958, repr., Milkwood, NY: Kraus Reprint, 1986), quote on 41 and 42.

Christine de Pizan, *The Book of the City of Ladies*, trans. Rosalind Brown-Grant (London: Penguin Books, 1999), 12–13.

1. A book-length account of his life and travels with bibliography is Ross E. Dunn, *The Adventures of Ibn Battuta: A Muslim Traveler of the 14th Century*, rev. ed. (Berkeley: University of California Press, 2012).

2. There is a growing body of scholarship on Christine de Pizan. For an introduction, see

Nadia Margolis, *An Introduction to Christine de Pizan* (Gainesville: University Press of Florida, 2011).

3. See esp. Chris Wickham, *Framing the Early Middle Ages: Europe and the Mediterranean, 400–800* (Oxford: Oxford University Press, 2005), 591–596. Peregrine Horden and Nicholas Purcell, *The Corrupting Sea: A Study of Mediterranean History* (Oxford: Blackwell, 2000), 89–108, have abandoned cities as an analytic category and use "settlement" instead. For a more global perspective, see esp. Shadreck Chirikure, "Shades of Urbanism(s) and Urbanity in Pre-Colonial Africa," *Journal of Urban Archaeology* 1 (2020): 49–66; and Joyce Marcus and Jeremy A. Sabloff, "Introduction," in *The Ancient City: New Perspectives on Urbanism in the Old and New World*, ed. Joyce Marcus and Jeremy A. Sabloff (Santa Fe: School for Advanced Research Press, 2008), 3–26.

4. Classic studies are Spiro Kostof, *The City Shaped: Urban Patterns and Meanings through History* (Boston: Little, Brown, 1991); and Kostof with Greg Castillo, *The City Assembled: The Elements of Urban Form through History* (New York: Thames & Hudson, 2005). The bibliography on medieval and early modern urbanism is too large to cite here. It includes numerous works focused on particular cities; examples are Sarah Bassett, ed., *The Cambridge Companion to Constantinople* (Cambridge: Cambridge University Press, 2022); and Balázs Nagy et al., eds., *Medieval Buda in Context* (Leiden: Brill, [2016]).

5. Fabrizio Nevola, *Street Life in Renaissance Italy* (New Haven: Yale University Press, 2020), quotes on 19. Miri Rubin, *Cities of Strangers: Making Lives in Medieval Europe* (Cambridge: Cambridge University Press, 2020); and see the essays relevant to the chronology and geography of the present book in Zeynep Çelik, Diane Favro, and Richard Ingersoll, *Streets: Critical Perspectives on Public Space* (Berkeley: University of California Press, 1994).

6. See esp. Jan Gadeyne, "Short Cuts: Observations on the Formation of the Medieval Street System in Rome," and Lila Yawn, "Public Access, Action, and Display in Rome of the Later *Anni Mille*," both in *Perspectives on Public Space in Rome, from Antiquity to the Present Day*, ed. Gregory Smith and Jan Gadeyne (Farnham, Surrey: Ashgate, 2013), 67–83 and 85–105, respectively; and see, for the history of a single street, Maurizio Caperna, *La Lungara: Storia e vicende edilizie dell'area tra il Gianicolo e il Tevere* (Rome: Edizioni Quasar, 2013); and for street cleaning and sanitation, Pamela O. Long, *Engineering the Eternal City: Infrastructure, Topography, and the Culture of Knowledge in Late Sixteenth-Century Rome* (Chicago: University of Chicago Press, 2018), 43–62 and 163–188 with further bibliography.

7. In addition to those mentioned in note 4 above, other examples include Bárbara Boloix-Gallardo, ed., *A Companion to Islamic Granada* (Leiden: Brill, 2022); and Susana Zapke and Elisabeth Gruber, eds., *A Companion to Medieval Vienna* (Leiden: Brill, 2021).

8. See James Crow, "The Infrastructure of a Great City: Earth, Walls, and Water in Late Antique Constantinople," in *Technology in Transition, 300–650*, ed. Luke Lavan et al. (Leiden: Brill, 2007), 251–285, esp. 262–268 on the walls of Constantinople; and Michael Wolfe, *Walled Towns and the Shaping of France: From the Medieval to the Early Modern Era* (New York: Palgrave Macmillan, 2009).

9. Jessica Maier, *Rome Measured and Imagined: Early Modern Maps of the Eternal City* (Chicago: University of Chicago Press, 2015); and Ferdinand Opll, "The Heritage of Maps and City Views," in Zapke and Gruber, *Companion to Medieval Vienna*, 135–159.

10. Nicola Camerlenghi, *St. Paul's Outside the Walls: A Roman Basilica, from Antiquity to the Modern Era* (Cambridge: Cambridge University Press, 2018); and Finbarr Barry Flood, *The Great Mosque of Damascus: Studies on the Makings of an Umayyad Visual Culture* (Leiden: Brill, 2001).

11. Iñigo Almela, "Religious Architecture as an Instrument for Urban Renewal: Two Religious Complexes from the Saadian Period in Marrakesh," *Al-Masāq* 31, no. 3 (2019): 272–302;

Niall Brady, "The Gothic Barn of England: Icon of Prestige and Authority," in *Technology and Resource Use in Medieval Europe: Cathedrals, Mills, and Mines*, ed. Elizabeth Bradford Smith and Michael Wolfe (Aldershot: Ashgate, 1997), 76–105; and Maureen C. Miller, *The Bishop's Palace: Architecture and Authority in Medieval Italy* (Ithaca: Cornell University Press, 2000).

12. See esp. Anthony Gerbino and Stephen Johnston, *Compass and Rule: Architecture as Mathematical Practice in England* (Oxford: Museum of the History of Science, 2009); Ann C. Huppert, *Becoming an Architect in Renaissance Italy: Art, Science, and the Career of Baldassarre Peruzzi* (New Haven: Yale University Press, 2015); and Gül Kale, "From Measuring to Estimation: Definitions of Geometry and Architect-Engineer in Early Modern Ottoman Architecture," *Journal of the Society of Architectural Historians* 79 (June 2020): 132–151.

13. For settlement patterns in many locales in the early medieval centuries, see Niall Brady and Claudia Theune, eds., *Settlement Change across Medieval Europe: Old Paradigms and New Vistas* (Leiden: Sidestone Press, 2019); and John Schofield and Heiko Steuer, "Urban Settlement," in *The Archaeology of Medieval Europe*, vol. 1, *Eighth to Twelfth Centuries AD*, ed. James Graham-Campbell with Magdalena Valor (Aarhus: Aarhus University Press, 2007), 111–153.

14. For the plague, see Lester K. Little, ed., *Plague and the End of Antiquity: The Pandemic of 541–750* (Cambridge: Cambridge University Press, 2007); and see Albrecht Berger, "Urban Development and Decline: Fourth–Fifteenth Centuries," in *Cambridge Companion to Constantinople*, 33–49; Gilbert Dagron, "The Urban Economy, Seventh–Twelfth Centuries," Klaus-Peter Matschke, "The Late Byzantine Urban Economy, Thirteenth–Fifteenth Centuries," Paul Magdalino, "Medieval Constantinople: Built Environment and Urban Development," and Charalambos Bouras, "Master Craftsmen, Craftsmen, and Building Activities in Byzantium," all in *The Economic History of Byzantium from the Seventh through the Fifteenth Century*, ed. Angeliki E. Laiou, 3 vols. (Washington, DC: Dumbarton Oaks, 2002), 2: 393–462, 464–496, 529–537, and 539–554, respectively; and André Raymond, "The Ottoman Conquest and the Development of the Great Arab Towns," in Raymond, *Arab Cities in the Ottoman Period: Cairo, Syria and the Maghreb* (Aldershot: Ashgate 2002), 17–34.

15. Wickham, *Framing the Early Middle Ages*, 613–620.

16. Corisande Fenwick, "From Africa to Ifrīqiya: Settlement and Society in Early Medieval North Africa (650–800)," *Al-Masāq* 25, no. 1 (2013): 9–33.

17. Wickham, *Framing the Early Middle Ages*, 591–692.

18. Good introductions to urbanism in western Europe include Hans Andersson and Barbara Scholkmann, "Towns," in *The Archaeology of Medieval Europe*, vol. 2, *Twelfth to Sixteenth Centuries*, ed. Martin Carver and Jan Klápště (Aarhus: Aarhus University Press, 2011), 370–407; David Nicholas, *The Growth of the Medieval City: From Late Antiquity to the Early Fourteenth Century* (London: Longman, 1990); Nicholas, *The Later Medieval City, 1300–1500* (London: Longman, 1997); D. M. Palliser, *The Cambridge Urban History of Britain*, vol. 1, *600–1540* (Cambridge: Cambridge University Press, 2000); and Wickham, *Framing the Early Middle Ages*, 591–692, esp. 685–688 (for coastal emporia).

19. Caroline Goodson, *Cultivating the City in Early Medieval Italy* (Cambridge: Cambridge University Press, 2021); D. Fairchild Ruggles, *Islamic Gardens and Landscapes* (Philadelphia: University of Pennsylvania Press, 2008); and Anatole Tchikine, "'L'anima del giardino': Water, Gardens, and Hydraulics in Sixteenth-Century Florence and Naples," in *Technology and the Garden*, ed. Michael G. Lee and Kenneth I. Helphand (Washington, DC: Dumbarton Oaks, 2014), 129–153.

20. George Dameron, "Feeding the Medieval Italian City-State," *Speculum* 92 (October 2017): 976–1019. Other relevant studies include William Caferro, "City and Countryside in Siena in the Second Half of the Fourteenth Century," *Journal of Economic History*, 54 (March 1991): 85–103, who outlines the debate about exploitation for Florence and Siena in Italy; and

Jane Grenville, "Urban and Rural Houses and Households in the Late Middle Ages: A Case Study from Yorkshire," in *Medieval Domesticity: Home, Housing, and Household in Medieval England*, ed. Maryanne Kowaleski and P.J.P. Goldberg (Cambridge: Cambridge University Press, 2008), 92–123, who discusses reciprocal influences of town and country in the construction of houses.

21. James A. Galloway, Derek Keene, and Margaret Murphy, "Fueling the City: Production and Distribution of Firewood and Fuel in London's Region, 1290–1400," *Economic History Review* 49 (August 1996): 447–472.

22. Hana Taragan, "Constructing a Visual Rhetoric: Images of Craftsmen and Builders in the Umayyad Palace at Qusayr 'Amra," *Al-Masāq* 20, no. 2 (2008): 141–160.

23. Jonathan Bardill, "Building Materials and Techniques," in *The Oxford Handbook of Byzantine Studies*, ed. Elizabeth Jeffreys with John Haldon and Robin Cormack (Oxford: Oxford University Press, 2008), 335–352; Crow, "Infrastructure of a Great City"; Robert Ousterhout, *Master Builders of Byzantium* (Princeton: Princeton University Press, 1999), 128–137; Enrico Zanini, "Constantinople: Building and Maintenance," in Bassett, *Cambridge Companion to Constantinople*, 102–116; and Zanini, "Technology and Ideas: Architects and Master-Builders in the Early Byzantine World," in Lavan et al., *Technology in Transition*, 381–405.

24. Ousterhout, *Master Builders*, 58–84; and Zanini, "Technology and Ideas."

25. See esp. Ken Dark and Jan Kostenec, *Hagia Sophia in Context: An Archaeological Reexamination of the Cathedral of Byzantine Constantinople* (Oxford: Oxbow Books, 2019), which shows the importance of the larger complex of ecclesiastical buildings in which the church was situated; for the building's subsequent influence, see Metin Ahunbay and Zeynep Ahunbay, "Structural Influence of Hagia Sophia on Ottoman Mosque Architecture," in *Hagia Sophia from the Age of Justinian to the Present*, ed. Robert Mark and Ahmet S. Çakmak (Cambridge: Cambridge University Press, 1992), 179–194; and Gülru Necipoğlu, *The Age of Sinan: Architectural Culture in the Ottoman Empire* (London: Reaktion Books, 2005), 88–89, 139–140, and passim.

26. Wickham, *Framing the Early Middle Ages*, 609–612.

27. See esp. Louise Cooke, "Approaches to the Conservation and Management of Earthen Architecture in Archaeological Contexts," 2 vols. (PhD diss., University College London, 2008), esp. 1: 67, which is concerned with conservation but also gives a good description of the process; and Barry Kemp, "Soil (including Mudbrick Architecture)," in *Ancient Egyptian Materials and Technology*, ed. Paul T. Nicholson and Ian Shaw (Cambridge: Cambridge University Press, 2000), 78–103.

28. Ignacio Arce, "Umayyad Building Techniques and the Merging of Roman-Byzantine and Partho-Sassanian Traditions: Continuity and Change," in Lavan et al., *Technology in Transition*, 491–537.

29. Sheila S. Blair and Jonathan M. Bloom, *The Art and Architecture of Islam, 1250–1800* (New Haven: Yale University Press, 1994); Jonathan M. Bloom, *Architecture of the Islamic West: North Africa and the Iberian Peninsula, 700–1800* (New Haven: Yale University Press, 2020), esp. 18–34; Richard Ettinghausen, Oleg Grabar, and Marilyn Jenkins-Madina, *Islamic Art and Architecture, 650–1250*, 2nd ed. (New Haven: Yale University Press, 2001); Robert Hillenbrand, *Islamic Art and Architecture* (London: Thames & Hudson, 1999); and Julio Navarro Palazón and Pedro Jiménez Castillo, "Southerners: House and Garden in Al-Andalus," pt. 3 of "Housing," in Carver and Klápště, *Archaeology of Medieval Europe*, vol. 2, *Twelfth to Sixteenth Centuries*, 176–188. And see Olga Bush, *Reframing the Alhambra: Architecture, Poetry, Textiles and Court Ceremonial* (Edinburgh: Edinburgh University Press, 2018), for the intrinsic importance of ornamentation, poetic inscriptions, and textiles at the Alhambra.

30. See esp. Ronald Lewcock, "Materials and Techniques," in *Architecture of the Islamic*

World: Its History and Social Meaning, ed. George Michell (London: Thames & Hudson, 1978), 129–143, esp. 133–134.

31. Lewcock, "Materials and Techniques," 136 and 141–142.

32. Bush, *Reframing the Alhambra*, esp. 17–109 and 166–207; Antonio Fernández-Puertas, *The Alhambra*, 2 vols. (London: Saqi Books, 1997); and Ieva Rėklaitytė, "The Rumor of Water: A Key Element of Morrish Granada," in Boloix-Gallardo, *Companion to Islamic Granada*, 441–462.

33. Necipoğlu, *Age of Sinan*, esp. 12–23 and 71–124.

34. Necipoğlu, *Age of Sinan*, 21 and 208–222, for a detailed description of the Süleymaniye complex.

35. Marco Valenti, "Architecture and Infrastructure in the Early Medieval Village: The Case of Tuscany," in Lavan et al., *Technology in Transition*, 451–489; and more generally, Jan Klápště and Nissen Jaubert, "Rural Settlement," in Graham-Campbell with Valor, *Archaeology of Medieval Europe*, 1: 76–110.

36. Helena Hamerow, *Early Medieval Settlements: The Archaeology of Rural Communities in Northwest Europe, 400–900* (Oxford: Oxford University Press, 2002), esp. 12–26; and Klápště and Jaubert, "Rural Settlement," 89–92.

37. Hamerow, *Early Medieval Settlements*, 25–26.

38. See Kelly DeVries and Robert Douglas Smith, *Medieval Military Technology*, 2nd ed. (Toronto: University of Toronto Press, 2012), 213–222; and Johnny De Meulemeester and Kieran O'Conor, "Fortifications," in Graham-Campbell with Valor, *Archaeology of Medieval Europe*, vol. 1, *Eighth to Twelfth Centuries*, 316–341, esp. 325–331.

39. DeVries and Smith, *Medieval Military Technology*, 223–259; De Meulemeester and O'Conor, "Fortifications," 331–334; and Werner Meyer et al., "Archaeologies of Coercion," in Carver and Klápště, *Archaeology of Medieval Europe*, vol. 2, *Twelfth to Sixteenth Centuries*, 230–276.

40. A good introduction is Robert Bork, *Late Gothic Architecture: Its Evolution, Extinction, and Reception* (Turnhout: Brepols, 2018), 21–52.

41. See esp. Lynn T. Courtenay, ed., *The Engineering of Medieval Cathedrals* (Aldershot: Ashgate, 1997); Robert Mark, "Technological Innovation in High Gothic Architecture," in Smith and Wolfe, *Technology and Resource Use*, 11–25; and Nancy Y. Wu, ed., *Ad Quadratum: The Practical Application of Geometry in Medieval Architecture* (Aldershot: Ashgate, 2002).

42. See esp. Philippe Jansen, "La mobilité des maîtres-maçons en Italie au Moyen Âge: Une Mobilité technique ou culturelle?," in *Les systèmes de mobilité de la préhistoire au Moyen Âge*, ed. Nicholas Naudinot et al. (Antibes: Éditions APDCA, 2015), 305–318; and Lon R. Shelby and Jacques Heyman, "Mason (i)," in *Grove Art on Line* (2003), https://doi.org/10.1093/gao/978188 4446054.article.T054911.

43. For the transition from Gothic to classical architecture, see Bork, *Late Gothic Architecture*. For Vitruvius, see Vitruvius, *Ten Books on Architecture*, ed. and trans. Ingrid D. Rowland, commentary and illustrations by Thomas Noble Howe (Cambridge: Cambridge University Press, 1999); and for a discussion of Vitruvian influence in the Renaissance, Pamela O. Long, *Artisan/Practitioners and the Rise of the New Sciences, 1400–1600* (Corvallis, OR: Oregon State University Press, 2011), 62–126. For the practice of drawing in Renaissance architecture, see esp. Huppert, *Becoming an Architect*. Renaissance humanism is a vast field of study. Good entryways include Christopher S. Celenza, *The Intellectual World of the Italian Renaissance* (Cambridge: Cambridge University Press, 2018); Jill Kray, ed., *The Cambridge Companion to Renaissance Humanism* (Cambridge: Cambridge University Press, 2004); and Anthony Grafton, *Commerce with the Classics* (Ann Arbor: University of Michigan Press, 1997).

44. Hamerow, *Early Medieval Settlements*, 3–38; and Sarah Rees Jones, "Building Domesticity in the City: English Urban Housing before the Black Death," in Kowaleski and Goldberg, eds., *Medieval Domesticity*, 66–91.

45. See esp. Brady, "Gothic Barn of England"; Lynn T. Courtenay, "Scale and Scantling: Technological Issues in Large-Scale Timberwork of the High Middle Ages," in Smith and Wolfe, *Technology and Resource Use*, 42–75; and Rees Jones, "Building Domesticity in the City."

46. See Mark Gardiner, "Timber Churches in Medieval England: A Preliminary Study," in *Historic Wooden Architecture in Europe and Russia: Evidence, Study and Restoration*, ed. Evgeny Khodakovsky and Siri Skjold Lexau (Basel: Birkhäuser, 2016), 28–41; Andrine Nilsen, *Vernacular Buildings and Urban Social Practice: Wood and People in Early Modern Swedish Society* (Oxford: Archaeopress, 2020).

47. Evgeny Khodakovsky, "Introduction: Wood in the Architecture of Europe and Russia: National Specifics and International Research," Jergen H. Jensenius, "Wooden Churches in Viking and Medieval Norway: Two Geometric and Static Strategies," and Leif Anker, "Stave Church Research and the Norwegian Stave Church Programme: New Findings—New Questions," all in Khodakovsky and Lexau, *Historic Wooden Architecture in Europe and Russia*, 6–17, 20–27, and 94–109, respectively.

48. See esp. Mark Girouard, *Elizabethan Architecture: Its Rise and Fall, 1540–1640* (New Haven: Yale University Press, 2009), 256–298 for windows; and Else Roesdahl and Frans Verhaeghe, "Material Culture—Artifacts and Daily Life," in Carver and Klápště, *Archaeology of Medieval Europe*, vol. 2, *Twelfth to Sixteenth Centuries*, 189–225.

49. Anatole Tchikine, "Technology of Grandeur: Early Modern Aqueducts in Portugal," in Ana Duarte Rodrigues and Carmen Toribio Marín, eds., *The History of Water Management in the Iberian Peninsula between the 16th and 19th Centuries* (Cham: Birkhäuser, 2020), 139–158.

50. See esp. James Crow, Jonathan Bardill, and Richard Bayliss, *The Water Supply of Byzantine Constantinople* (London: Society for the Promotion of Roman Studies, 2008); Crow, "Ruling the Waters: Managing the Water Supply of Constantinople, AD 330–1204," *Water History* 20 (2012): 35–55; Crow, "Waters for a Capital: Hydraulic Infrastructure and Use in Byzantine Constantinople," in Bassett, *Cambridge Companion to Constantinople*, 67–86; and Brooke Shilling and Paul Stephenson, eds., *Fountains and Water Culture in Byzantium* (Cambridge: Cambridge University Press, 2016).

51. Crow, "Ruling the Waters." For Ottoman work on the city's hydraulic infrastructure, see esp. Federica Broilo, "A Dome for the Water: Canopied Fountains and Cypress Trees in Byzantine and Early Ottoman Constantinople," and Johan Mårtelius, "Sinan's Ablution Fountains," both in Shilling and Stephenson, *Fountains and Water Culture in Byzantium*, 314–323 and 324–340, respectively.

52. Deniz Karakaş, "Water for the City: Builders, Technology, and Private Initiative," in *A Companion to Early Modern Istanbul*, ed. Shirine Hamadeh and Çiğdem Kafescioğlu (Leiden: Brill, 2022), 308–338.

53. Amalia Levanoni, "Water Supply in Medieval Middle Eastern Cities: The Case of Cairo," *Al-Masāq* 20, no. 2 (2008): 179–205.

54. See esp. Long, *Engineering the Eternal City*, 63–91 and 102–112; Katherine W. Rinne, *The Waters of Rome: Aqueducts, Fountains, and the Birth of the Baroque City* (New Haven: Yale University Press, 2010), 38–55, 83–108, 122–134, and 138–154; and Paolo Squatriti, *Water and Society in Early Medieval Italy, AD 400–1000* (Cambridge: Cambridge University Press, 1998), 14–16.

55. Long, *Engineering the Eternal City*, 22–24 and 43–62.

56. David Gentilcore, "The Cistern-System of Early Modern Venice: Technology, Politics, and Culture in a Hydraulic Society," *Water History* 13 (October 2021): 1–32.

57. See esp. Jaime-Chaim Shulman, *A Tale of Three Thirsty Cities: The Innovative Water Supply Systems of Toledo, London and Paris in the Second Half of the Sixteenth Century* (Leiden: Brill, 2018), 99–165; and Cristiano Zanetti, *Janello Torriani and the Spanish Empire: A Vitruvian Artisan at the Dawn of the Scientific Revolution* (Leiden: Brill, 2017), esp. 338–401.

58. Tchikine, "Technology of Grandeur,"150; and André Teixeira and Rodrigo Banha da Silva, "The Water Supply and Sewage Networks in Sixteenth Century Lisbon: Drawing the Renaissance City," in Rodrigues and Toribio Marín, *History of Water Management*, 3–24.

59. Teixeira and Silva, "Water Supply," 14–21.

60. Derek Keene, "Issues of Water in Medieval London to c. 1300," *Urban History* 28 (August 2001): 161–179; and Roberta Magnusson, "Public and Private Urban Hydrology: Water Management in Medieval London," in *Wind and Water in the Middle Ages: Fluid Technologies from Antiquity to the Renaissance*, ed. Stephen A. Walton. (Tempe: Arizona Center for Medieval and Renaissance Studies, 2006), 171–187, esp. 172–183.

61. Keene, "Issues of Water in Medieval London."

62. Shulman, *A Tale of Three Thirsty Cities*, 166–231; and Leslie Tomory, *The History of the London Water Industry, 1580–1820* (Baltimore: Johns Hopkins University Press, 2017), esp. 1–41 for the earlier period.

63. Elisabeth Gruber, "Meeting Water Needs as a Major Challenge in an Urban Context: Examples from the Danube Region (1300–1600)," in *The Power of Urban Water: Studies in Premodern Urbanism*, ed. Nicola Chiarenza, Annette Haug, and Ulrich Müller (Berlin: De Gruyter, 2020), 179–195; and for Vienna itself, Heike Krause, Paul Mitchell, and Christoph Sonnlechner, "The Urban Waterscape in Medieval Vienna," in Zapke and Gruber, *Companion to Medieval Vienna*, 222–264.

64. Betty Arndt, "Medieval and Post-Medieval Urban Water Supply and Sanitation: Archaeological Evidence from Göttingen and North German Towns," in Chiarenza, Haug, and Müller, *Power of Urban Water*, 213–227.

65. [Jaime-]Chaim Shulman, "The Groundbreaking Water Supply Systems of Central and Eastern European Cities, 1300–1580," *Technology and Culture* 60 (July 2019): 726–769.

CHAPTER 4: Transportation and Communication

Epigraph: Pamela O. Long, David McGee, and Alan M. Stahl, *The Book of Michael of Rhodes: A Fifteenth Century Maritime Manuscript*, 3 vols. (Cambridge, MA: MIT Press, 2009), 1: 211–217 and 2: 273–281, a facsimile edition (vol. 1), transcription and English translation of the text (vol. 2), and book of studies by specialists on the topics Michael treats (vol. 3). Michael names the positions that he was assigned for each voyage, but the exact duties associated with these positions often are not precisely known.

1. See esp. Pamela O. Long, "Introduction: The World of Michael of Rhodes, Venetian Mariner," and Alan M. Stahl, "Michael of Rhodes: Mariner in Service to Venice," both in Long, McGee, and Stahl, *Book of Michael of Rhodes*, 3: 1–33 and 35–98, respectively; and see also, for the larger context of building ships and writing about them, Frederick M. Hocker and John M. McManamon, "Mediaeval Shipbuilding in the Mediterranean and Written Culture at Venice," *Mediterranean Historical Review* 21, no. 1 (2006): 1–37.

2. Raffaella Franci, "Mathematics in the Manuscript of Michael of Rhodes," in Long, McGee, and Stahl, *Book of Michael of Rhodes*, 3: 115–146, and for the will (the testament of the sailor), 1:504–505 and 2: 604–605.

3. Michael McCormick, *Origins of the European Economy: Communications and Commerce, AD 300–900* (Cambridge: Cambridge University Press, 2001); and the useful review by Julia M. H. Smith in *Speculum* 78, no. 3 (July 2003): 956–959. See also Johannes Preiser-Kapeller, Lucian Reinfandt, and Yannis Stouritis, eds. "Migration History of the Afro-Urasian Transition

Zone, c. 300–1500: An Introduction (with a Chronological Table of Selected Events of Political and Migration History)," in *Migration Histories of the Medieval Afroeurasian Transition Zone: Aspects of Mobility between Africa, Asia, and Europe, 300–1500 C.E.*, ed. Preiser-Kapeller, Reinfandt, and Stouratis (Leiden: Brill, 2020), 1–48, as well as the other detailed studies in this volume. For slavery, see note 37 of the introduction to the present book.

4. See Martha C. Howell, *Commerce before Capitalism in Europe, 1300–1600* (Cambridge: Cambridge University Press, 2010); McCormick, *Origins of the European Economy*; Preiser-Kapeller, Reinfandt, and Stouritis, "Migration History of the Afro-Eurasian Transition Zone"; and Chris Wickham, *Framing the Early Middle Ages: Europe and the Mediterranean, 400–800* (Oxford: Oxford University Press, 2005), 693–824. A classic study is Robert S. Lopez, *The Commercial Revolution of the Middle Ages, 950–1350* (Cambridge: Cambridge University Press, 1976); and two useful resources, John Block Friedman and Kristen Mossler Figg, eds., *Trade, Travel, and Exploration in the Middle Ages* (New York: Garland, 2000); and Paul B. Newman, *Travel and Trade in the Middle Ages* (Jefferson, NC: McFarland, 2011).

5. George F. Bass, "The Development of Maritime Archaeology," in *The Oxford Handbook of Maritime Archaeology*, ed. Alexis Catsambis, Ben Ford, and Donny L. Hamilton (Oxford: Oxford University Press, 2011), 3–22; and Ruthy Gertwagen, "Nautical Technology," in *A Companion to Mediterranean History*, ed. Peregrine Horden and Sharon Kinoshita (Chichester, West Sussex: John Wiley and Sons, 2014), 154–169.

6. See esp. John H. Pryor and Elizabeth M. Jeffreys, *The Age of the Dromōn: The Byzantine Navy ca. 500–1204* (Leiden: Brill, 2006), 146–152; and Eric Rieth, "Mediterranean Ship Design in the Middle Ages," in Catsambis, Ford, and Hamilton, *Oxford Handbook of Maritime Archaeology*, 406–425. See also Horst Nowacki and Matteo Valleriani, eds., *Shipbuilding Practice and Ship Design Methods from the Renaissance to the 18th Century: A Workshop Report* (Berlin: Max Planck Institute for the History of Science, 2003), Preprint 245; Pryor, "The Mediterranean Round Ship," in *Cogs, Caravels and Galleons: The Sailing Ship, 1000–1650*, ed. Robert Gardiner and Richard W. Unger (London: Conway Maritime Press, 1994), 59–77, esp. 65–67; and Richard W. Unger, "Ships and Sailing Routes in Maritime Trade around Europe, 1300–1600," in *The Routledge Handbook of Maritime Trade around Europe, 1300–1600*, ed. Wim Blockmans, Mikhail Krom, Justyna Wubs-Mrozewicz (New York: Routledge, 2017), 17–35, esp. 22–26.

7. See esp. George F. Bass et al., *Serçe Limanı: An Eleventh-Century Shipwreck*, 2 vols. (College Station: Texas A&M University Press, 2004 and 2009).

8. Renard Gluzman, *Venetian Shipping from the Days of Glory to Decline, 1453–1571* (Leiden: Brill, 2021), 47–72; and Pryor, "Mediterranean Round Ship," 59–77.

9. For Byzantine galleys, see esp. Pryor and Jeffreys, *Age of the Dromōn*; and see John H. Pryor, "Shipping and Seafaring," in *The Oxford Handbook of Byzantine Studies*, ed. Elizabeth Jeffreys with John Haldon and Robin Cormack (Oxford: Oxford University Press, 2008), 482–491. See also Stefania Montemezzo, "Ships and Trade: The Role of Public Navigation in Renaissance Venice," in *Reti maritime come fattori dell'integrazione europea / Maritime Networks as a Factor in European Integration*, ed. Giampiero Nigro (Florence: Firenze University Press, 2019), 473–484. For the merchant-patricians that controlled Venetian shipping, see Monique O'Connell, "Venice: City of Merchants or City for Merchandise?," in Blockmans, Krom, and Wubs-Mrozewicz, *Routledge Handbook of Maritime Trade*, 103–120.

10. Julian Whitewright, "Technological Continuity and Change: The Lateen Sail of the Medieval Mediterranean," *Al-Masāq* 24, no. 1 (2012): 1–19; and see Pryor, "Mediterranean Round Ship," 67–69; and Pryor and Jeffreys, *Age of the Dromōn*, 153–161.

11. See esp. Dionisius A. Agius, *Classic Ships of Islam: From Mesopotamia to the Indian Ocean* (Leiden: Brill, 2008), 141–168 and 277–357; and Hikmat Homsi, "Navigation and Ship-Building," in *Science and Technology in Islam*, pt. 2, *Technology and Applied Sciences*, ed. A. Y.

al-Hassan, Maqbul Ahmed, and A. Z. Iskandar (Paris: UNESCO, 2001), 217–246; and Christophe Picard, *Sea of the Caliphs: The Mediterranean in the Medieval Islamic World*, trans. Nicholas Elliot (Cambridge, MA: Harvard University Press, 2018).

12. Lawrence V. Mott, *Sea Power in the Medieval Mediterranean: The Catalan-Aragonese Fleet in the War of the Sicilian Vespers* (Gainesville: University Press of Florida, 2003), 185–209. For Roger of Lauria, see Charles D. Stanton, *Roger of Lauria (c. 1250–1305): Admiral of Admirals* (Woodbridge, Suffolk: Boydell Press, 2019).

13. Jan Bill, "The Oseberg Ship and Ritual Burial," and Peter Pentz, "Ships and the Vikings," both in *Vikings: Life and Legend*, ed. Gareth Williams, Peter Pentz, and Matthias Wemhoff (Ithaca, NY: Cornell University Press, 2014), 200–201 and 202–233, respectively; and Jan Bill, "Ships and Seamanship," in *The Oxford Illustrated History of the Vikings*, ed. Peter Sawyer (Oxford: Oxford University Press, 1997), 182–201.

14. See esp. Bill, "Ships and Seamanship"; Kelly DeVries and Robert Douglas Smith, *Medieval Military Technology*, 2nd ed. (Toronto: University of Toronto Press, 2012), 291–294; Owain T. P. Roberts, "Descendants of Viking Boats," in Gardiner and Unger, *Cogs, Caravels and Galleons*, 11–28; and Susan Rose, "Medieval Ships and Seafaring," in Catsambis, Ford, and Hamilton, *Oxford Handbook of Maritime Archaeology*, 426–444, esp. 427–428. Two detail-filled studies of (primarily) English ships are Ian Friel, *The Good Ship: Ships, Shipbuilding and Technology in England, 1200–1520* (Baltimore: Johns Hopkins University Press, 1995); and Gillian Hutchinson, *Medieval Ships and Shipping* (Rutherford, PA: Fairleigh Dickinson University Press, 1994).

15. Detlev Ellmers, "The Cog as Cargo Carrier," in Gardiner and Unger, *Cogs, Caravels and Galleons*, 29–46; Newman, *Travel and Trade*, 127–129; and Rose, "Medieval Ships and Seafaring," 428–430. For an introduction to the Hanseatic League, see Donald J. Herreld, *A Companion to the Hanseatic League* (Leiden: Brill, 2015).

16. Jan Bill, "Ship Construction Tools and Techniques," in Gardiner and Unger, *Cogs, Caravels and Galleons*, 151–159; and Friel, *Good Ship*, 39–67.

17. For some of the broader implications of the Atlantic turn in the Iberian world, see esp. Francesco Contente Domingues, "Science and Technology in Portuguese Navigation: The Idea of Experience in the Sixteenth Century," in *Portuguese Oceanic Expansion, 1400–1800*, ed. Francisco Bethencourt and Diogo Ramada Curto (Cambridge: Cambridge University Press, 2007), 460–479; Henrique Leitão, "Instruments and Artisanal Practices in Long Distance Oceanic Voyages," *Centaurus* 60 (2018): 189–202; Antonio Sánchez, "Charts for an Empire: A Global Trading Zone in Early Modern Portuguese Cartography," *Centaurus* 60 (2018): 173–188; Sánchez, "Cosmography, Maritime Culture, and Practical Knowledge in the Early Modern Spanish Empire," in *The Routledge Hispanic Studies Companion to Early Modern Spanish Literature and Culture*, ed. Rodrigo Cacho Casal and Caroline Egan (Abingdon: Routledge, 2022), 79–92; and Filipe Themudo Barata, "Portugal and the Mediterranean Trade: A Prelude to the Discovery of the New World," *Al-Masāq* 17, no. 2 (2005): 205–219. For Spain's shipbuilding program and its relationship to timber resources, see John T. Wing, *Roots of Empire: Forests and State Power in Early Modern Spain, c. 1500–1750* (Leiden: Brill, 2015), 44–84.

18. Ian Friel, "The Carrack: The Advent of the Full Rigged Ship," and Carla Rahn Phillips, "The Caravel and the Galleon," both in Gardiner and Unger, *Cogs, Caravels and Galleons*, 77–90 and 91–114, respectively; and see Fred Hocker, "Postmedieval Ships and Seafaring in the West," in Catsambis, Ford, and Hamilton, *Oxford Handbook of Maritime Archaeology*, 445–472, esp 450–453; and Gluzman, *Venetian Shipping*, 47–72.

19. Wickham, *Framing the Early Middle Ages*, 800–801; and James Bond, "Canal Construction in the Early Middle Ages: An Introductory Review," and Stephen Rippon, "Waterways and Water Transport on Reclaimed Coastal Marshlands: The Somerset Levels and Beyond," both in *Waterways and Canal-Building in Medieval England*, ed. John Blair (Oxford: Oxford University

Press, 2007), 153–206 and 207–227, respectively. See also McCormick, *Origins of the European Economy*, esp. 644–669 for river transport in early medieval commerce.

20. John Langdon, "The Efficiency of Inland Water Transport in Medieval England," in Blair, *Waterways and Canal-Building*, 110–130; and see also, for Wales, Robert Weeks, "Transport and Trade in South Wales c. 1100–c1400: A Study in Historical Geography" (PhD diss.: University of Wales College, 2003), 171–192.

21. Homsi, "Navigation and Ship-Building," esp. 226–229; Mott, *Sea Power*, 210–215; and David Goodman, *Power and Penury: Government, Technology, and Science in Philip II's Spain* (Cambridge: Cambridge University Press, 1988), 88–150.

22. Scholarship on the Venetian arsenal and Venetian ships includes Robert C. Davis, *Shipbuilders of the Venetian Arsenal: Workers and Workplace in the Preindustrial City* (Baltimore: Johns Hopkins University Press, 1991); Paola Lanaro, "Le donne velere nell'arsenale di Venezia: Donne e lavoro operaio in una società preindustriale," in *L'Arsenale di Venezia: Da grande complesso industrial a risorsa patrimoniale*, ed. Paola Lanaro and Christophe Austruy (Venice: Marsilio, 2020), 57–82; Franco Rossi, "L'Arsenale: I quadri direttivi," and Giovanni Caniato, "L'Arsenale: Maestranze e organizzazione del lavoro," both in in *Storia di Venezia: Alle origini alle caduta della Serenissima*, vol. 5, *Il Rinascimento: Società ed economia*, ed. Alberto Tenenti and Ugo Tucci (Rome: Istituto della Enciclopedia Italiana fondata da Giovanni Treccani, 1996), 593–639 and 641–677, respectively; and Matteo Valleriani, "Structure and Epistemology of the Practical Knowledge of the Venetian Arsenal," in *Les ingénieurs, des intermédiaires? Transmission et cooperation à l'épreuve du terrain (Europe, XV–XVIII siècle)*, ed. Stéphane Blond et al. (Toulouse: Presses Universitaires du Midi, 2021), 31–52.

23. Gluzman, *Venetian Shipping*, esp. 198–203.

24. The harbors of Europe and the Mediterranean are the focus of a series of databases that contain much information about the hundreds of harbors and landing places that functioned in the ancient and medieval worlds. See as an example, Alkiviadis Ginalis et al., *Harbours and Landing Places on the Balkan Coasts of the Byzantine Empire (4th to 12th Centuries)*, European Harbour Data Repository, vol. 4, ed. L. Werther, H. Müller, and M. Foucher (Jena: Friedrich-Schiller-Universität Jena, 2019). See also Wim Blockmans, Mikhail Krom, Justyna Wubs-Mrozewicz, "Maritime Trade around Europe, 1300–1600: Commercial Networks and Urban Autonomy," and Unger, "Ships and Sailing Routes," both in Blockmans, Krom, and Wubs-Mrozewicz, *Routledge Handbook of Maritime Trade*, 1–14 and 17–35, respectively (the handbook also contains many articles on specific ports on the Mediterranean and the Atlantic). See also Myrto Veikou, "Mediterranean Byzantine Ports and Harbours in the Complex Interplay between Environment and Society: Spatial, Socioeconomic and Cultural Considerations based on Archaeological Evidence from Greece, Cyprus and Asia Minor," Flora Karagianni, "Networks of Medieval City-Ports on the Black Sea (7th–15th Century): The Archaeological Evidence," Søren M. Sindbaek, "Northern Emporia and Maritime Networks: Modelling Past Communication Using Archaeological Network Analysis," and Johannes Preiser-Kapeller, "Harbours and Maritime Mobility: Networks and Entanglements," all in *Harbours and Maritime Networks as Complex Adaptive Systems*, ed. Johannes Preiser-Kapeller and Falko Daim (Mainz: Verlag des Römisch-Germanisches Zentralmuseum, 2015), 39–60, 83–104, 105–117, and 119–139, respectively. See also Angela Orlandi, "Between the Mediterranean and the North Sea: Networks of Men and Ports (14th–15th Centuries)," in Nigro, *Reti maritime / Maritime Networks*, 49–67; and Giovanni Tarantino and Paola von Wyss-Giacosa, eds., *Twelve Cities—One Sea: Early Modern Mediterranean Port Cities and Their Inhabitants* (Naples: Edizione Scientifiche Italiana, 2023).

25. Çağla Caner Yüksel, "A Tale of Two Port Cities: Ayasuluk (Ephesus) and Balat (Miletus) during the Beyliks Period," *Al-Masāq* 31, no. 3 (2019): 338–365; and see Unger, "Ships and Sailing Routes," 26–30. For detailed studies of numerous other Byzantine harbors and landing places,

see the studies in Falko Daim and Ewald Kilsinger, eds., *The Byzantine Harbours of Constan-tinople*, trans. Leo Ruickie and Antje Bosselmann-Ruickbie (Mainz: Verlag des Römisch-Germanisches Zentralmuseum, 2021); and Johannes Preiser-Kapeller, Taxiarchis G. Kolias, and Falko Dam, eds., *Seasides of Byzantium: Harbours and Anchorages of a Mediterranean Empire* (Mainz: Verlag des Römisch-Germanisches Zentralmuseum, 2021).

26. George Christ, "Collapse and Continuity: Alexandria as a Declining City with a Thriv-ing Port (Thirteenth to Sixteenth Centuries)," in Blockmans, Krom, and Wubs-Mrozewicz, *Routledge Handbook of Maritime Trade*, 121–140; Miriam Frenkel, "Medieval Alexandria—Life in a Port City," *Al-Masāq* 26, no. 1 (2014): 5–35; and John P. Cooper, " 'Fear God; Fear the *Bogaze*': The Nile Mouths and the Navigational Landscape of the Medieval Nile Delta, Egypt," *Al-Masāq* 24, no. 1 (April 2012): 53–73.

27. Mark Gardiner, "Hythes, Small Ports, and Other Landing Places in Later Medieval England," in Blair, *Waterways and Canal-Building*, 85–109; Maryanne Kowaleski, "The Mari-time Trade Networks of Late Medieval London," in Blockmans, Krom, and Wubs-Mrozewicz, *Routledge Handbook of Maritime Trade*, 383–410; and Gustaf Milne, *The Port of Medieval Lon-don* (Stroud, Gloucestershire: Tempus, 2003); and for Wales, Weeks, "Transport and Trade in South Wales," 182–188.

28. Eric H. Ash, *Power, Knowledge, and Expertise in Elizabethan England* (Baltimore: Johns Hopkins University Press, 2004), 55–86.

29. Olivia Remie Constable, *Housing the Stranger in the Mediterranean World: Lodging, Trade, and Travel in Late Antiquity and the Middle-Age*s (Cambridge: Cambridge University Press, 2003).

30. The career of Michael of Rhodes gives an idea of the many tasks required to rig and sail a ship—see esp. Stahl, "Michael of Rhodes: Mariner"; and concerning the training of ship crews and navigators, mostly in a later period but with much relevant material, Margaret E. Schotte, *Sailing School: Navigating Science and Skill, 1550–1800* (Baltimore: Johns Hopkins University Press, 2019).

31. For Roman roads, see esp. Lorenzo Quilici, "Land Transport, Part I: Roads and Bridges," in *The Oxford Handbook of Engineering and Technology in the Classical World*, ed. John Peter Oleson (New York: Oxford University Press, 2008), 551–579. And see Klaus Belke, "Communi-cations: Roads and Bridges," in Jeffreys with Haldon and Cormack, *Oxford Handbook of Byzan-tine Studies*, 295–307; McCormick, *Origins of the European Economy*, 67–77 and 394–402; and Newman, *Travel and Trade*, 12–63.

32. Magdolna Szilágyi, *On the Road: The History and Archaeology of Medieval Communica-tion Networks in East-Central Europe* (Budapest: Archaeolingua, 2014). See also McCormick, *Origins of the European Economy*, esp. 394–402, 445–450, and 548–564; Rainer Christoph Schwinges, ed., *Strassen- und Verkehrswesen in hohen und späten Mittelalt* (Ostfildern: Jan Thorbecke Verlag, 2007); and Thomas Szabó, ed., *Die Welt der europäischen Strassen: Von der Antike bis in die Frühe Neuzeit* (Cologne: Böhlau Verlag, 2009).

33. Paul Hindle, *Medieval Roads and Tracks*, 3rd ed. (Buckinghamshire: Shire Publications, 1998), 8–19; and, for roads in medieval Wales, Weeks, "Transport and Trade in South Wales," 131–170.

34. Oana Toda, "Evidence on the Engineering and Upkeep of Roads in Late Medieval Transylvania," *Annales Universitatis Apulensis, Series Historica* 17, no. 2 (2013): 173–200; and Toda, "Economic and Material Aspects of the Late Medieval Bridges from Transylvania: The Written Sources," *Banatica* 27 (2017): 361–397.

35. See esp. Nicholas Brooks, *Communities and Warfare, 700–1400* (London: Hambledon Press, 2000), 1–31; Giovanni Coppola, *Ponti medievali in legno* (Bari: Laterza, 1996); Marc Guyon, *Les fondations des ponts en France: Sabot métalliques des pieux de fondation, de l'antiquité à*

l'époque modern (Montagnac: Éditions Monique Mergoil, 2000); Donald R. Hill, *Islamic Science and Engineering* (Edinburgh: Edinburgh University Press, 1993), 149–158; David Harrison, *The Bridges of Medieval England: Transport and Society 400–1800* (Oxford: Clarendon Press, 2004); Fügen İlter,"The Main Features of the Seljuk, the Beylik and the Ottoman Bridges of the Turkish Anatolian Architecture from the XIIth to the XVIth Centuries," *Belleten* 57, no. 219 (1994): 481–494; Danièle James-Raoul and Claude Thomasset, eds., *Le Ponts au Moyen Âge* (Paris: Presses de l'Université-Sorbonne, 2006); Newman, *Travel and Trade*, 48–63; Andrew Petersen, "Medieval Bridges of Palestine," in *Egypt and Syria in the Fatimid, Ayyubid and Mamluk Eras*, vol. 6, ed. U. Vermeulen and K. D'Hulster (Leuven: Peeters, 2010), 291–306; and Petersen, "Roman, Medieval or Ottoman: Historic Bridges of the Lebanon Coast," in *Bridge of Civilizations: The Near East and Europe c. 1100–1300*, ed. Peter Edbury, Denys Pringle, and Balázs Major (Oxford: Archaeopress Publishing, 2019), 175–203.

36. Les Hamil, *Bridge Hydraulics* (Boca Raton, FL: CRC Press, 1999), esp. 9–23 and 61–102; Martin Cook, *Medieval Bridges* (Buckinghamshire: Shire Publications, 1998); Harrison, *Bridges of Medieval England*, 80–82.

37. Harrison, *Bridges of Medieval England*, esp. 99–135.

38. Harrison, *Bridges of Medieval England*, esp. 155–167; Pamela O. Long, *Engineering the Eternal City: Infrastructure, Topography, and the Culture of Knowledge in Sixteenth-Century Rome* (Chicago: University of Chicago Press, 2018), 43–62, 93–102, 163–188; and Andrew L. Russell and Lee Vinsel, "After Innovation, Turn to Maintenance," *Technology and Culture* 29 (January 2018): 1–25.

39. Richard W. Bulliet, *The Camel and the Wheel*, Morningside ed. (New York: Columbia University Press, 1990); Pekka Masonen, "The Sahara as Highway for Trade and Knowledge," in *Highways, Byways, and Road Systems in the Pre-Modern World*, ed. Susan E. Alcock, John Bodel, and Richard J. A. Talber (Chichester, West Sussex: Wiley-Blackwell, 2012), 168–184; and Daniel E. Schafer, "Caravans," in Friedman and Figg, *Trade, Travel, and Exploration*, 94–96.

40. Jonathan M. Bloom, *Paper before Print: The History and Impact of Paper in the Islamic World* (New Haven: Yale University Press, 2001), 20–23; Michelle P. Brown, *Understanding Illuminated Manuscripts: A Guide to Technical Terms*, revised by Elizabeth C. Teviotdale and Nancy K Turner, rev. ed. (Malibu, CA: J. Paul Getty Museum, 2018), 80 and 93.

41. Carlo Bertelli, "The Production and Distribution of Books in Late Antiquity," in *The Sixth Century: Production, Distribution, and Demand*, ed. Richard Hodges and William Bowden (Leiden: Brill, 1998), 41–60; Bloom, *Paper before Print*, 20–23; Brown, *Understanding Illuminated Manuscripts*, 30; Barbara Crostini, "Byzantium," in *The Oxford Illustrated History of the Book*, ed. James Raven (Oxford: Oxford University Press, 2020), 54–83; Anthony Grafton, "Premodern Regimes and Practices," and James Raven, "Book," both in *Information: A Historical Companion*, ed. Ann Blair et al. (Princeton: Princeton University Press, 2021), 3–20 and 333–338, respectively; and Erik Kwakkel, *Books before Print* (Leeds: Arc Humanities Press, 2018).

42. Bloom, *Paper before Print*, 25–29; Brown, *Understanding Illuminated Manuscripts*, 80–81; Kwakkel, *Books before Print*; and David Rundle, "Medieval Western Europe," in Raven, *Oxford Illustrated History of the Book*, 113–136, esp. 117–125.

43. Bloom, *Paper before Print*, 29–45, 65–70 (for molds), and 204–206; Brown, *Understanding Illuminated Manuscripts*, 79–80; Daven Christopher Chamberlain, "Paper," in *The Oxford Companion to the Book*, ed. Michael F. Suarez and H. R. Woudhuysen (Oxford: Oxford University Press, 2010), 1: 79–87; and Paul M. Dover, *The Information Revolution in Early Modern Europe* (Cambridge: Cambridge University Press, 2021), 38–55.

44. Bloom, *Paper before Print*, 107; Brown, *Understanding Illuminated Manuscripts*, 60 and 74.

45. Bloom, *Paper before Print*, 204–213; and Brown, *Understanding Illuminated Manuscripts*, 79–80.

46. John Lowden, "Book Production," in Jeffreys with Haldon and Cormack, *Oxford Handbook of Byzantine Studies*, 462–471; Judith Waring, "Byzantine Book Culture," in *A Companion to Byzantium*, ed. Liz James (Chichester, West Sussex: Wiley-Blackwell, 2010), 275–288; and N. G. Wilson, "The History of the Book in Byzantium," in Suarez and Woudhuysen, *Oxford Companion to the Book*, 1: 35–37.

47. Sheila S. Blair and Jonathan M. Bloom, "The Islamic World," in Raven, *Oxford Illustrated History of the Book*, 195–220, esp. 195–200; Jonathan M. Bloom and Sheila Blair, *Islamic Arts* (New York: Phaidon Press, 1997), 58–78; Bloom, *Paper before Print*, 99–101; Elias Muhanna, "Realms of Information in the Medieval Islamic World," in Blair et al., *Information*, 21–37; Geoffrey Roper, "The History of the Book in the Muslim World," in Suarez and Woudhuysen, *Oxford Companion to the Book*, 1: 321–339; and J. Sourdel-Thomine, "Khat (A) Writing," in *The Encyclopaedia of Islam*, ed. Bearman et al., new ed., 11 vols. (Leiden: Brill, 1978–2001), 4: 1113–1122.

48. Blair and Bloom, "Islamic World," 200–205; Bloom, *Paper before Print*, 106–111; Dimitri Gutas, *Greek Thought, Arabic Culture: The Graeco-Arabic Translation Movement in Baghdad and Early Abbasid Society (2nd–4th / 8th–10th Centuries)* (London: Routledge, 1998); Beatrice Gruendler, "Aspects of Craft in the Arabic Book Revolution," in *Globalization of Knowledge in the Post-Antique Mediterranean, 700–1500*, ed. Sonja Brentjes and Jürgen Renn (London: Routledge, 2016), 31–65; Geoffrey Khan, "Arabic Documents from the Early Islamic Period," and Esther-Miriam Wagner, "Scribal Practice in the Jewish Community of Medieval Egypt," both in *Scribal Practices and the Social Construction of Knowledge in Antiquity, Late Antiquity and Medieval Islam*, ed. Myriam Wissa (Leuven: Peeters, 2017), 69–90 and 91–110, respectively.

49. Bloom, *Paper before Print*, 113–123.

50. See esp. Eltjo Buringh, *Medieval Manuscript Production in the Latin West: Explorations with a Global Database* (Leiden: Brill, 2011); Christopher de Hamel, "The European Medieval Book," in Suarez and Woudhuysen, *Oxford Companion to the Book*, 1: 38–51; and Andrew Pettegree, *The Book in the Renaissance* (New Haven: Yale University Press, 2010), 3–20.

51. Michele P. Brown, *The British Library Guide to Writing and Scripts: History and Techniques* (Toronto: University of Toronto Press, 2001), 72–78; Brown, *Understanding Illuminated Manuscripts*, 82; and Kwakkel, *Books before Print*.

52. See esp. Jonathan J. G. Alexander, *Medieval Illuminators and their Methods of Work* (New Haven: Yale University Press, 1992); Brown, *Understanding Illuminated Manuscripts*; Christopher de Hamel, *The British Library Guide to Manuscript Illumination: History and Techniques* (Toronto: University of Toronto Press, 2001), esp. 39–81; and Susan L'Engle and Gerald B. Guest, eds., *Tributes to Jonathan J. G. Alexander: The Making and Meaning of Illuminated Medieval and Renaissance Manuscripts, Art and Architecture* (London: Harvey Miller, 2006).

53. Kai-Wing Chow, "Reinventing Gutenberg: Woodblock and Movable-Type Printing in Europe and China," in *Agent of Change: Print Culture Studies after Elizabeth L. Eisenstein*, ed. Sabrina Alcorn Baron, Eric N. Lindquist, and Eleanor E. Shevlin (Amherst: University of Massachusetts Press; Washington, DC: Center for the Book, Library of Congress, 2007), 169–192; Lucien Febvre and Henri-Jean Martin, *The Coming of the Book: The Impact of Printing, 1480–1800*, trans. David Gerard (London: Verso, 1976), 45–49; and David Landau and Peter Parshall, *The Renaissance Print, 1470–1550* (New Haven: Yale University Press, 1994), esp. 21–23, 33–102, and 169–259. For xylography in Asia, see Cynthia Brokaw, "Medieval and Early Modern East Asia," in Raven, *Oxford Illustrated History of the Book*, 84–112; and Bruce Rusk, "Xylography," in Blair et al., *Information*, 828–831.

54. See esp. Ann Blair, "Information in Early Modern Europe," in Blair et al., *Information*, 61–85; Cristina Dondi, "The European Printing Revolution," and James Mosely, "The Technologies of Print," both in Suarez and Woudhuysen, *Oxford Companion to the Book*, 1: 53–67 and

89–104, respectively; Febvre and Martin, *Coming of the Book*, 49–51; and Pettegree, *The Book in the Renaissance*, esp. 21–42.

55. James Raven and Joran Proot, "Renaissance and Reformation," in Raven, *Oxford Illustrated History of the Book*, 137–168, esp. 137–151; Febvre and Martin, *Coming of the Book*, 49–76; and Pettegree, *The Book in the Renaissance*, 21–42.

56. Dover, *Information Revolution*, esp. 38–55 and 190; Elizabeth L. Eisenstein, *The Printing Press as an Agent of Change: Communications and Cultural Transformations in Early Modern Europe*, 2 vols. (Cambridge: Cambridge University Press, 1979). For critiques and discussions, see esp. Sabrina Alcorn Baron, Eric N. Lindquist, and Eleanor F. Shevlin, "Introduction," in Baron, Lindquist, and Shevlin, *Agent of Change*, 1–22, as well as the other essays in this volume; and Adrian Johns, *The Nature of the Book: Print and Knowledge in the Making* (Chicago: University of Chicago Press, 1998), esp. 1–57. See also Ann Blair, *Too Much to Know: Managing Scholarly Information before the Modern Age* (New Haven: Yale University Press, 2010); Blair, "Information in Early Modern Europe"; and James Raven, *What Is the History of the Book?* (Cambridge: Polity Press, 2018).

57. See esp. Paul Goldman, "The History of Illustration and Its Technologies," in Suarez and Woudhuysen, *Oxford Companion to the Book*, 1: 137–145; Landau and Parshall, *Renaissance Print*, esp. 23–30, 103–168, and 310–358; Evelyn Lincoln, *The Invention of the Italian Renaissance Printmaker* (New Haven: Yale University Press, 2000); and Sachiko Kusukawa, *Picturing the Book of Nature: Image, Text, and Argument in Sixteenth-Century Human Anatomy and Medical Botany* (Chicago: University of Chicago Press, 2012).

58. See esp. Ricardo Córdoba, *Craft Treatises and Handbooks: The Dissemination of Technical Knowledge in the Middle Ages* (Turnhout: Brepols, 2013); Dover, *Information Revolution*; William Eamon, *Science and the Secrets of Nature: Books of Secrets in Medieval and Early Modern Culture* (Princeton: Princeton University Press, 1994); Anthony Grafton, *Inky Fingers: The Making of Books in Early Modern Europe* (Cambridge, MA: Harvard University Press, 2020), 29–55; Allison Kavey, *Books of Secrets: Natural Philosophy in England, 1550–1600* (Urbana: University of Illinois Press, 2007); Elaine Yen Tien Leong, *Recipes and Everyday Knowledge: Medicine, Science, and the Household in Early Modern England* (Chicago: University of Chicago Press, 2018); Elaine Leong and Alisha Rankin, eds., *Secrets and Knowledge in Medicine and Science, 1500–1800* (Farnham, Surrey: Ashgate, 2011); Pamela O. Long, *Openness, Secrecy, Authorship: Technical Arts and the Culture of Knowledge from Antiquity to the Renaissance* (Baltimore: Johns Hopkins University Press, 2001); Long, "Manuals," in Blair et al., *Information*, 589–593; and Pamela H. Smith, *From Lived Experience to the Written Word: Reconstructing Practical Knowledge in the Early Modern World* (Chicago: University of Chicago Press, 2022).

59. McCormick, *Origins of the European Economy*; and Peregrine Horden and Nicholas Purcell, *The Corrupting Sea: A Study of Mediterranean History* (Oxford: Blackwell, 2000), 53–88 and 123–172. See also Blockmans, Krom, and Wubs-Mrozewicz, "Maritime Trade around Europe, 1300–1600"; Olivia Remie Constable, *Trade and Traders in Muslim Spain: The Commercial Realignment of the Iberian Peninsula, 900–1500* (Cambridge: Cambridge University Press, 1994); Ádela Fábregas García, "Other Markets: Complementary Commercial Zones in the Nasrid World of the Western Mediterranean (Seventh/Thirteenth to Ninth/Fifteenth Centuries)," *Al-Masāq* 25, no. 1 (2013): 135–153; and Yuen-Gen Liang et al., "Unity and Disunity across the Strait of Gibraltar," *Medieval Encounters* 19 (2013): 1–40; For piracy, see esp. Clifford R. Bachman, "Piracy," in Horden and Kinoshita, *Companion to Mediterranean History*, 170–183; Travis Bruce, "Piracy as Statecraft: The Mediterranean Policies of the Fifth/Eleventh-Century Taifa of Denia," *Al-Masāq* 22, no. 5 (2010): 235–248; and Thomas Heebøll-Holm, Philipp Höhn, and Gregor Rohmann, eds., *Merchants, Pirates, and Smugglers: Criminalization, Economics and the Transformation of the Maritime World (1200–1600)* (Frankfurt: Campus Verlag, 2019).

60. Foundational studies were carried out by Shelomo Dov Goitein, *A Mediterranean Society: The Jewish Communities of the Arab World as Portrayed in the Documents of the Cairo Geniza*, 6 vols. (Berkeley: University of California Press, 1967–1993). More recently, focusing on eleventh-century commercial correspondence, see esp. Jessica L. Goldberg, *Trade and Institutions in the Medieval Mediterranean: The Geniza Merchants and their Business World* (Cambridge: Cambridge University Press, 2012), 1–29 for the discovery of the Geniza and the historiographic context of the collection.

61. Goldberg, *Trade and Institutions*, esp. 93–179 and 187–210; Goitein, *Mediterranean Society*, 1: 27–28, 42–59, and 272–352.

62. Goldberg, *Trade and Institutions*, 213–215 (for markets and emporia); and see also Ivana Ait, "Donne in affari: Il caso di Roma (secoli XIV-XV)," in *Donne del Rinascimento a Roma e dintorni*, ed. Anna Esposito (Rome: Roma nel Rinascimento, 2013), 53–83; Anna Bellavitis, *Il lavoro delle donne nelle città dell'Europa moderna* (Rome: Viella, 2016), esp. 97–159; Madeleine Pelner Cosman, *Women at Work in Medieval Europe* (New York: Checkmark Books, 2000), esp. 75–93; Barbara A. Hanawalt and Anna Dronzek, "Women in Medieval Urban Society," and Benjamin R. McRee and Trisha K. Dent, "Working Women in the Medieval City," both in *Women in Medieval Western European Culture*, ed. Linda E. Mitchell (London: Routledge, 1999), 30–45, esp. 41 and 241–256, respectively; Kathryn Reyerson, "Urban Economies," in *The Oxford Handbook of Women and Gender in Medieval Europe*, ed. Judith Bennett and Ruth Karras (Oxford: Oxford University Press, 2013), 295–307; Maya Shatzmiller, *Labour in the Medieval Islamic World* (Leiden: Brill, 1994), 348–351; Lucienne Thys-Şenocak, "Women in the City," in *A Companion to Early Modern Istanbul*, ed. Shirine Hamadeh and Çiğdem Kafescioğlu (Leiden: Brill, 2022), 86–113; Merry E. Wiesner-Hanks, *Women and Gender in Early Modern Europe*, 4th ed. (Cambridge: Cambridge University Press, 2019), 130–153; and Maria Paola Zanoboni, *Donne al lavoro nell'Italia e nell'Europa medievali (secoli XIII-XV)* (Milan: Editoriale Jouvence, 2016), 39–78. See also Terry G. Wilfong, *Women of Jeme: Lives in a Coptic Town in Late Antique Egypt* (Ann Arbor: University of Michigan Press, 2002), 117–149, who shows that seventh- and eighth-century women in the southern Egyptian town of Jeme played important economic roles as moneylenders and as participants in the economic life of Jeme and nearby towns. For the importance of local markets in regions of the North Sea, see the essays for the earlier centuries in Leen Van Molle and Yves Segers, eds., *The Agro-Food Market: Production, Distribution and Consumption*, vol. 2 in *Rural Economy and Society in North-Western Europe, 500–2000* (Turnhout: Brepols, 2013).

63. Sonja Brentjes, *Travelers from Europe in the Ottoman and Safavid Empires, 16th–17th Centuries: Seeking, Transforming, Discarding Knowledge* (Farnham, Surrey: Ashgate, 2010); and Sonja Brentjes, Alexander Fidora, and Matthias M. Tischler, "Towards a New Approach to Medieval Cross-Cultural Exchanges," *Journal of Transcultural Medieval Studies* 1 (2014): 9–50.

CHAPTER 5: Crafts and Industries

Epigraphs: Ikhwān al-Ṣafā, "Epistle 8 on the Practical Crafts," trans. Nader El-Bizri, in *Epistles of the Brethren of Purity: On Composition and the Arts: An Arabic Critical Edition and English Translation of Epistles 6–8*, by Ikhwān al-Ṣafā, ed. and trans. Nader El-Bizri and Godefroid de Callataÿ (New York: Oxford University Press, 2018), 137–165, quote on 140.

Theophilus, *On Divers Arts: The Foremost Medieval Treatise on Painting, Glassmaking and Metalwork*, trans. John G. Hawthorne and Cyril Stanley Smith (New York: Dover, 1979), 78.

1. John G. Hawthorne and Cyril Stanley Smith, "Introduction," in Theophilus, *On Divers Arts*, xv–xxxv; and Pamela O. Long, *Openness, Secrecy, Authorship: Technical Arts and the Culture of Knowledge from Antiquity to the Renaissance* (Baltimore: Johns Hopkins University Press, 2001), 85–88.

2. For an insightful discussion, see Margaret S. Graves, *Arts of Allusion: Object, Ornament, and Architecture in Medieval Islam* (New York: Oxford University Press, 2018), 38–42.

3. Theophilus, *On Divers Arts*, 77–80; and Long, *Openness, Secrecy, Authorship*, 85–88.

4. Studies focused on objects and materiality for the period covered by this book include Caroline Walker Bynum, *Christian Materiality: An Essay on Religion in Late Medieval Europe* (Brooklyn: Zone Books, 2011); Paula Findlen, ed., *Early Modern Things: Objects and their Histories, 1500–1800* (London: Routledge, 2013); and Paula Hohti Erichsen, *Artisans, Objects, and Everyday Life in Renaissance Italy: The Material Culture of the Middling Class* (Amsterdam: Amsterdam University Press, 2020). See also, for a wide-ranging collection of studies on the production of things, Philippe Braunstein and Luca Molá, eds., *Produzione e techniche*, vol. 3 in *Il Rinascimento Italiano e Europa* (Treviso: Foundazione Cassamarca, 2007).

5. Sharon Farmer, *The Silk Industries of Medieval Paris: Artisanal Migration, Technological Innovation, and Gendered Experience* (Philadelphia: University of Pennsylvania Press, 2017). See also Anna Bellavitis, *Il lavoro delle donne nelle città dell'Europa moderna* (Rome: Viella, 2016), 134–146; Barbara A. Hanawalt and Anna Dronzek, "Women in Medieval Urban Society," and Benjamin R. McRee and Trisha K. Dent, "Working Women in the Medieval City," both in *Women in Medieval Western European Culture*, ed. Linda E. Mitchell (London: Routledge, 1999), 30–45, esp. 41, and 241–256, respectively; Madeleine Pelner Cosman, *Women at Work in Medieval Europe* (New York: Checkmark Books, 2000), 75–102; Merry E. Wiesner-Hanks, *Women and Gender in Early Modern Europe*, 4th ed. (Cambridge: Cambridge University Press, 2019), esp. 112–156; and Maria Paola Zanoboni, *Donne al lavoro nell'Italia e nell'Europa medievali (secoli XIII–XV)* (Milan: Editorale Jouvence, 2016), 69–78.

6. Alice-Mary Talbot, "Women," and Nicholas Oikonomidès, "Entrepreneurs," both in *The Byzantines*, ed. Guglielmo Cavallo, trans. Thomas Dunlap, Teresa Lavender Fagan, and Charles Lambert (Chicago: University of Chicago Press, 1997), 117–143, esp. 130–131, and 144–171.

7. See esp. Gilbert Dagron, "The Urban Economy, Seventh–Twelfth Centuries," in *The Economic History of Byzantium from the Seventh through the Fifteenth Century*, ed. Angeliki E. Laiou, 3 vols. (Washington, DC: Dumbarton Oaks, 2002), 2: 393–461; George C. Maniatis, *Guilds, Price Formation, and Market Structures in Byzantium* (Farnham, Surrey: Ashgate/Variorum, 2009); and Anna Muthesius, "Silk in the Medieval World," in *The Cambridge History of Western Textiles*, ed. David Jenkins, 2 vols. (Cambridge: Cambridge University Press, 2003), 1: 325–354, esp. 328–330. For an English translation of the *Book of the Eparch*, see Edwin H. Freshfield, *Ordinances of Leo VI, c. 897: From the Book of the Eparch* (1938, repr., London: Variorum Reprints, 1970); and see Johannes Koder, "The Authority of the *Eparchos* in the Markets of Constantinople (according to the *Book of the Eparch*)," in *Authority in Byzantium*, ed. Pamela Armstrong (Farnham, Surrey: Ashgate, 2013), 83–110. I thank Petros Bouras-Vallianatos for references and advice on this topic.

8. Shelomo Dov Goitein, *A Mediterranean Society: The Jewish Communities of the Arab World as Portrayed in the Documents of the Cairo Geniza*, vol. 1, *Economic Foundations* (Berkeley: University of California Press, 1967), 80–92.

9. Goitein, *Mediterranean Society*, 1: 91–100.

10. Maya Shatzmiller, *Labour in the Medieval Islamic World* (Leiden: Brill, 1994), 348–359; and see also Goitein, *Mediterranean Society*, 1: 127–131.

11. Steven A. Epstein, "Urban Society," in *The New Cambridge Medieval History*, vol. 5, *c. 1198–c. 1300*, ed. David Abulafia (Cambridge: Cambridge University Press, 1999), 26–37.

12. Theophilus, *On Divers Arts*; and see C. R. Dodwell, "Introduction," in *Theophilus: The Various Arts, De Diversis Artibus* (Oxford: Clarendon, 1986), ix–lxxvi, esp. xxxiii–xliv.

13. Introductions to a vast literature include John Blair and Nigel Ramsay, eds., *English Medieval Industries: Craftsmen, Techniques, Products* (London: Hambledon Press, 1991); Ricardo

Córdoba et al., "Technology, Craft and Industry," in *The Archaeology of Medieval Europe*, vol. 1, *Eighth to Twelfth Centuries AD*, ed. James Graham-Campbell with Magdalena Valor (Aarhus: Aarhus University Press, 2006), 208–236; and Ricardo Córdoba and Ulrich Müller, "Manufacture and Production," in *The Archaeology of Medieval Europe*, vol. 2, *Twelfth to Sixteenth Centuries*, ed. Martin Carver and Jan Klápště (Aarhus: Aarhus University Press, 2011), 277–327.

14. Steven A. Epstein, *The Economic and Social History of Later Medieval Europe, 1000–1500* (Cambridge: Cambridge University Press, 2009), 111–122; S. R. Epstein, "Craft Guilds, Apprenticeship, and Technological Change in Preindustrial Europe," *Journal of Economic History* 58 (September 1998): 684–713; S. R. Epstein and Maarten Prak, eds., *Guilds, Innovation and the European Economy, 1400–1800* (Cambridge: Cambridge University Press, 2008); Sheilagh Ogilvie, *The European Guilds: An Economic Analysis* (Princeton: Princeton University Press, 2019); and Gervase Rosser, *The Art of Solidarity in the Middle Ages: Guilds in England, 1250–1550* (Oxford: Oxford University Press, 2015). See also the articles collected in a special issue of the *International Review of Social History*, especially, Maarten Prak, "Preface: S. R. Epstein (1960–2007) and the Guilds," Jan Lucassen, Tine De Moor, and Jan Luiten van Zanden, "The Return of the Guilds: Towards a Global History of the Guilds in Pre-Industrial Times," Clare Crowston, "Women, Gender, and Guilds in Early Modern Europe: An Overview of Recent Research," and Luca Mocarelli, "Guilds Reappraised: Italy in the Early Modern Period," *International Review of Social History*, supplement 16, *The Return of the Guilds*, 53 (2008): 1–3, 5–18, 19–44, and 159–178, respectively.

15. For these complex issues, see as a start, Carlo M. Belfanti, "Guilds, Patents, and the Circulation of Technical Knowledge: Northern Italy during the Early Modern Age," *Technology and Culture* 45 (2004): 569–589; Mario Biagioli, "Patent Republic: Representing Inventions, Constructing Rights and Authors," *Social Research* 73, no. 4 (Spring 2006): 1129–1172; Marius Buning, *Knowledge, Patents, Power: The Making of a Patent System in the Dutch Republic* (Leiden: Brill, 2022); Karel Davids, "Craft Secrecy in Europe in the Early Modern Period: A Comparative View," *Early Science and Medicine* 10 (2005): 341–348; Liliane Hilaire-Pérez and Catherine Verna, "Dissemination of Technical Knowledge in the Middle Ages and the Early Modern Era: New Approaches and Methodological Issues," *Technology and Culture* 47 (July 2006): 536–565; Long, *Openness, Secrecy, Authorship*, esp. 93–101; Luca Molà, "Stato e impresa: Privilegi per l'introduzione di nuove arti e brevetti," in Braunstein and Molá, *Produzione e techniche*, 533–572; Dagmar Schäfer, Annapurna Mamidipudi, and Marius Buning, eds., *Ownership of Knowledge: Beyond Intellectual Property* (Cambridge, MA: MIT Press, 2023); and Koen Vermeir, "Openness versus Secrecy? Historical and Historiographical Remarks," *British Journal for the History of Science* 45 (June 2012): 165–188.

16. For detailed treatments of mining and metallurgy in specific regions, see Ian Blanchard, *Mining, Metallurgy and Minting in the Middle Ages*, 3 vols. (Stuttgart: Franz Steiner, 2001); and Robert Bork et al., eds., *De Re Metallica: The Uses of Metal in the Middle Ages* (Aldershot: Ashgate, 2005). For a summary of metalworking techniques, see Antonella Fuga, *Artists' Techniques and Materials*, trans. Rosanna M. Giammanco Frongia (Los Angeles: J. Paul Getty Museum, 2006), 298–359. For Byzantium see esp. Marlia Mundell Mango, "Metalwork," in *The Oxford Handbook of Byzantine Studies*, ed. Elizabeth Jeffreys with John Haldon and Robin Cormack (Oxford: Oxford University Press, 2008), 444–451; Klaus-Peter Matschke, "Mining," and Maria K. Papathanassiou, "Metallurgy and Metalworking Techniques," both in Laiou, *Economic History of Byzantium*, 1: 115–120 and 121–127, respectively; and Gerald W. R. Ward, ed., *The Grove Encyclopedia of Materials and Techniques in Art* (Oxford: Oxford University Press, 2008), 371–390.

17. Anthony Cutler, "The Industries of Art," in Laiou, *Economic History of Byzantium*, 1: 569–575; and Fuga, *Artists' Materials and Techniques*, 160–165.

18. Gene W. Heck, "Gold Mining in Arabia and the Rise of the Islamic State," *Journal of the Economic and Social History of the Orient* 42, no. 3 (1999): 364–395.

19. Ahmad Y. al-Hassan, "Mining and Metallurgy," in *Science and Technology in Islam*, ed. Ahmad Y. al-Hassan, Maqbul Ahmed, and A. Z. Iskandar, pt. 2, *Technology and Applied Science* (Paris: United Nations Educational, Scientific and Cultural Organization, 2001), 85–106, esp. 85–88.

20. Al-Hassan, "Mining and Metallurgy," pt. 2, 89–96; Martha E. Morgan, "Reconstructing Early Islamic Maghribi Metallurgy" (PhD diss., University of Arizona, 2009), esp. 283–314, an archaeological investigation of metal production at al-Basra, Morocco; and Rachel Ward, *Islamic Metalwork* (New York: Thames & Hudson, 1993). See also Graves, *Arts of Illusion*, 78–83; and for a detailed study of one mint, Alan Stahl, *Zecca: The Mint of Venice in the Middle Ages* (Baltimore: Johns Hopkins University Press, 2000).

21. Córdoba et al., "Technology, Craft and Industry," 210; Córdoba and Müller, "Manufacture and Production," esp. 321–324; Al-Hassan, "Mining and Metallurgy," pt. 2, 97–104; and Morgan, "Reconstructing," 295–296.

22. Henriette Lyngstrøm, "Farmers, Smelters, Smiths: Relations between Production, Consumption, and Distribution of Iron in Denmark, 500 BC–AD 1500," in *Prehistoric and Medieval Direct Iron Smelting in Scandinavia and Europe: Aspects of Technology and Society*, ed. Lars Christian Nørbach, *Acta Jutlandica* 76, special issue (2003): 21–25.

23. Maria Elena Cortese, "Medieval Ironworking on Mount Amiata (Siena, Italy): Economy, Society, Technology," and Vasco La Salvia and L'ubonír Mihok, "The Smithy Workshop of the Medieval Site of Leopoli-Cencelle (Viterbo, Italy)," both in Nørbach, *Prehistoric and Medieval Direct Iron Smelting*, 55–59 and 125–128, respectively. See also Patrice Beck, Philippe Braunstein, and Michel Philippe, "Wood, Iron, and Water in the Othe Forest in the Late Middle Ages: New Findings and Perspectives," in *Technology and Resource Use in Medieval Europe: Cathedrals, Mills, and Mines*, ed. Elizabeth Bradford Smith and Michael Wolfe (Aldershot: Ashgate, 1997), 173–184; and Radomír Pleiner, *Iron in Archaeology: The European Bloomery Smelters* (Prague: Archeologický Ústav AV ČR, 2000); and Pleiner, *Iron in Archaeology: Early European Blacksmiths* (Prague: Archeologický Ústav AV ČR, 2006).

24. Enzo Baraldi, "Una nuova etá del ferro: Macchine e processi della siderugia," in Braunstein and Molá, *Produzione e techniche*, 199–216; C. Cucini and M. Tizzoni, "The Lombard Iron Masters Migrations and the Spread of the Blast Furnace in Europe, with a Focus on the 16th–17th Centuries," *Metalla* 26, no. 1 (2022): 37–66; Richard C. Hoffmann, *An Environmental History of Medieval Europe* (Cambridge: Cambridge University Press, 2014), esp. 223–225; Pleiner, *Iron in Archaeology*, 283–284; and Alan Williams, *The Knight and the Blast Furnace: A History of the Metallurgy of Armour in the Middle Ages and the Early Modern Period* (Leiden: Brill, 2003).

25. See esp. Tina Asmussen and Pamela O. Long, "Introduction: The Cultural and Material Worlds of Mining in Early Modern Europe," in special issue, *The Cultural and Material Worlds of Mining in Early Modern Europe*, ed. Tina Asmussen, *Renaissance Studies* 34 (February 2020): 8–30, which includes further essential bibliography; Baraldi, "Una nuova etá del ferro"; Philippe Braunstein, ed., *La sidérurgie Alpine en Italie (XIIᵉ–XVIIᵉ siècle)* (Rome: École Française de Rome, 2001); Jeannette Graulau, *The Underground Wealth of Nations: On the Capitalist Origins of Silver Mining, A.D. 1150–1450* (New Haven: Yale University Press, 2019); and Raffaello Vergani, "L'attivitá mineraria e metallurgica: Argento e rame," in Braunstein and Molá, *Produzione e techniche*, 217–233. For mine shares, see esp. Tina Asmussen, "The *Kux* as a Site of Mediation: Economic Practices and Material Desires in the Early Modern German Mining Industry," in *Sites of Mediation—Connected Histories of Places, Processes, and Objects: Europe and Beyond, 1450–1650*, ed. Susanna Burghartz, Lucas Burkart, and Christine Göttler (Leiden: Brill, 2017), 159–182.

26. See esp. Vergani, "L'attivitá mineraria e metallurgica," 230–233, with further bibliography.

27. See Marian Campbell, "Gold, Silver and Precious Stones," in Blair and Ramsay, *English Medieval Industries*, 117–126; Franz Kirchweger, "Goldsmiths in Medieval Vienna (1150–1527): Their Trade and Their Art," in *A Companion to Medieval Vienna*, ed. Susana Zapke and Elisabeth Gruber (Leiden: Brill, 2021), 469–496; Pamela H. Smith, *The Body of the Artisan: Art and Experience in the Scientific Revolution* (Chicago: University of Chicago Press, 2004), esp. 74–80 and 106–107; and Smith, *From Lived Experience to the Written Word: Reconstructing Practical Knowledge in the Early Modern World* (Chicago: University of Chicago Press, 2022), 43–63.

28. Claude Blair and John Blair, "Copper Alloys," in Blair and Ramsay, *English Medieval Industries*, 81–106, esp. 89–93 (for bell founding); Richard Ettinghausen, Oleg Grabar, and Marilyn Jenkins-Madina, *Islamic Art and Architecture, 650–1250*, 2nd ed. (New Haven: Yale University Press, 2001), 66–67 and 97–98; and Vannoccio Biringuccio, *The Pirotechnia of Vannoccio Biringuccio: The Classic Sixteenth-Century Treatise on Metals and Metallurgy*, trans. and ed. Cyril Stanley Smith and Martha Teach Gnudi (1959, repr., New York: Dover, 1990), 213–277, for casting bells and guns. For Biringuccio, see also Andrea Bernardoni, *La conoscenza del fare: Ingegneria, arte, scienza nel* De la pirotechnia *di Vannoccio Biringuccio* (Rome: L'Erma di Bretschneider, 2011); and Long, *Openness, Secrecy, Authorship*, 178–182.

29. For cogent pathways into a vast subject, see esp. Allison Margaret Bigelow, *Mining Language: Racial Thinking, Indigenous Knowledge, and Colonial Metallurgy in the Early Modern Iberian World* (Williamsburg, VA: Omohundro Institute of Early American History and Culture; Chapel Hill: University of North Carolina Press, 2020); and Renée Raphael, "Toward a Critical Transatlantic History of Early Modern Mining: Depiction, Reality, and Readers' Expectations in Álvaro Alonso Barba's 1640 *El arte de los metals*," *Isis* 14 (June 2023): 341–358.

30. Thomas Morel, *Underground Mathematics: Craft Culture and Knowledge Production in Early Modern Europe* (Cambridge: Cambridge University Press, 2023), esp. 1–117 for the fifteenth and sixteenth centuries.

31. Hoffmann, *Environmental History*, 219–227. See also Peter Anreiter et al., eds., *Mining in European History and Its Impact on Environment and Human Societies* (Innsbruck: Innsbruck University Press, 2010); and Frank Uekoetter, ed., "Mining in Central Europe: Perspectives from Environmental History," *Rachel Carson Center Perspectives* 2012, no. 10, doi.org/10.5282/rcc/5600.

32. For an introduction to European textiles with a regional emphasis, *Cambridge History of Western Textiles*, vol. 1; and Ward, *Grove Encyclopedia of Materials and Techniques*, 674–709. For Islamic textiles, see Patricia L. Baker, *Islamic Textiles* (London: British Museum Press, 1995); Jonathan Bloom and Sheila Blair, *Islamic Arts* (New York: Phaidon Press, 1997), 81–98, 221–246, and 361–383; Ahmad Y. al-Hassan, "Textiles and Other Manufacturing Industries," in al-Hassan, Ahmed, and Iskandar, *Science and Technology in Islam*, pt. 2, 135–163; and for textile uses, Frances Pritchard, "The Uses of Textiles, c. 1000–1500," in Jenkins, *Cambridge History of Western Textiles*, 1: 355–377.

33. Bloom and Blair, *Islamic Arts*, 80–98, esp. 86–87; al-Hassan, "Textiles and Other Manufacturing Industries," pt. 2, 147–153; Lise Bender Jørgensen et al., "Textile Industries of the Early Medieval World," and Leslie Clarkson, "The Linen Industry in Early Modern Europe," both in Jenkins, *Cambridge History of Western Textiles*, 1: 118–175 and 473–492, respectively; and Ward, *Grove Encyclopedia of Materials*, 674–675.

34. Bloom and Blair, Islamic Arts, 80–98, esp. 87; al-Hassan, "Textiles and Other Manufacturing Industries," pt. 2, 139–143; Eliyahu Ashtor and Halil Inalcik, "Ḳuṭn, Ḳuṭun (A.), cotton," in *The Encyclopaedia of Islam*, new ed., ed. P. J. Bearman et al., 11 vols. (Leiden: Brill, 1978–2001), 5: 554–566; Maureen Fennell Mazzaoui, "The First European Cotton Industry: Italy and Germany, 1100–1800," in *The Spinning World: A Global History of Cotton Textiles, 1200–1850*, ed. Giorgio Riello and Prasannan Parthasarathi, 63–88 (Oxford: Oxford University Press, 2009);

and for a global perspective, Giorgio Riello, *Cotton: The Fabric That Made the Moden World* (Cambridge: Cambridge University Press, 2013).

35. See esp. Dagron, "Urban Economy," 438–444; David Jacoby, "Silk Production," in Jeffreys with Haldon and Cormack, *Oxford Handbook of Byzantine Studies*, 421–428; Anna Muthesius, *Byzantine Silk Weaving, A.D. 400 to A.D. 1200* (Vienna: Verlag Fassbaender, 1997); and Muthesius, *Studies in Medieval and Islamic Silk Weaving* (London: Pindar Press, 1995).

36. Muthesius, *Byzantine Silk Weaving*, 5–33; and Muthesius, *Studies*, 122–134.

37. See Farmer, *Silk Industries of Medieval Paris*; Luca Molà, *The Silk Industry of Renaissance Venice* (Baltimore: Johns Hopkins University Press, 2000); Luca Molà, Reinhold C. Mueller, and Claudio Zainer, *La seta in Italia dal Medioevo al Seicento: Dal baco al drappo* (Venice: Marsilio, 2000); Muthesius, "Silk in the Medieval World," 1: 326–333; and Shatzmiller, *Labour in the Medieval Islamic World*, 348–354.

38. Córdoba and Müller, "Manufacture and Production," 315–317; al-Hassan, "Textiles and Other Manufacturing Industries," pt. 2, 144–147; John S. Lee, *The Medieval Clothier* (Woodbridge: Boydell Press, 2018); and John H. Munro, "Medieval Woollens: Textiles, Textile Technology and Industrial Organisation, c. 800–1500," in Jenkins, *Cambridge History of Western Textiles*, 1: 181–227, esp. 182–185.

39. Lee, *Medieval Clothier*, 45–46; and Munro, "Medieval Woollens," 197–200.

40. Lee, *Medieval Clothier*, 47–50; and Munro, "Medieval Woollens," 200–204.

41. Córdoba et al., "Technology, Craft, and Industry"; Lee, *Medieval Clothier*, 50–53; and Munro, "Medieval Woollens," 192–197.

42. Lee, *Medieval Clothier*, 53–56; and Munro, "Medieval Woollens," 204–209.

43. Lee, *Medieval Clothier*, 56–59; Munro, "Medieval Woollens," 209–211; and Penelope Walton, "Textiles," in Blair and Ramsay, *English Medieval Industries*, 332–337.

44. Lee, *Medieval Clothier*, 61–64; and Munro, "Medieval Woollens," 211–215.

45. See Lee, *Medieval Clothier*, esp.76–114; Munro, "Medieval Woollens," 217–222; and Walton, "Textiles," 332–337. For the complexity of the wool industry in one locale, see Andrea Mozzato, "The Production of Woollens in Fifteenth- and Sixteenth-Century Venice," in *At the Centre of the Old World: Trade and Manufacturing in Venice and the Venetian Mainland, 1400–1800*, ed. Paola Lanaro (Toronto: Centre for Reformation and Renaissance Studies, 2006), 73–107.

46. Carole Collier Frick, *Dressing Renaissance Florence: Families, Fortunes, and Fine Clothing* (Baltimore: Johns Hopkins University Press, 2002), esp. 32–56 and 228–230. See also Ulinka Rublack, *Dressing Up: Cultural Identity in Renaissance Europe* (Oxford: Oxford University Press, 2010). These two books, taken together, constitute an excellent introduction to the burgeoning field of clothing studies.

47. Tehmina Goskar, "Material Worlds: The Shared Cultures of Southern Italy and Its Mediterranean Neighbours in the Tenth to Twelfth Centuries," *Al-Masāq* 23, no. 3 (2011): 189–204.

48. For the use of ceramics to identify exchange (by whatever means, such as commerce, gift, warfare, or theft), see, for example, Claire Déléry, "Using *Cuerda Seca* Ceramics as a Historical Source to Evaluate Trade and Cultural Relations between Christian Ruled Lands and Al-Andalus, from the Tenth to Thirteenth Centuries," *Al-Masāq* 21, no. 1 (2009): 31–58; Margaret S. Graves, "Ceramics: The Art of Being Human," in *Ceramic Art*, ed. Margaret S. Graves et al. (Princeton: Princeton University Press, 2023), 19–46, esp. 20–26; Michael McCormick, *Origins of the European Economy: Communications and Commerce, AD 300–900* (Cambridge: Cambridge University Press, 2001), 656–663 and passim; and Chris Wickham, *Framing the Early Middle Ages: Europe and the Mediterranean, 400–800* (Oxford: Oxford University Press, 2005), esp. 700–706.

49. Sequoia Miller, "Ceramic Art: An Introduction," in Graves et al., *Ceramic Art*, 1–8, esp. 6–14.

50. Miller, "Ceramic Art," 6–7.

51. Véronique François and Jean-Michel Spieser, "Pottery and Glass in Byzantium," In Laiou, *Economic History of Byzantium*, 2: 593–609, esp. 598–609.

52. See Pamela Armstrong, "Ceramics," in Jeffreys with Haldon and Cormack, *Oxford Handbook of Byzantine Studies*, 429–443; Bloom and Blair, *Islamic Arts*, 81–86; Ettinghausen, Grabar, and Jenkins-Madina, *Islamic Art and Architecture*, 249; Fuga, *Artists' Materials and Techniques*, 210–247; Ahmad Y. al-Hassan, "Alchemy, Chemistry and Chemical Technology," in al-Hassan, Ahmed, and Iskandar, *Science and Technology in Islam*, pt. 2, *Technology and Applied Science*, 41–83, esp. 77–80; Catherine Hess, "Brilliant Achievements: The Journey of Islamic Glass and Ceramics to Renaissance Italy," and Linda Komaroff, "Color, Precious Metal, and Fire: Islamic Ceramics and Glass," both in *The Arts of Fire: Islamic Influences on Glass and Ceramics of the Italian Renaissance*, ed. Catherine Hess (Los Angeles: J. Paul Getty Museum, 2004), 1–32 and 34–53, respectively; and Ward, *Grove Encyclopedia of Materials and Techniques*, 92–101.

53. Córdoba et al., "Technology, Craft, and Industry," 214; Córdoba and Müller, "Manufacture and Production," 279–280 and 288–307; Ettinghausen, Grabar, and Jenkins-Madina, *Islamic Art and Architecture*, 67–71, 93–94, and 249–252; and Hess, "Brilliant Achievements," 11–16.

54. See esp. Martin Schönfield, "Was There a Western Inventor of Porcelain?," *Technology and Culture* 39 (October 1998): 716–727; John Cherry, "Pottery and Tile," in Blair and Ramsay, *English Medieval Industries*, 189–209; and Hess, "Brilliant Achievements," 8–10.

55. For the use of relatively scarce tin in the pottery of al-Andalus, see Anna McSweeney, "The Tin Trade and Medieval Ceramics: Tracing the Sources of Tin and Its Influence on Mediterranean Ceramics Production," *Al-Masāq* 23, no. 3 (2011): 155–169; and for the influence of Islamic techniques on Italian maiolica, Hess, "Brilliant Achievements," 12–19.

56. See R. J. Charleston, "Vessel Glass," in Blair and Ramsay, *English Medieval Industries*, 237–264; Córdoba et al., "Technology, Craft and Industry," 213–214; Córdoba and Müller, "Manufacture and Production," 280–281; François and Spieser, "Pottery and Glass in Byzantium," esp. 2: 594–598; Fuga, *Artists' Materials and Techniques*, 248–291; al-Hassan, "Alchemy, Chemistry and Chemical Technology," pt. 2, esp. 74–77; Michel Philippe, *Naissance de la verrerie moderne, xiie–xvie siècles: Apects économiques, techniques et humains* (Turnhout: Brepols, 1998); Sebastian Strobl, *Glastechnik des Mittelalters* (Stuttgart: Gentner, 1990); and Ward, *Grove Encyclopedia of Materials and Techniques*, 239–252.

57. Ettinghausen, Grabar, and Jenkins-Madina, *Islamic Art and Architecture*, 71–72, 206–209, and 252–253; Hess, "Brilliant Achievements," 3–7 and 16–17; and Komaroff, "Color, Precious Metal, and Fire."

58. Marco Beretta, "Glassmaking Goes Public: The Cultural Background to Antonio Nari's *L'Arte Vetraria* (1612)," *Technology and Culture* 58 (October 2017): 1046–1070; Camilla Bertini, Julian Henderson, and Simon Chenery, "Seventh to Eleventh Century CE Glass from Northern Italy: Between Continuity and Innovation," *Archaeological and Anthropological Sciences* 12 (2020), article no. 120, http://doi.org/10.1007/s12520-020-01048-8; R. J. Charlston, "Vessel Glass," in Blair and Ramsay, *English Medieval Industries*, 237–264; Corine Maitte, "L'arte del vetro: Innovazione e trasmissione delle techniche," in Braunstein and Molá, *Produzione e techniche*, 235–259; W. Patrick McCray, *Glassmaking in Renaissance Venice: The Fragile Craft* (Aldershot: Ashgate, 1999); Francesca Trivellato, *Fondamenta dei vetrai: Lavoro, tecnologia e mercato a Venezia tra Sei e Settecento* (Rome: Donzelli Editore, 2000), 171–187, for women glassworkers; and Trivellato, "Murano Glass, Continuity and Transformation (1400–1800)," in Lanaro, *At the Centre of the Old World*, 143–184.

59. Sarah Brown and David O'Connor, *Medieval Craftsmen: Glass-Painters* (London: British Museum Press, 1991), esp. 46–64; Fuga, *Artists' Materials and Techniques*, 292–297; Lech

Kalinowski, "Medieval Stained Glass in Poland," in *Ars Vitrea: Collected Writings on Mediaeval Stained Glass*, by Lech Kalinowski and Helena Małkiewiczówna, ed. Dobrosława Horzela and Joanna Utzig, trans. Piotr Mizia and Zbigniew Suszczyński (Kraków: Polish Academy of Arts and Sciences, 2016), 69–126; and Richard Marks, "Window Glass," in Blair and Ramsay, *English Medieval Industries*, 265–294.

60. Ward, *Grove Encyclopedia of Materials and Techniques*, 338–342; John Cherry, "Leather," in Blair and Ramsay, *English Medieval Industries*, 295–318; Quita Mould, Ian Carlisle, and Esther Cameron, *Craft, Industry and Everyday Life: Leather and Leatherworking in Anglo-Scandinavian and Medieval York* (York: Council for British Archaeology, 2003), 3222–3227; and Roy Thomson, "Leather Working Process," in *Leather and Fur: Aspects of Early Medieval Trade and Technology*, ed. Esther Cameron (London: Archetype Publications, 1998), 1–9.

61. Cherry, "Leather"; Ricardo Cordoba de la Llave, "Textes techniques médiévaux sur le tannage et la teinture du cuir," and Eva Halasz-Csiba, "Peaux et cuirs: Méthode d'investigation de la dimension historique du tannage en France (XIVᵉ–XVIIIᵉ siècles)," both in *Le travail du cuir de la Préhistoire à nos jours*, ed. Frédérique Audoin-Rouzeau and Sylvie Beyries (Antibes: Éditions APDCA, 2002), 351–365 and 387–398, respectively; and Thomson, "Leather Working Process."

62. See Olaf Goubitz, Carol van Driel-Murray, and Willy Groenman-van Waateringe, *Stepping through Time: Archaeological Footwear from Prehistoric Times until 1800* (Zwolle: Stichting Promotie Archeologie, 2001); Janne Harjula, *Sheaths, Scabbards and Grip Coverings: The Use of Leather for Portable Personal Objects in the 14th–16th Century Turku* (Turku: Archaeologia Medii Aevi Finlandiae, 2005); and Harjula, *Before the Heels: Footwear and Shoemaking in Turku in the Middle Ages and the Beginning of the Early Modern Period* (Turku: Archaeologia Medii Aevi Finlandiae, 2008), 138–141.

63. Harjula, *Before the Heels*, 138–141; Mould, Carlisle, and Cameron, *Craft, Industry and Everyday Life*, 3235–3244; and Michelle O'Malley, "A Pair of Little Gilded Shoes: Commission, Cost, and Meaning in Renaissance Footwear," *Renaissance Quarterly* 63 (Spring 2010): 45–83.

64. Robin Cormack, *Byzantine Art*, 2nd ed. (Oxford: Oxford University Press, 2018), esp. 107–117, 147–151, 167–169, and 187–193; Cormack, "Wall-Paintings and Mosaics," in Jeffreys, Haldon, and Cormack, *Oxford Handbook of Byzantine Studies*, 385–406, esp. 387–393; Cutler, "Industries of Art," 1: 557–561; Fuga, *Artists' Materials and Techniques*, 183–191; Dale Kinney, "Mosaics at Ravenna," in *Making Medieval Art*, ed. Phillip Lindley (Donington, Lincolnshire: Shaun Tyas, 2003), 81–90; and Ward, *Grove Encyclopedia of Materials and Techniques*, 403–409.

65. Cormack, "Wall Paintings and Mosaics," 387–388; Cutler, "Industries of Art," 1: 561–565; Fuga, *Artists' Materials and Techniques*, 99–107; and Ward, *Grove Encyclopedia of Materials and Techniques*, 223–225.

66. Robin Cormack, *Icons* (London: British Museum, 2007), esp. 31–45; and Cutler, "Industries of Art," 1: 565–569.

67. Anthony Cutler, *The Hand of the Master: Craftsmanship, Ivory, and Society in Byzantium (9th–11th Centuries)* (Princeton: Princeton University Press, 1994), esp. 70–153; Cutler, "Industries of Art," 1: 578–581; Ettinghausen, Grabar, and Jenkins-Medina, *Art and Architecture*, 95–96 and 277; and Sarah M. Guérin, "Forgotten Routes? Italy, Ifrīqiya and the Trans-Saharan Ivory Trade," *Al-Masāq* 29, no. 1 (2013): 70–91.

68. Ettinghausen, Grabar, and Jenkins-Medina, *Islamic Art and Architecture*, 59–66 and 95–96.

69. Ettinghausen, Grabar, and Jenkins-Medina, *Islamic Art and Architecture*, 65–66, 200–203, and 253–256.

70. Suzanne B. Butters, *The Triumph of Vulcan: Sculptors' Tools, Porphyry, and the Prince in Ducal Florence*, 2 vols. (Florence: Leo S. Olschki, 1996); Phillip Lindley, "Gothic Sculpture:

Studio and Workshop Practice," in Lindley, *Making Medieval Art*, 54–80; and Ward, *Grove Encyclopedia of Materials and Techniques*, 620–635.

71. Anabel Thomas, *The Painter's Practice in Renaissance Tuscany* (Cambridge: Cambridge University Press, 1995), 149–152; Ann Massing, "A Short History of Tempera Painting," Achim Timmermann, "The Workshop Practice of Medieval Painters," and Dagmar Eichberger, "The Winged Altarpiece in Early Netherlandish Art," all in Lindley, *Making Medieval Art*, 30–41, 42–53 and 152–172, respectively.

72. Thomas, *Painter's Practice*, 155–157.

73. See esp. Carmen C. Bambach, *Drawing and Painting in the Italian Renaissance Workshop: Theory and Practice, 1300–1600* (Cambridge: Cambridge University Press, 1999); Fuga, *Artists' Techniques and Materials*, 98–111; Roger Ling and Lesley Ling, "Fresco and Other Forms of Wall Painting," in Lindley, *Making Medieval Art*, 4–29; Timmermann, "Workshop Practice of Medieval Painters"; and Ward, *Grove Encyclopedia of Materials and Techniques*, 223–225.

74. Daniel V. Thompson Jr., trans., *The Craftsman's Handbook: "Il Libro dell'Arte" Cennino d'Andrea Cennini* (1933, repr., New York: Dover, 1954). See also Fuga, *Artists' Techniques and Materials*, 112–117; Massing, "Short History of Tempera Painting"; and Ward, *Grove Encyclopedia of Materials and Techniques*, 668–671.

75. Fuga, *Artists' Techniques and Materials*, 121–131; and Ward, *Grove Encyclopedia of Materials and Techniques*, 421–425.

76. For the growing value of ingenuity in this period, see Richard J. Oosterhoff, José Ramón Marcaida, and Alexander Marr, eds., *Ingenuity in the Making: Matter and Technique in Early Modern Europe* (Pittsburgh: University of Pittsburgh Press, 2021).

CHAPTER 6: Instruments and Machines Including Weapons

Epigraph: Ibn al Razzāz al-Jazarī, *The Book of Knowledge of Ingenious Mechanical Devices*, trans. and annotated by Donald R. Hill (Dordrecht: D. Reidel, 1974), 15.

1. See Ahmad Dallal, "Science, Medicine, and Technology: The Making of a Scientific Culture," in *The Oxford History of Islam*, ed. John L. Esposito (Oxford: Oxford University Press, 1999), 155–213, esp. 193–196; Dimitri Gutas, *Greek Thought, Arabic Culture: The Graeco-Arabic Translation Movement in Baghdad and Early Abbasid Society (2nd–4th / 8th–10th Centuries)* (London: Routledge, 1998); Robert Hall, "Mechanics," in *Science and Technology in Islam*, pt. 1, *The Exact and Natural Sciences*, ed. A. Y. al-Hassan, Maqbul Ahmed, and A. Z. Iskandar (Paris: UNESCO, 2001), 297–336; and Donald R. Hill, *Islamic Science and Engineering* (Edinburgh: Edinburgh University Press, 1993).

2. See esp. Vitruvius, *Ten Books on Architecture*, trans. Ingrid D. Rowland, commentary and illustrations by Thomas Noble Howe (Cambridge: Cambridge University Press, 1999), 119–134 and 292–317 for book ten; and Vitruve, *De l'architecture, Livre X*, ed. and trans. Louis Callebat, commentary by Philippe Fleury (Paris: Les Belles Lettres, 1986). For the Vitruvian tradition in later centuries, see Pamela O. Long, *Artisan/Practitioners and the Rise of the New Sciences* (Corvallis: Oregon State University Press, 2011), 62–93.

3. Richard L. Hills, *Power from the Wind: A History of Windmill Technology* (Cambridge: Cambridge University Press, 1994), 11–17; and Vaclav Smil, *Energy and Civilization: A History* (Cambridge, MA: MIT Press, 2017), 157–163.

4. Hills, *Power from the Wind*, 24–214; Andreas Ney, *Wasser- und Windmühlen in Europa in der Spätantike und dem Mittelalter nach archäologischen, bildlichen und schriftlichen Quellen* (Detmold: Verlag Moritz Schäfer, [2019]), 63–83; Tim Sistrunk, "The Right to Wind in the Later Middle Ages," in *Wind and Water in the Middle Ages: Fluid Technologies from Antiquity to the Renaissance*, ed. Steven A. Walton (Tempe: Arizona Center for Medieval and Renaissance Studies, 2006), 153–167; and Smil, *Energy and Civilization*, 157–163.

5. See esp. Luisa Dolza, "*Utilitas et delectatio*: Libri di tecniche e teatri di macchine," in *Produzione e techniche*, ed. Philippe Braunstein and Luca Molà, vol. 3 in *Il Rinascimento Italiano e l'Europa* (Treviso: Fondazione Cassamarca, 2007), 115–143; Paolo Galluzzi, *The Italian Renaissance of Machines* (Cambridge, MA: Harvard University Press, 2020); and Pamela O. Long, "Picturing the Machine: Francesco di Giorgio and Leonardo da Vinci in the 1490s," in *Picturing Machines, 1400–1700*, ed. Wolfgang Lefèvre (Cambridge, MA: MIT Press, 2004), 117–141.

6. Dolza, "*Utilitas et delectatio*"; Luisa Dolza and Hélène Vérin, "Illustrating Mechanics: The Enigma of Theater of Machines during the Renaissance," *Revue d'Histoire Modern et Contemporaine* 51/52, no. 2 (2004): 7–37; Galluzzi, *Italian Renaissance of Machines*; Pamela O. Long, "Power, Patronage, and the Authorship of *Ars*: From Mechanical Know-How to Mechanical Knowledge in the Last Scribal Age," *Isis* 88 (March 1997): 1–41; Marcus Popplow, "Why Draw Pictures of Machines? The Social Contexts of Early Modern Machine Drawings," in Lefèvre, *Picturing Machines*, 17–48; and Jonathan Sawday, *Engines of the Imagination: Renaissance Culture and the Rise of the Machine* (London: Routledge, 2007).

7. Banū Mūsā bin Shakir, *The Book of Ingenious Devices*, trans. Donald R. Hill (Dordrecht: Reidel, 1979); Donald R. Hill, *A History of Engineering in Classical and Medieval Times* (London: Routledge, 1996), 202–203; Hill, "Mechanical Technology," in *Science and Technology in Islam*, pt. 2, *Technology and Applied Sciences*, ed. A. Y. al-Hassan, Maqbul Ahmed, and A. Z. Iskandar (Paris: UNESCO, 2001), 165–192, esp. 181–183; and Elly R. Truitt, *Medieval Robots: Mechanism, Magic, Nature, and Art* (Philadelphia: University of Pennsylvania Press, 2015), esp. 19–20.

8. See esp. François Charette, *Mathematical Instrumentation in Fourteenth-Century Egypt and Syria: The Illustrated Treatise of Najm al-Dīn al-Miṣrī* (Leiden: Brill, 2003), 49–112; Dallal, "Science, Medicine, and Technology," esp. 179–182; Muammar Dizer, "Observatories and Astronomical Instruments," in al-Hassan, Ahmed, and Iskandar, *Science and Technology in Islam*, pt. 1, 235–266; Hill, *History of Engineering*, 190–195; Hill, *Islamic Science and Engineering*, 48–52; Hill, "Mechanical Technology," 189–190; David A. King, *In Synchrony with the Heavens: Studies in Astronomical Timekeeping and Instrumentation in Medieval Islamic Civilization*, vol. 2, *Instruments of Mass Calculation* (Leiden: Brill, 2005–2014); Josefina Rodríguez-Arribas et al., eds., *Astrolabes in Medieval Cultures* (Leiden: Brill, 2019); and George Saliba, *Islamic Science and the Making of the European Renaissance* (Cambridge, MA: MIT Press, 2011), esp. 221–232.

9. Charette, *Mathematical Instrumentation*, esp. 3–21; and Dallal, "Science, Medicine, and Technology," esp. 177–183.

10. al-Jazarī, *Book of Knowledge of Ingenious Mechanical Devices*; and see Lamia Balafrej, "Automated Slaves, Ambivalent Images, and Noneffective Machines in al-Jazari's Compendium of Mechanical Arts, 1206," *Inquiries into Art, History, and the Visual* 4 (2022): 737–774; Vivek Gupta, "Routes of Translation: Connected Book Histories and al-Jazari's Robotic Wonders from the Mamluks to Mandu," *South Asian Studies* 39, no. 2 (2023): 1–22; Hill, *History of Engineering*, 146–152; Hill, *Islamic Science and Engineering*, 97–105 and 124–135; Meekyung MacMurdie, "The Manuscript Machine: Assemblages and Divisions in Jazari's *Compendium*," in *Destroyed—Disappeared—Lost—Never Were*, ed. Beate Fricke and Aden Kumler (University Park: Penn State University Press, 2022), 113–128; and Angela Vanhaelen, "Strange Things for Strangers: Transcultural Automata in Early Modern Amsterdam," *Art Bulletin* 103, no. 3 (2021): 42–68.

11. Hill, *History of Engineering*, 199–222; Hill, *Islamic Engineering and Science*, 140–148; and Truitt, *Medieval Robots*, 1–39.

12. Truitt, *Medieval Robots*, 122–137.

13. Lily Filson, "Magical and Mechanical Evidence: The Late-Renaissance Automata of Francesco I de' Medici," in *Evidence in the Age of the New Sciences*, ed. James A. T. Lancaster and Richard Raiswell (Cham: Springer, 2018), 177–201; and Matteo Valleriani, "The Garden of Pratolino: Ancient Technology Breaks through the Barriers of Modern Iconology," in Ludi

Naturae: *Spiele der Natur in Kunst und Wissenschaft*, ed. Natascha Adamowsky, Hartmut Böhme, and Robert Felfe (Munich: Weilhelm Fink, 2011), 121–141.

14. For sundials, see esp. Gerhard Dohrn-Van Rossum, *History of the Hour: Clocks and Modern Temporal Orders*, trans. Thomas Dunlap (Chicago: University of Chicago Press, 1996), esp. 20–21; and René R. J. Rohr, *Sundials: History, Theory, and Practice*, trans. G. Godin (1970; repr., New York: Dover, 1996).

15. See esp. Antonio Fernández-Puertas, *Clepsidras y Relojes musulmanes / Muslim Water Clocks and Mechanical Time Pieces* (Granada: Fundación Pública Andaluza El Legado Andalusi, 2010); Hill, *History of Engineering*, 224–242; Hill, *Islamic Science and Engineering*, 126–135; and Truitt, *Medieval Robots*, 142–145.

16. Dohrn-Van Rossum, *History of the Hour*, 5–6, 117–118, 242–245, 270, and 276–278.

17. Dohrn-Van Rossum, *History of the Hour*, 125–172; Hill, *History of Engineering*, 242–245; and Truitt, *Medieval Robots*, esp. 141–53.

18. See esp. Günther Oestmann, *The Astronomical Clock of Strasbourg Cathedral: Function and Significance*, trans. Bruce W. Irwin (Leiden: Brill, 2020).

19. For a cogent summary of a complex subject, see Nancy Mason Bradbury and Carolyn P. Colette, "Changing Times: The Mechanical Clock in Late Medieval Literature," *Chaucer Review* 43 (2009): 351–375.

20. For a lucid introduction to the topic of navigation, see Jim Bennett, *Navigation: A Very Short Introduction* (Oxford: Oxford University Press, 2017); and see Eric H. Ash, "Navigation Techniques and Practice in the Renaissance," in *The History of Cartography*, vol. 3, *Cartography in the Renaissance*, pt. 1, ed. David Woodward (Chicago: University of Chicago Press, 2007), 509–527; Ruthy Gertwagen, "Nautical Technology," in *A Companion to Mediterranean History*, ed. Peregrine Horden and Sharon Kinoshita (Chichester, West Sussex: Wiley Blackwell, 2014), 154–169; and Robert D. Hicks, "Nautical Astronomy and Celestial Navigation," in *The Oxford Encyclopedia of Maritime History*, ed. John B. Hattendorf, 4 vols. (Oxford: Oxford University Press, 2007), 2: 619–638.

21. See Dionisius A. Agius, *Classic Ships of Islam: From Mesopotamia to the Indian Ocean* (Leiden: Brill, 2008), esp. 171–202; Bennett, *Navigation*, 22–33; Hikmat Homsi, "Navigation and Shipbuilding," in al-Hassan, Ahmed, and Iskandar, *Science and Technology in Islam*, pt. 2, 217–246, esp. 243–246; Richard A. Paselk, "Navigational Instruments: Measurement of Altitude," and Jobst Broelmann, "Navigational Instruments: Measurement of Direction," both in Hattendorf, *Oxford Encyclopedia of Maritime History*, 3: 29–42 and 3: 42–52, respectively; and María M. Portuondo, *Secret Science: Spanish Cosmography and the New World* (Chicago: University of Chicago Press, 2009), esp. 48–50. For portolan charts, see esp. Tony Campbell, "Portolan Charts from the Late Thirteenth Century to 1500," in *The History of Cartography*, vol. 1, *Cartography in Prehistoric, Ancient, and Medieval Europe and the Mediterranean*, ed. J. B. Harley and David Woodward (Chicago: University of Chicago Press, 1987), 371–463; and Campbell, "A Critical Re-examination of Portolan Charts with a Reassessment of Their Replication and Seaboard Function," Map History / History of Cartography (website), 2024, https://www.maphistory.info /portolan.html, a collection of numerous further studies and revisions up to (as of now) 2024.

22. Bennett, *Navigation*, 36–41; Henrique Leitão, "Instruments and Artisanal Practices in Long Distance Ocean Voyages," *Centaurus* 60 (2018): 189–202; Antonio Sánchez [Martínez], "Charts for an Empire: A Global Trading Zone in Early Modern Portuguese Nautical Cartography," *Centaurus* 60 (2018): 173–188; Carla Rahn Phillips, "Americas: Exploration Voyages, 1492–1509," Louis De Vorsey, "Americas: Exploration Voyages, 1500–1620 (Eastern Coast)," and John Hemming, "Americas: Exploration Voyages, 1500–1616 (South America)," all in Hattendorf, *Oxford Encyclopedia of Maritime History*, 1: 23–28, 1: 28–36, and 1: 36–42, respectively.

23. Bruno Almeida, "Transmitting Nautical and Cosmographical Knowledge in the 16th

and 17th Centuries: The Case of Pedro Nuñes," *Centaurus* 60 (2018): 216–229; Bennett, *Navigation*, 54–66; Norman J. W. Thrower, "Mercator, Gerardus (1512–1594)," and António Estácio dos Reis, "Nuñes, Pedro (1502–1578)," both in Hattendorf, *Oxford Encyclopedia of Maritime History*, 2: 555–556 and 3: 116–117, respectively.

24. See esp. Almeida, "Transmitting Nautical and Cosmographical Knowledge"; Bennett, *Navigation*, 53–66; Leitão, "Instruments and Artisanal Practices"; Edward Collins, "Interactions of Portuguese Artisanal Culture in the Maritime Enterprise of 16th-Century Seville," *Centaurus* 60 (2018): 203–215; Portuondo, *Secret Science*, 1–102; Sánchez [Martínez], "Charts for an Empire"; Antonio Sánchez Martínez, "Los artifices del Plus Ultra: Pilotos, cartógrafos y cosmógrafos en la Casa de la Contratación de Sevilla durante el siglo XVI," *Hispania* 70 (September–December 2010): 607–632; Alison Sandman, "Mirroring the World: Sea Charts, Navigation, and Territorial Claims in Sixteenth-Century Spain," in *Merchants and Marvels: Commerce, Science, and Art in Early Modern Europe*, ed. Pamela H. Smith and Paula Findlen (New York: Routledge, 2002), 83–108; and Margaret E. Schotte, *Sailing School: Navigating Science and Skill, 1550–1800* (Baltimore: Johns Hopkins University Press, 2019).

25. Lesley B. Cormack, Steven A. Walton, and John A. Schuster, eds., *Mathematical Practitioners and the Transformation of Natural Knowledge in Early Modern Europe* (Cham: Springer, 2017); Serafina Cuomo, "Shooting by the Book: Notes on Niccolò Tartaglia's Nova Scientia," *History of Science* 35 (June 1997): 155–188; and for Galileo's compass, Galileo Galilei, *Operations of the Geometric and Military Compass*, trans. and intro. Stillman Drake (Washington, DC: Smithsonian Institution Press, 1978).

26. James A. Bennett, "Shopping for Instruments in Paris and London," in Smith and Findlen, *Merchants and Marvels*, 370–395; Anthony Gerbino and Stephen Johnston, *Compass and Rule: Architecture as Mathematical Practice in England* (Oxford: Museum of the History of Science, 2009), esp. 45–64; Cesare S. Maffioli, "A Fruitful Exchange/Conflict: Engineers and Mathematicians in Early Modern Italy," *Annals of Science* 70, no. 2 (2013): 197–228; Alexander Marr, *Between Raphael and Galileo: Mutio Oddi and the Mathematical Culture of Late Renaissance Italy* (Chicago: University of Chicago Press, 2011); and Steven A. Walton, "Technologies of Pow(d)er: Military Mathematical Practitioners' Strategies and Self-Presentation," in Cormack, Walton, and Schuster, *Mathematical Practitioners*, 87–113.

27. See esp. Edgar Zilsel, *The Social Origins of Modern Science*, ed. Diederick Raven, Wolfgang Krohn, and Robert S. Cohen (Dordrecht: Kluwer Academic, 2000), which includes his articles relevant to the "Zilsel thesis." For recent assessments, see Lesley B. Cormack, "Handwork and Brainwork: Beyond the Zilsel Thesis," in Cormack, Walton, and Schuster, *Mathematical Practitioners*, 11–35; William Eamon, "Corn, Cochineal, and Quina: The 'Zilsel Thesis' in a Colonial Iberian Setting," *Centaurus* 60 (2018): 141–158; Henrique Laitão and Antonio Sánchez [Martínez], "Zilsel's Thesis, Maritime Culture, and Iberian Science in Early Modern Europe," *Journal of the History of Ideas* 78 (April 2017): 191–210; Long, *Artisan/Practitioners*, esp. 10–29; and Donata Romizi, Monika Wulz, and Elisabeth Nemeth, eds., *Edgar Zilsel: Philosopher, Historian, Sociologist* (Cham: Springer, 2022).

28. Cormack, "Handwork and Brainwork," esp. 15–18 and 23–26; Laitão and Sánchez, "Zilsel's Thesis, Maritime Culture, and Iberian Science"; Long, *Artisan/Practitioners*, 94–106; Pamela O. Long, "Trading Zones in Early Modern Europe," *Isis* 106 (December 2015): 840–847; Pamela H. Smith, *The Body of the Artisan: Art and Experience in the Scientific Revolution* (Chicago: University of Chicago Press, 2004); and Smith, ed., "The Making and Knowing Project," https://www.makingandknowing.org/, accessed January 2024. The manuscript, Ms. Bn FMs. Fr. 640, is housed in the Bibliothèque Nationale in Paris.

29. For an introduction that demonstrates the complexity of the topic, see Bernard S.

Bachrach and David S. Bachrach, *Warfare in Medieval Europe, c. 400–c. 1453*, 2nd ed. (Abingdon: Routledge, 2022).

30. See esp. Christopher Allmand, "War and the Non-Combatant in the Middle Ages," in *Medieval Warfare: A History*, ed. Maurice Keen (Oxford: Oxford University Press, 1999), 253–272; John Haldon, *Warfare, State and Society in the Byzantine World* (London: University College London Press, 1999), 234–279; and John A. Lynn II, *Women, Armies, and Warfare in Early Modern Europe* (Cambridge: Cambridge University Press, 2008).

31. Bachrach and Bachrach, *Warfare in Medieval Europe*, 237–239; Kelly DeVries and Robert Douglas Smith, *Medieval Military Technology*, 2nd ed. (Toronto: University of Toronto Press, 2012), 34–41; John Haldon, "Some Aspects of Early Byzantine Arms and Armour," and Michael Gorelik, "Arms and Armour in South-Eastern Europe in the Second Half of the First Millennium AD," both in *A Companion to Medieval Arms and Armour*, ed. David Nicolle (Woodbridge: Boydell Press, 2002), 65–79 and 127–147, respectively; and Donald R. Hill, "Military Technology," in al-Hassan, Ahmed, and Iskandar, *Science and Technology in Islam*, pt. 2, 247–270, esp. 249–251 and 255–260.

32. Bachrach and Bachrach, *Warfare in Medieval Europe*, 237–239; Devries and Smith, *Medieval Military Technology*, 38–41; Hill, "Military Technology," 260; and Matthew Strickland and Robert Hardy, *The Great Warbow: From Hastings to the Mary Rose* (Thrupp, Stroud: Sutton Publishers, 2005), 34–48 for the ancient antecedents of the longbow.

33. Bachrach and Bachrach, *Warfare in Medieval Europe*, 239–242; DeVries and Smith, *Medieval Military Technology*, 41–46; Nicolas Prouteau, "L'artilleur et l'artillerie avant le temps des canons," in *Artillerie et fortification, 1200–1600*, ed. Nicolas Prouteau, Emmanuel de Crouy-Chanel, and Nicolas Faucherre (Rennes: Presses Universitaires de Rennes, 2011), 23–32; and Strickland and Hardy, *Great Warbow*, 113–135.

34. Bachrach and Bachrach, *Warfare in Medieval Europe*, 223–229 and 235–236; DeVries and Smith, *Medieval Military Technology*, 5–41; Paddy Griffith, *The Viking Art of War* (London: Greenhill Books, 1998), 162–166 and 173; Haldon, "Some Aspects of Early Byzantine Arms," 65–71; Hill, "Military Technology," 253–255; Anne Pedersen, "Scandinavian Weaponry in the Tenth Century: The Example of Denmark," and Shihab al-Sarraf, "Close Combat Weapons in the Early 'Abbāsid Period: Maces, Axes and Swords," both in Nicolle, *Companion to Medieval Arms and Armour*, 25–35 and 149–178, respectively.

35. Bachrach and Bachrach, *Warfare in Medieval Europe*, 226–233; DeVries and Smith, *Medieval Military Technology*, 18–24; Pedersen, "Scandinavian Weaponry in the Tenth Century," 28–29; and Alan Williams, "The Metallurgy of Medieval Arms and Armour," in Nicolle, *Companion to Medieval Arms and Armour*, 45–54, esp. 45–51.

36. Lynn White Jr., *Medieval Technology and Social Change* (Oxford: Oxford University Press, 1962), 1–38; and see DeVries and Smith, *Medieval Military Technology*, 11–15, for the debate concerning the onset of mounted shock combat.

37. White, *Medieval Technology and Social Change*, 1–38.

38. A good summary is DeVries and Smith, *Medieval Military Technology*, 99–114; and see Bernard S. Bachrach, *Early Carolingian Warfare: Prelude to Empire* (Philadelphia: University of Pennsylvania Press, 2001); and Alex Roland, "Once More in the Stirrups: Lynn White Jr., Medieval Technology, and Social Change," *Technology and Culture* 44 (July 2003): 574–585.

39. An early, highly influential article was Elizabeth A. R. Brown, "The Tyranny of a Construct: Feudalism and Historians of Medieval Europe," *American Historical Review* 79 (October 1974): 1063–1088. See also Susan Reynolds, *Fiefs and Vassals: The Medieval Evidence Reinterpreted* (Oxford: Oxford University Press, 1994); Bachrach and Bachrach, *Warfare in Medieval Europe*, 119–121; and Devries and Smith, *Medieval Military Technology*, 99–114, who discuss the

various meanings given to the term and the role of the concept of feudalism in White's thesis. It is beyond the scope of this book to detail the ongoing discussions concerning medieval feudalism. For a good account of the complexities of the topic and its implications for periodization, see Charles West, *Reframing the Feudal Revolution: Political and Social Transformation between the Marne and Moselle, c. 800–c. 1100* (Cambridge: Cambridge University Press, 2013), esp. 1–16 and 255–263.

40. See esp. Andrew Ayton, "Arms, Armour, and Horses," in Keen, *Medieval Warfare*, 188–208; Bachrach and Bachrach, *Warfare in Medieval Europe*, 223–226, 231, and 233–235; Devries and Smith, *Medieval Military Technology*, 53–98; and Williams, "Metallurgy of Medieval Arms and Armour," 49–54.

41. Bachrach and Bachrach, *Warfare in Medieval Europe*, 247–252; and DeVries and Smith, *Medieval Military Technology*, 165–181 (siege machines), and 117–128 (trebuchets and other non-gunpowder artillery).

42. Bachrach and Bachrach, *Warfare in Medieval Europe*, 243–245; Paul E. Chevedden, "The Invention of the Counterweight Trebuchet: A Study in Cultural Diffusion," in *Dumbarton Oaks Papers*, No. 54, ed. Alice-Mary Talbot (Washington, DC: Dumbarton Oaks, 2000), 71–116; Paul E. Chevedden et al., "The Traction Trebuchet: A Triumph of Four Civilizations," *Viator* 31 (January 2000): 433–486; DeVries and Smith, *Medieval Military Technology*, 117–128; Hill, "Military Technology," 260–266; and W.T.S. Tarver, "The Traction Trebuchet: A Reconstruction of an Early Medieval Siege Engine," *Technology and Culture* 36 (January 1995): 136–167.

43. Bachrach and Bachrach, *Warfare in Medieval Europe*, 243–245; and Chevedden, "Invention of the Counterweight Trebuchet." See also George T. Dennis, "Byzantine Heavy Artillery: The Helepolis," *Greek, Roman, and Byzantine Studies* 39 (1998): 99–115; and Hill, "Military Technology," 260–266.

44. Bert S. Hall, *Weapons and Warfare in Renaissance Europe* (Baltimore: Johns Hopkins University Press, 1997), 34–40.

45. See esp. Bachrach and Bachrach, *Warfare in Medieval Europe*, 245–247; Ahmad Y. al-Hassan, "Military Fires, Gunpowder and Firearms," in al-Hassan, Ahmed, and Iskandar, eds., *Science and Technology in Islam*, pt. 2, 107–134; and Alex Roland, "Secrecy, Technology, and War: Greek Fire and the Defense of Byzantium, 678–1204," *Technology and Culture* 33 (October 1992): 655–679.

46. Bachrach and Bachrach, *Warfare in Medieval Europe*; 245–247; and Hall, *Weapons and Warfare*, 41–104. For a contextual study of gun manufacture (in Florence, Italy), see Fabrizio Ansani, "The Life of a Renaissance Gunmaker: Bonaccorso Ghiberti and the Development of Florentine Artillery in the Late Fifteenth Century," *Technology and Culture* 58 (July 2017): 749–789.

47. Brice Cossart, "Producing Skills for an Empire: Theory and Practice in the Seville School of Gunners during the Golden Age of the Carrera de Indias," *Technology and Culture* 58 (April 2017): 459–486; Cossart, *Les Artilleurs et la Monarchie Hispanique (1560–1610)* (Paris: Classiques Garnier, 2021); Rainer Leng, *Ars belli: Deutsche taktische und kriegstechnische Bilderhanschriften und Traktate im 15. Und 16. Jahrhundert*, 2 vols. (Wiesbaden: Reichert, 2002); Walton, "Proto-Scientific Revolution or Cookbook Science? Early Gunnery Manuals in the Craft Treatise Tradition," in *Craft Treatises and Handbooks: The Dissemination of Technical Knowledge in the Middle Ages*, ed. Ricardo Córdoba (Turnhout: Brepols, 2013), 221–236; and Walton, "Technologies of Pow(d)er." And see Clifford J. Rogers, "The Artillery and Artillery Fortress Revolutions Revisited," in Prouteau, Crouy-Chanel, and Faucherre, *Artillerie et fortification, 1200–1600*, 75–80, for the thesis that the greater efficiency of artillery in the fifteenth century and the greater effectiveness of fortification in the sixteenth both constitute revolutions.

48. See esp. Bachrach and Bachrach, *Warfare in Medieval Europe*, 254–262; and DeVries and Smith, *Medieval Military Technology*, 272–278.

49. See esp. Geoffrey Parker, *The Military Revolution: Military Innovation and the Rise of the West, 1500–1800*, 2nd ed. (Cambridge: Cambridge University Press, 1996); Hall, *Weapons and Warfare*, 201–234; Christos G. Makrypoulias, "'Our Engines Are Better Than Yours': Perception and Reality of Late Byzantine Military Technology," in *Byzantium and the West: Perception and Reality (11th–15th c.)*, ed. Nikolaos G. Chrissis, Athina Kolia-Dermitzaki, and Angeliki Papageorgiou (Abingdon: Routledge, 2019), 306–315; Clifford J. Rogers, *The Military Revolution Debate: Readings on the Military Transformation of Early Modern Europe* (Boulder, CO: Westview Press, 1995); and David Eltis, *The Military Revolution in Sixteenth-Century Europe* (London: Tauris Academic Studies, 1995). For an astute global analysis, see Hyeok Hweon Kang, "Difference in an Age of Parity: Technology and Global Military History," in *The Military Revolution and Revolutions in Military Affairs*, ed. Mark C. Fissel (Berlin: De Gruyter, 2022), 29–64.

50. See esp. Bachrach and Bachrach, *Warfare in Medieval Europe*, 164–221; Lynn, *Women, Armies, and Warfare*; and Mary E. Wiesner-Hanks, *Women and Gender in Early Modern Europe*, 4th ed. (Cambridge: Cambridge University Press, 2019), 323–325. For Joan of Arc as a military leader, see Kelly DeVries, *Joan of Arc: A Military Leader* (Stroud, Gloucestershire: History Press, 2011).

Conclusion

Epigraph: Cited in Constant Méheut, "Facing a Future of Drought, Spain Turns to Medieval Solutions and 'Ancient Wisdom,'" *New York Times*, July 20, 2023, https://www.nytimes.com /2023/07/19/world/europe/spain-drought-acequias.html.

1. Méheut, "Facing a Future of Drought"; and Stephen Burgen, "Spring Time: Why an Ancient Water System Is Being Brought Back to Life in Spain," *The Guardian*, April 11, 2022, https://www.theguardian.com/environment/2022/apr/11/ancient-water-system-restore-spain -sierra-nevada-aoe. The efforts to recover the medieval canals is being undertaken in different areas and is led in part by the laboratory of biocultural archaeology of the University of Granada (the MEMOLab project).

2. David E. Nye, *Technology Matters: Questions to Live With* (Cambridge, MA: MIT Press, 2006), quote on ix.

3. Maarten Prak, "Preface: S. R. Epstein (1960–2007) and the Guilds," *International Review of Social History*, supplement 16, *The Return of the Guilds*, 53 (2008): 1–3.

4. Pamela O. Long, "Trading Zones in Early Modern Europe," *Isis* 106 (December 2015): 840–847; Lissa Roberts, Simon Schaffer, and Peter Dear, eds., *The Mindful Hand: Inquiry and Invention in the Late Renaissance to Early Industrialisation* (Amsterdam: Knoinklijke Nederlandse Akademie van Wetenschappen, 2007); and Pamela H. Smith, *The Body of the Artisan: Art and Experience in the Scientific Revolution* (Chicago: University of Chicago Press, 2004).

5. Marcus Popplow, "Formalization and Interaction," *Isis* 106 (December 2015): 848–856.

6. For Zilsel, see Edgar Zilsel, *The Social Origins of Modern Science*, ed. Diederick Raven, Wolfgang Krohn, and Robert S. Cohen (Dordrecht: Kluwer Academic, 2000), which includes his articles relevant to the "Zilsel thesis." For recent assessments, see Lesley B. Cormack, "Handwork and Brainwork: Beyond the Zilsel Thesis," in Cormack, Walton, and Schuster, *Mathematical Practitioners*, 11–35; William Eamon, "Corn, Cochineal, and Quina: The 'Zilsel Thesis' in a Colonial Iberian Setting," *Centaurus* 60 (2018): 141–158; Henrique Laitão and Antonio Sánchez [Martínez], "Zilsel's Thesis, Maritime Culture, and Iberian Science in Early Modern Europe, *Journal of the History of Ideas* 78 (April 2017): 191–210; Long, *Artisan/Practitioners*, esp. 10–29; and Donata Romizi, Monika Wulz, and Elisabeth Nemeth, eds., *Edgar Zilsel: Philosopher, Historian, Sociologist* (Cham: Springer, 2022).

7. Good introductions include Katharine Park and Lorraine Daston, eds., *The Cambridge History of Science*, vol. 3, *Early Modern Science* (Cambridge: Cambridge University Press, 2006); and Steven Shapin, *The Scientific Revolution*, 2nd ed. (Chicago: University of Chicago Press, 2018). See also Jorge Cañizares-Esguerra, "On Ignored Global 'Scientific Revolutions,'" *Journal of Early Modern History* 21 (2017): 420–432; Kapil Raj, "Thinking without the Scientific Revolution: Global Interactions and the Construction of Knowledge," *Journal of Early Modern History* 21 (2017): 445–458; and Antonio Sánchez, "Practical Knowledge and Empire in the Early Modern Iberian World: Towards an Artisanal Turn," *Centaurus* 61 (2019): 268–281.

8. For scholarship in other regions of the world that suggest such a comparison, see, for example, Hyeok Hweon Kang, "Crafting Knowledge: Artisan, Officer, and the Culture of Making in Chosŏn Korea, 1392–1910" (PhD diss., Harvard University, 2020), esp. 7–17; and Dorothy Ko, *The Social Life of Inkstones: Artisans and Scholars in Early Qing China* (Seattle: University of Washington Press, 2017), esp. 198–201.

Bibliography

Aberth, John. *An Environmental History of the Middle Ages: The Crucible of Nature*. London: Routledge, 2013.

Abulafia, David. *The Great Sea: A Human History of the Mediterranean*. Oxford: Oxford University Press, 2011.

Agius, Dionisius A. *Classic Ships of Islam: From Mesopotamia to the Indian Ocean*. Leiden: Brill, 2008.

Ågren, Maria, ed. *Making a Living, Making a Difference: Gender and Work in Early Modern European Society*. Oxford: Oxford University Press, 2017.

Ahmet, K. "A Middle Byzantine Olive Press Room at Aphrodisias." *Anatolian Studies* 51 (2001): 150–167.

Ahunbay, Metin and Zeynep Ahunbay. "Structural Influence of Hagia Sophia on Ottoman Mosque Architecture." In *Hagia Sophia from the Age of Justinian to the Present*, edited by Robert Mark and Ahmet S. Çakmak, 179–194. Cambridge: Cambridge University Press, 1992.

Ait, Ivana. "Donne in affari: Il caso di Roma (secoli XIV–XV)." In Esposito, *Donne del Rinascimento a Roma e dintorni*, 53–83.

Alexander, Jonathan J. G. *Medieval Illuminators and Their Methods of Work*. New Haven: Yale University Press, 1992.

Allmand, Christopher. "War and the Non-Combatant in the Middle Ages." In Keen, *Medieval Warfare*, 253–272.

Almeida, Bruno. "Transmitting Nautical and Cosmographical Knowledge in the 16th and 17th Centuries: The Case of Pedro Nuñes." *Centaurus* 60 (2018): 216–229.

Almela, Iñigo. "Religious Architecture as an Instrument for Urban Renewal: Two Religious Complexes from the Saadian Period in Marrakesh." *Al-Masāq* 31, no. 3 (2019): 272–302.

Anderson, Christy, Anne Dunlop, and Pamela H. Smith. *The Matter of Art: Materials, Practices, Cultural Logics, c. 1250–1750*. Manchester: Manchester University Press, 2014.

Andersson, Hans and Barbara Scholkmann. "Towns." In Carver and Klápště, *Archaeology of Medieval Europe*, vol. 2, *Twelfth to Sixteenth Centuries*, 370–407.

Andreolli, Bruno and Massimo Montanari, eds. *Il bosco nel medioevo*. Bologna: Cooperativa Libraria Universitaria Editrice Bologna, 1988.

Angelos, Mark. "Urban Women, Investment, and the Commercial Revolution of the Middle Ages." In Mitchell, *Women in Medieval Western European Culture*, 257–272.

Anker, Leif. "Stave Church Research and the Norwegian Stave Church Programme: New

Findings—New Questions." In Khodakovsky and Lexau, *Historic Wooden Architecture in Europe and Russia*, 94–109.

Anreiter, Peter, et al., eds. *Mining in European History and Its Impact on Environment and Human Societies*. Innsbruck: Innsbruck University Press, 2010.

Ansani, Fabrizio. "The Life of a Renaissance Gunmaker: Bonaccorso Ghiberti and the Development of Florentine Artillery in the Late Fifteenth Century." *Technology and Culture* 58 (July 2017): 749–789.

Appuhn, Karl. *A Forest on the Sea: Environmental Expertise in Renaissance Venice*. Baltimore: Johns Hopkins University Press, 2009.

Arce, Ignacio. "Umayyad Building Techniques and the Merging of Roman-Byzantine and Partho-Sassanian Traditions: Continuity and Change." In Lavan et al., *Technology in Transition*, 491–537.

Armstrong, Pamela. "Ceramics." In Jeffreys with Haldon and Cormack, *Oxford Handbook of Byzantine Studies*, 429–443.

Arndt, Betty. "Medieval and Post-Medieval Urban Water Supply and Sanitation: Archaeological Evidence from Göttingen and North German Towns." In Chiarenza, Haug, and Müller, *Power of Urban Water*, 213–227.

Ash, Eric H. *The Draining of the Fens: Projectors, Popular Politics, and State Building in Early Modern England*. Baltimore: Johns Hopkins University Press, 2017.

———. "Navigation Techniques and Practice in the Renaissance." In *The History of Cartography*, vol. 3, *Cartography in the Renaissance*, pt. 1, edited by David Woodward, 509–527. Chicago: University of Chicago Press, 2007.

———. *Power, Knowledge, and Expertise in Elizabethan England*. Baltimore: Johns Hopkins University Press, 2004.

Ashburner, Walter. "The Farmer's Law." *Journal of Hellenic Studies* 32 (1912): 68–95.

Ashtor, Eliyahu and Halil Inalcik. "Ḳuṭn, Ḳuṭun (A.), Cotton." In Bearman et al., *Encyclopaedia of Islam*, 5: 554–566.

Asmussen, Tina. "The *Kux* as a Site of Mediation: Economic Practices and Material Desires in the Early Modern German Mining Industry," 159–182. In *Sites of Mediation—Connected Histories of Places, Processes, and Objects: Europe and Beyond, 1450–1650*, edited by Susanna Burghartz, Lucas Burkart, and Christine Göttler. Leiden: Brill, 2017.

Asmussen, Tina and Pamela O. Long. "Introduction: The Cultural and Material Worlds of Mining in Early Modern Europe." In special issue, *The Cultural and Material Worlds of Mining in Early Modern Europe*, edited by Tina Asmussen, *Renaissance Studies* 34 (February 2020): 8–30.

Astill, Grenville and John Langdon, eds. *Medieval Farming and Technology: The Impact of Agricultural Change in Northwest Europe*. Leiden: Brill, 1997.

Aston, Michael, ed. *Medieval Fish, Fisheries and Fishponds in England*. 2 parts. Oxford: BAR Publishing, 1988.

Audoin-Rouzeau, Frédérique and Sylvie Beyries, eds. *Le travail du cuir de la Préhistoire à nos jours*." Antibes: Éditions APDCA, 2002.

Avni, Gideon. "Early Islamic Irrigated Farmsteads and the Spread of Qanats in Eurasia." *Water History* 10 (2018): 313–338.

Ayalon, Etan, Rafael Frankel, and Amoss Kloner, eds. *Oil and Wine Presses in Israel from the Hellenistic, Roman and Byzantine Periods*. BAR International Series 1972. Oxford: Archaeopress, 2009.

Ayton, Andrew. "Arms, Armour, and Horses." In Keen, *Medieval Warfare*, 188–208.

Bachman, Clifford R. "Piracy." In Horden and Kinoshita, *Companion to Mediterranean History*, 170–183.

Bachrach, Bernard S. *Early Carolingian Warfare: Prelude to Empire.* Philadelphia: University of Pennsylvania Press, 2001.

Bachrach, Bernard S. and David S. Bachrach. *Warfare in Medieval Europe, c.400–c.1453.* 2nd ed. Abingdon: Routledge, 2022.

Baker, Patricia L. *Islamic Textiles.* London: British Museum Press, 1995.

Balafrej, Lamia. "Automated Slaves, Ambivalent Images, and Noneffective Machines in al-Jazari's Compendium of Mechanical Arts, 1206." *Inquiries into Art, History, and the Visual* 4 (2022): 737–774.

Bambach, Carmen C. *Drawing and Painting in the Italian Renaissance Workshop: Theory and Practice, 1300–1600.* Cambridge: Cambridge University Press, 1999.

Banū Mūsā bin Shakir. *The Book of Ingenious Devices.* Translated by Donald R. Hill. Dordrecht: Reidel, 1979.

Baraldi, Enzo. "Una nuova etá del ferro: Macchine e processi della siderugia." In Braunstein and Molá, *Produzione e techniche,* 199–216.

Barceló, Miquel. "The Missing Water-Mill: A Question of Technological Diffusion in the High Middle Ages." In Barceló and Sigaut, *Making of Feudal Agricultures?,* 255–314.

Barceló, Miquel and François Sigaut, eds. *The Making of Feudal Agricultures?,* Leiden: Brill, 2004.

Bardill, Jonathan. "Building Materials and Techniques." In Jeffreys with Haldon and Cormack *Oxford Handbook of Byzantine Studies,* 335–352.

Barker, Graeme. "Castles in the Desert." In Barker, *Farming the Desert,* 1–20.

———, ed. *Farming the Desert: The UNESCO Libyan Valleys Archaeological Survey.* Vol. 1, *Synthesis.* London: UNESCO, the Department of Antiquities, Tripoli, and the Society for Libyan Studies, 1996.

Barker, Graeme with David D. Gilbertson. "Farming the Desert: Retrospect and Prospect." In Barker, *Farming the Desert,* 343–363.

Barker, Graeme, David D. Gilbertson with Chris O. Hunt and David Mattingly. "Romano-Libyan Agriculture: Integrated Models." In Barker, *Farming the Desert,* 265–290.

Barker, Hannah. *That Most Precious Merchandise: The Mediterranean Trade in Black Sea Slaves, 1260–1500.* Philadelphia: University of Pennsylvania Press, 2019.

Baron, Sabrina Alcorn, Eric N. Lindquist, and Eleanor E. Shevlin, eds. *Agent of Change: Print Culture Studies after Elizabeth L. Eisenstein.* Amherst: University of Massachusetts Press; Washington, DC: Center for the Book, Library of Congress, 2007.

———. "Introduction," In Baron, Lindquist, and Shevlin, *Agent of Change,* 1–22.

Barrett, James H. "Medieval Sea Fishing, AD 500–1500: Chronology, Causes and Consequences." In Barrett and Orton, *Cod and Herring,* 250–272.

Barrett, James H. and David R. Orton, eds. *Cod and Herring: The Archaeology and History of Medieval Sea Fishing.* Oxford: Oxbow Books, 2016.

Bass, George F. "The Development of Maritime Archaeology." In Catsambis, Ford, and Hamilton, *Oxford Handbook of Maritime Archaeology,* 3–22.

Bass, George F., et al. *Serçe Limani: An Eleventh-Century Shipwreck.* 2 vols. College Station: Texas A&M University Press, 2004 and 2009.

Bassett, Sarah, ed. *The Cambridge Companion to Constantinople.* Cambridge: Cambridge University Press, 2022.

Bauch, Martin and Gerrit Jasper Schenk, eds. *The Crisis of the 14th Century: Teleconnections between Environmental and Societal Change?* Berlin: De Gruyter, 2020.

———. "Teleconnections, Correlations, Causalities between Nature and Society? An Introductory Comment on the 'Crisis of the Fourteenth Century.'" In Bauch and Schenk, *Crisis of the 14th Century,* 1–16.

Bavel, Bas J. P. van. *Manors and Markets: Economy and Society in the Low Countries, 500–1600.* Oxford: Oxford University Press, 2010.

Bavel, Bas J. P. van and Richard W. Hoyle. "Introduction: Social Relations, Property and Power in the North Sea Area, 500–2000." In Bavel and Hoyle, *Social Relations: Property and Power*, 1–47

——, eds. *Social Relations: Property and Power.* Vol 4 in *Rural Economy and Society in North-Western Europe, 500–2000,* ed. Erik Thoen. Turnhout: Brepols, 2010.

Bearman, P. J., et al., eds. *The Encyclopaedia of Islam.* New ed., 11 vols. Leiden: Brill, 1978–2001.

Béaur, Gérhard and Larent Feller. "Northern France, 1000–1750." In Vanhaute, Devos, and Lambrecht, *Making a Living*, 98–125.

Beck, Patrice, Philippe Braunstein, and Michel Philippe. "Wood, Iron, and Water in the Othe Forest in the Late Middle Ages: New Findings and Perspectives." In Smith and Wolfe, *Technology and Resource Use in Medieval Europe*, 173–184.

Bekker-Nielsen, Tønnes and Ruthy Gertwagen, eds. *The Inland Seas: Towards an Ecohistory of the Mediterranean and the Black Sea.* Stuttgart: Franz Steiner, [2016].

Belfanti, Carlo M. "Guilds, Patents, and the Circulation of Technical Knowledge: Northern Italy during the Early Modern Age." *Technology and Culture* 45 (2004): 569–589.

Belich, James. "The Black Death and the Spread of Europe." In Belich et al., *Prospect of Global History*, 93–105.

Belich, James, et al., eds. *The Prospect of Global History.* Oxford: Oxford University Press, 2016.

Belich, James, John Darwin, and Chris Wickham. "Introduction: The Prospect of Global History." In Belich et al., *Prospect of Global History*, 3–22.

Belke, Klaus. "Communications: Roads and Bridges." In Jeffreys with Haldon and Cormack, *Oxford Handbook of Byzantine Studies*, 295–307.

Bellavitis, Anna. *Il lavoro delle donne nelle città dell'Europa moderna.* Rome: Viella, 2016.

Bennett, James A. *Navigation: A Very Short Introduction.* Oxford: Oxford University Press, 2017.

——. "Shopping for Instruments in Paris and London." In Smith and Findlen, *Merchants and Marvels*, 370–395.

Bennett, Judith M. and Ruth Mazo Karras, eds. *The Oxford Handbook of Women and Gender in Medieval Europe.* Oxford: Oxford University Press, 2013.

Bensch, Stephen P. "From Prizes of War to Domestic Merchandise: The Changing Face of Slavery in Catalonia and Aragon, 1000–1400." *Viator* 25 (1994): 63–93.

Beretta, Marco. "Glassmaking Goes Public: The Cultural Background to Antonio Nari's *L'Arte Vetraria* (1612)." *Technology and Culture* 58 (October 2017): 1046–1070.

Berger, Albrecht. "Urban Development and Decline: Fourth–Fifteenth Centuries." In Bassett, *Cambridge Companion to Constantinople*, 33–49.

Bernardoni, Andrea. *La conoscenza del fare: Ingegneria, arte, scienza nel De la pirotechnia di Vannoccio Biringuccio.* Rome: L'Erma di Bretschneider, 2011.

Bertelli, Carlo. "The Production and Distribution of Books in Late Antiquity." Hodges and Bowden, *Sixth Century*, 41–60.

Bertini, Camilla, Julian Henderson, and Simon Chenery. "Seventh to Eleventh Century CE Glass from Northern Italy: Between Continuity and Innovation." *Archaeological and Anthropological Sciences* 12 (2020), article no. 120, http://doi.org/10.1007/s12520-020-01048-8.

Biagioli, Mario. "Patent Republic: Representing Inventions, Constructing Rights and Authors." *Social Research* 73 (Spring 2006): 1129–1172.

Bigelow, Allison Margaret. *Mining Language: Racial Thinking, Indigenous Knowledge, and Colonial Metallurgy in the Early Modern Iberian World.* Williamsburg, VA: Omohundro Institute of Early American History and Culture; Chapel Hill: University of North Carolina Press, 2020.

Bill, Jan. "The Oseberg Ship and Ritual Burial." In Williams, Pentz, and Wemhoff, *Vikings: Life and Legend*, 200–201.

———. "Ship Construction Tools and Techniques." In Gardiner and Unger, *Cogs, Caravels and Galleons*, 151–159.

———. "Ships and Seamanship." In *The Oxford Illustrated History of the Vikings*, edited by Peter Sawyer, 182–201. Oxford: Oxford University Press, 1997.

Biringuccio, Vannoccio. *The Pirotechnia of Vannoccio Biringuccio: The Classic Sixteenth-Century Treatise on Metals and Metallurgy*. Translated and edited by Cyril Stanley Smith and Martha Teach Gnudi. 1959. Reprint, New York: Dover, 1990.

Blair, Ann. "Information in Early Modern Europe." In Blair et al., *Information*, 61–85.

———. *Too Much to Know: Managing Scholarly Information before the Modern Age*. New Haven: Yale University Press, 2010.

Blair, Ann, et al., eds. *Information: A Historical Companion*. Princeton: Princeton University Press, 2021.

Blair, Claude and John Blair. "Copper Alloys." In Blair and Ramsay, *English Medieval Industries*, 81–89.

Blair, John, ed. *Waterways and Canal Building in Medieval England*. Oxford: Oxford University Press, 2007.

Blair, John and Nigel Ramsay, eds. *English Medieval Industries: Craftsmen, Techniques, Products*. London: Hambledon Press, 1991.

Blair, Sheila S. and Jonathan M. Bloom. *The Art and Architecture of Islam, 1250–1800*. New Haven: Yale University Press, 1994.

———. "The Islamic World." In Raven, *Oxford Illustrated History of the Book*, 195–220.

Blanchard, Ian. *Mining, Metallurgy and Minting in the Middle Ages*. 3 vols. Stuttgart: Franz Steiner, 2001.

Bloch, Marc. "Avènement et conquête du Moulin à eau." *Annales d'histoire économique et histoire* 7 (1935): 538–563. Translated as "The Advent and Triumph of the Watermill," in *Land and Work in Mediaeval Europe: Selected Papers by Marc Bloch*, translated by J. E. Anderson, 136–168. New York: Harper & Row, 1969.

Blockmans, Wim, Mikhail Krom, and Justyna Wubs-Mrozewicz. "Maritime Trade around Europe, 1300–1600: Commercial Networks and Urban Autonomy." In Blockmans, Krom, and Wubs-Mrozewicz, *Routledge Handbook of Maritime Trade*, 1–14.

———, eds. *The Routledge Handbook of Maritime Trade around Europe, 1300–1600*. New York: Routledge, 2017.

Bloom, Jonathan M. *Architecture of the Islamic West: North Africa and the Iberian Peninsula, 700–1800*. New Haven: Yale University Press, 2020.

———. *Paper before Print: The History and Impact of Paper in the Islamic World*. New Haven: Yale University Press, 2001.

Bloom, Jonathan and Sheila Blair. *Islamic Arts*. New York: Phaidon Press, 1997.

Boloix-Gallardo, Bárbara, ed. *A Companion to Islamic Granada*. Leiden: Brill, 2022.

Bond, James, "Canal Construction in the Early Middle Ages: An Introductory Review." In Blair, *Waterways and Canal-Building*, 153–206.

Bonde, Sheila and Clark Maines. "Inside the Black Box: The Technology of Medieval Water Management at the Charterhouse of Bourgfontaine." *Technology and Culture* 53 (July 2012): 625–670.

Bork, Robert. *Late Gothic Architecture: Its Evolution, Extinction, and Reception*. Turnhout: Brepols, 2018.

Bork, Robert, et al., eds. De Re Metallica: *The Uses of Metal in the Middle Ages*. Aldershot: Ashgate, 2005.

Bouras, Charalambos. "Master Craftsmen, Craftsmen, and Building Activities in Byzantium." In Laiou, *Economic History of Byzantium*, 2: 539–554.

Bowman, Alan K. and Eugene Rogan. "Agriculture in Egypt from Pharaonic to Modern Times." In Bowman and Rogan, *Agriculture in Egypt*, 1–32.

———, eds. *Agriculture in Egypt from Pharaonic to Modern Times*. Oxford: Oxford University Press, 1999.

Bradbury, Nancy Mason and Carolyn P. Colette. "Changing Times: The Mechanical Clock in Late Medieval Literature." *Chaucer Review* 43 (2009): 351–375.

Brady, Niall. "The Gothic Barn of England: Icon of Prestige and Authority." In Smith and Wolfe, *Technology and Resource Use*, 76–105.

Brady, Niall and Claudia Theune, eds. *Settlement Change across Medieval Europe: Old Paradigms and New Vistas*. Leiden: Sidestone Press, 2019.

Braemer, Frank, et al. "Conquest of New Lands and Water Systems in the Western Fertile Crescent (Central and Southern Syria)." *Water History* 2 (2010): 91–114.

Braudel, Fernand. *The Mediterranean and the Mediterranean World in the Age of Philip II*. Translated by Siân Reynolds. 2 vols. New York: Harper & Row, 1972.

Braunstein, Philippe, ed. *La sidérurgie Alpine en Italie (XIIe–XVIIe siècle)*. Rome: École Française de Rome, 2001.

Braunstein, Philippe and Luca Molá, eds. *Produzione e techniche*. Vol. 3, *Il Rinascimento Italiano e Europa*. Treviso: Foundazione Cassamarca, 2007.

Bray, Francesca and Barbara Hahn. " 'The Goddess Technology is a Polyglot': A Critical Review of Eric Schatzberg, *Technology: Critical History of a Concept*." *History and Technology* 38, no. 4 (2022): 275–316.

Brentjes, Sonja. *Travelers from Europe in the Ottoman and Safavid Empires, 16th–17th Centuries: Seeking, Transforming, Discarding Knowledge*. Farnham, Surrey: Ashgate, 2010.

Brentjes, Sonja, Alexander Fidora, and Matthias M. Tischler. "Towards a New Approach to Medieval Cross-Cultural Exchanges." *Journal of Transcultural Medieval Studies* 1 (2014): 9–50.

Broelmann, Jobst. "Navigational Instruments: Measurement of Direction." In Hattendorf, *Oxford Encyclopedia of Maritime History*, 3: 42–52.

Broilo, Federica. "A Dome for the Water: Canopied Fountains and Cypress Trees in Byzantine and Early Ottoman Constantinople." In Shilling and Stephenson, *Fountains and Water Culture in Byzantium*, 314–323.

Brokaw, Cynthia. "Medieval and Early Modern East Asia." In Raven, *Oxford Illustrated History of the Book*, 84–112.

Brooks, George. "The 'Vitruvian Mill' in Roman and Medieval Europe." In Walton, *Wind and Water in the Middle Ages*, 1–38.

Brooks, Nicholas. *Communities and Warfare, 700–1400*. London: Hambledon Press, 2000.

Brown, Elizabeth A. R. "The Tyranny of a Construct: Feudalism and Historians of Medieval Europe." *American Historical Review* 79 (October 1974): 1063–1088.

Brown, Michele P. *The British Library Guide to Writing and Scripts: History and Techniques*. Toronto: University of Toronto Press, 2001.

———. *Understanding Illuminated Manuscripts: A Guide to Technical Terms*. Revised by Elizabeth C. Teviotdale and Nancy K. Turner, rev. ed. Malibu, CA: J. Paul Getty Museum, 2018.

Brown, Sarah and David O'Connor. *Medieval Craftsmen: Glass-Painters*. London: British Museum Press, 1991.

Bruce, Travis. "Piracy as Statecraft: The Mediterranean Policies of the Fifth/Eleventh-Century Taifa of Denia." *Al-Masāq* 22.5 (2010): 235–248.

Brunner, Karl. "Continuity and Discontinuity of Roman Agricultural Knowledge in the Early Middle Ages," 21–49. In *Agriculture in the Middle Ages: Technology, Practice, and Representation*, edited by Del Sweeney. Philadelphia: University of Pennsylvania Press, 1995.

Bryer, Anthony. "The Means of Agricultural Production: Muscles and Tools." In Laiou, *Economic History of Byzantium*, 1: 101–113.

Bulliet, Richard W. *The Camel and the Wheel*. Morningside ed. New York: Columbia University Press, 1990.

———. *Cotton, Climate, and Camels in Early Islamic Iran: A Moment in World History*. New York: Columbia University Press, 2009.

Buning, Marius. *Knowledge, Patents, Power: The Making of a Patent System in the Dutch Republic*. Leiden: Brill, 2022.

Burgen, Stephen. "Spring Time: Why an Ancient Water System Is Being Brought Back to Life in Spain." *The Guardian*, April 11, 2022, https://www.theguardian.com/environment/2022/apr/11/ancient-water-system-restore-spain-sierra-nevada-aoe.

Buringh, Eltjo. *Medieval Manuscript Production in the Latin West: Explorations with a Global Database*. Leiden: Brill, 2011.

Burke, Peter. *The French Historical Revolution: The Annales School, 1929–2014*. 2nd ed. Standford: Stanford University Press, 2015.

Bush, Olga. *Reframing the Alhambra: Architecture, Poetry, Textiles and Court Ceremonial*. Edinburgh: Edinburgh University Press, 2018.

Butters, Suzanne B. *The Triumph of Vulcan: Sculptors' Tools, Porphyry, and the Prince in Ducal Florence*. 2 vols. Florence: Leo S. Olschki, 1996.

Bynum, Carolyn Walker. *Christian Materiality: An Essay on Religion in Late Medieval Europe*. Brooklyn: Zone Books, 2011.

Caferro, William. "City and Countryside in Siena in the Second Half of the Fourteenth Century." *Journal of Economic History* 54 (March 1991): 85–103.

Camerlenghi, Nicola. *St. Paul's Outside the Walls: A Roman Basilica, from Antiquity to the Modern Era*. Cambridge: Cambridge University Press, 2018.

Campbell, Bruce M. S. *The Great Transition: Climate, Disease and Society in the Late-Medieval World*. Cambridge: Cambridge University Press, 2016.

Campbell, Marian. "Gold, Silver and Precious Stones." In Blair and Ramsay, *English Medieval Industries*, 117–126.

Campbell, Tony. "A Critical Re-examination of Portolan Charts with a Reassessment of Their Replication and Seaboard Function." Map History / History of Cartography (website), 2024, https://www.maphistory.info/portolan.html.

———. "Portolan Charts from the Late Thirteenth Century to 1500." In *The History of Cartography*, vol. 1, *Cartography in Prehistoric, Ancient, and Medieval Europe and the Mediterranean*, edited by J. B. Harley and David Woodward, 371–463. Chicago: University of Chicago Press, 1987.

Caner Yüksel, Çağla. "A Tale of Two Port Cities: Ayasuluk (Ephesus) and Balat (Miletus) during the Beyliks Period." *Al-Masāq* 31, no. 3 (2019): 338–365.

Caniato, Giovanni. "L'Arsenale: Maestranze e organizzazione del lavoro." In Tenenti and Tucci, *Storia di Venezia*, vol. 5, *Il Rinascimento*, 641–677.

Cañizares-Esguerra, Jorge. "On Ignored Global 'Scientific Revolutions.'" *Journal of Early Modern History* 21 (2017): 420–432.

Caperna, Maurizio. *La Lungara: Storia e vicende edilizie dell'area tra il Gianicolo e il Tevere*. Rome: Edizioni Quasar, 2013.

Carver, Martin and Jan Klápště. *The Archaeology of Medieval Europe*. Vol. 2, *Twelfth to Sixteenth Centuries*. Aarhus: Aarhus University Press, 2011.

Catsambis, Alexis, Ben Ford, and Donny L. Hamilton, eds. *The Oxford Handbook of Maritime Archaeology*. Oxford: Oxford University Press, 2011.

Cavaciocchi, Simonetta, ed. *L'uomo e la foresta secc. XIII–XVIII*. Prato: Istituto Internazionale di Storia Economica "F. Datini," 1995.

Cavallo, Guglielmo, ed. *The Byzantines*. Translated by Thomas Dunlap, Teresa Lavender Fagan, and Charles Lambert. Chicago: University of Chicago Press, 1997.

Celenza, Christopher S. *The Intellectual World of the Italian Renaissance*. Cambridge: Cambridge University Press, 2018.

Çelik, Zeynep, Diane Favro, and Richard Ingersoll. *Streets: Critical Perspectives on Public Space*. Berkeley: University of California Press, 1994.

Chamberlain, Daven Christopher. "Paper." In Suarez and Woudhuysen, *Oxford Companion to the Book*, 1: 79–87.

Charbonnier, Julien and Kristen Hopper. "The Qanāt: A Multidisciplinary and Diachronic Approach to the Study of Groundwater Catchment Systems in Archaeology." *Water History* 10 (2018): 3–11.

Charette, François. *Mathematical Instrumentation in Fourteenth-Century Egypt and Syria: The Illustrated treatise of Najm al-Dīn al-Mīṣrī*. Leiden: Brill, 2003.

Charleston, R. J. "Vessel Glass." In Blair and Ramsay, *English Medieval Industries*, 237–264.

Cherry, John. "Leather." In Blair and Ramsay, *English Medieval Industries*, 295–318.

———. "Pottery and Tile." In Blair and Ramsay, *English Medieval Industries*, 189–209.

Chevedden, Paul E. "The Invention of the Counterweight Trebuchet: A Study in Cultural Diffusion." In *Dumbarton Oaks Papers*, No. 54, edited by Alice-Mary Talbot, 71–116. Washington, DC: Dumbarton Oaks, 2000.

Chevedden, Paul E., et al. "The Traction Trebuchet: A Triumph of Four Civilizations." *Viator* 31 (January 2000): 433–486.

Chiarenza, Nicola, Annette Haug, and Ulrich Müller, eds. *The Power of Urban Water: Studies in Premodern Urbanism*. Berlin: De Gruyter, 2020.

Chirikure, Shadreck. "Shades of Urbanism(s) and Urbanity in Pre-Colonial Africa." *Journal of Urban Archaeology* 1 (2020): 49–66.

Chow, Kai-Wing. "Reinventing Gutenberg: Woodblock and Movable-Type Printing in Europe and China." In Baron, Lindquist, and Shevlin, *Agent of Change*, 169–192.

Christ, George. "Collapse and Continuity: Alexandria as a Declining City with a Thriving Port (Thirteenth to Sixteenth Centuries)." In Blockmans, Krom, and Wubs-Mrozewicz, *Routledge Handbook of Maritime Trade*, 121–140.

Christine de Pizan. *The Book of the City of Ladies*. Translated by Rosalind Brown-Grant. London: Penguin Books, 1999.

Clarkson, Leslie. "The Linen Industry in Early Modern Europe." In Jenkins, *Cambridge History of Western Textiles*, 1: 473–492.

Collins, Edward. "Interactions of Portuguese Artisanal Culture in the Maritime Enterprise of 16th-Century Seville." *Centaurus* 60 (2018): 203–215.

Comet, Georges. *Le paysan et son outile: Essai d'histoire technique des cereals (France, VIIIe–XVe siècle)*. Rome: École Française de Rome, 1992.

———. "Technology and Agricultural Expansion in the Middle Ages: The Example of France North of the Loire." In Astill and Langdon, *Medieval Farming and Technology*, 11–39.

Constable, Olivia Remie. *Housing the Stranger in the Mediterranean World: Lodging, Trade, and Travel in Late Antiquity and the Middle-Ages*. Cambridge: Cambridge University Press, 2003.

———. *Trade and Traders in Muslim Spain: The Commercial Realignment of the Iberian Peninsula, 900–1500*. Cambridge: Cambridge University Press, 1994.

Contente Domingues, Francesco. "Science and Technology in Portuguese Navigation: The Idea of Experience in the Sixteenth Century." In *Portuguese Oceanic Expansion, 1400–1800*, edited by Francisco Bethencourt and Diogo Ramada Curto, 460–479. Cambridge: Cambridge University Press, 2007.

Cook, Hadrian, et al. "The Origins of Water Meadows in England." *Agricultural History Review* 51 (2003): 155–162.

Cook, Martin. *Medieval Bridges*. Buckinghamshire, UK: Shire Publications, 1998.

Cooke, Louise. "Approaches to the Conservation and Management of Earthen Architecture in Archaeological Contexts." 2 vols. PhD diss., University College London, 2008.

Cooper, John P. "'Fear God; Fear the *Bogaze*': The Nile Mouths and the Navigational Landscape of the Medieval Nile Delta, Egypt." *Al-Masāq* 24, no. 1 (2012): 53–73.

Coppola, Giovanni. *Ponti medievali in legno*. Bari: Laterza, 1996.

Córdoba de la Llave, Ricardo, ed. *Craft Treatises and Handbooks: The Dissemination of Technical Knowledge in the Middle Ages*. Turnhout: Brepols, 2013.

———. "Some Reflections on the Use of Water Power in Al-Andalus." In *Economia e Energia Secc. XIII–XVIII*. Atti della Trentaquattresima Settimana di Studi, edited by Simonetta Cavaciocchi. Florence: Le Monnier, [2003].

———. "Textes techniques médiévaux sur le tannage et la teinture du cuir." In Audoin-Rouzeau and Beyries, *Le travail du cuir*, 351–365.

Córdoba [de la Llave], Ricardo, et al. "Technology, Craft and Industry." In Graham-Campbell with Valor, *Archaeology of Medieval Europe*, vol. 1, *Eighth to Twelfth Centuries AD*, 208–236.

Córdoba [de la Llave], Ricardo and Ulrich Müller. "Manufacture and Production." In Carver and Klápště, *Archaeology of Medieval Europe*, vol. 2, *Twelfth to Sixteenth Centuries*, 277–327.

Cormack, Lesley B. "Handwork and Brainwork: Beyond the Zilsel Thesis." In Cormack, Walton, and Schuster, *Mathematical Practitioners*, 11–35.

Cormack, Lesley B., Steven A. Walton, and John A. Schuster, eds. *Mathematical Practitioners and the Transformation of Natural Knowledge in Early Modern Europe*. Cham: Springer, 2017.

Cormack, Robin. *Byzantine Art*. 2nd ed. Oxford: Oxford University Press, 2018.

———. *Icons*. London: British Museum, 2007.

———. "Wall-Paintings and Mosaics." In Jeffreys with Haldon and Cormack, *Oxford Handbook of Byzantine Studies*, 385–406.

Cortese, Maria Elena. "Medieval Ironworking on Mount Amiata (Siena, Italy): Economy, Society, Technology." In Nørbach, *Prehistoric and Medieval Direct Iron Smelting*, 55–59.

Cosman, Madeleine Pelner. *Women at Work in Medieval Europe*. New York: Checkmark Books, 2000.

Cossart, Brice. *Les Artilleurs et la Monarchie Hispanique (1560–1610)*. Paris: Classiques Garnier, 2021.

———. "Producing Skills for an Empire: Theory and Practice in the Seville School of Gunners during the Golden Age of the Carrera de Indias." *Technology and Culture* 58 (April 2017): 459–486.

Courtenay, Lynn T., ed. *The Engineering of Medieval Cathedrals*. Aldershot: Ashgate, 1997.

———. "Scale and Scantling: Technological Issues in Large-Scale Timberwork of the High Middle Ages." In Smith and Wolfe, *Technology and Resource Use*, 42–75.

Cresswell, Robert. "Of Mills and Waterwheels: The Hidden Parameters of Technological Choice." In *Technological Choices: Transformation in Material Cultures since the Neolithic*, edited by Pierre Lemonnier, 181–213. London: Routledge, 1993.

Crossley, David. "The Archaeology of Water Power in Britain before the Industrial Revolution." In Smith and Wolfe, *Technology and Resource Use*, 109–124.

Crostini, Barbara. "Byzantium." In Raven, *Oxford Illustrated History of the Book*, 54–83.

Crow, James. "The Infrastructure of a Great City: Earth, Walls, and Water in Late Antique Constantinople." In Lavan et al., *Technology in Transition*, 251–285
———. "Ruling the Waters: Managing the Water Supply of Constantinople, AD 330–1204." *Water History* 20 (2012): 35–55.
———. "Waters for a Capital: Hydraulic Infrastructure and Use in Byzantine Constantinople." In Bassett, *Cambridge Companion to Constantinople*, 67–86.
Crow, James, Jonathan Bardill, and Richard Bayliss. *The Water Supply of Byzantine Constantinople*. London: Society for the Promotion of Roman Studies, 2008.
Crowston, Clare. "Women, Gender, and Guilds in Early Modern Europe: An Overview of Recent Research." *International Review of Social History*, supplement 16, *The Return of the Guilds* 53 (2008): 19–44.
Cucini, C. and M. Tizzoni, "The Lombard Iron Masters Migrations and the Spread of the Blast Furnace in Europe, with a Focus on the 16th–17th Centuries." *Metalla* 26, no. 1 (2022): 37–66.
Cunliffe, Barry. *Facing the Ocean: The Atlantic and Its Peoples*. Oxford: Oxford University Press, 2001.
Cuomo, Serafina. "Shooting by the Book: Notes on Niccolò Tartaglia's *Nova Scientia*." *History of Science* 35 (June 1997): 155–188.
Cutler, Anthony. *The Hand of the Master: Craftsmanship, Ivory, and Society in Byzantium (9th–11th Centuries)*. Princeton: Princeton University Press, 1994.
———. "The Industries of Art." In Laiou, *Economic History of Byzantium*, 1: 554–587.
Dagron, Gilbert. "The Urban Economy, Seventh–Twelfth Centuries," In Laiou, *Economic History of Byzantium*, 2: 393–462.
Daim, Falko and Ewald Kilsinger, eds. *The Byzantine Harbours of Constantinople*. Translated by Leo Ruickie and Antje Bosselmann-Ruickbie. Mainz: Verlag des Römisch-Germanisches Zentralmuseum, 2021.
Dalby, Andrew, trans. *Geoponika*. London: Prospect Books, 2011.
Dallal, Ahmad. "Science, Medicine, and Technology: The Making of a Scientific Culture." In *The Oxford History of Islam*, edited by John L. Esposito, 155–213. Oxford: Oxford University Press, 1999.
Dam, Petra J.E.M. van. "Ecological Challenges. Technical Innovations: The Modernization of Sluice Building in Holland, 1300–1600." *Technology and Culture* 43 (July 2002): 500–520.
———. "Sinking Peat Bogs: Environmental Change in Holland, 1350–1550." *Environmental History* 6 (January 2001): 32–45.
Dameron, George. "Feeding the Medieval Italian City-State." *Speculum* 92 (October 2017): 976–1019.
Dark, Ken and Jan Kostenec. *Hagia Sophia in Context: An Archaeological Re-examination of the Cathedral of Byzantine Constantinople*. Oxford: Oxbow Books, 2019.
Davids, Karel. "Craft Secrecy in Europe in the Early Modern Period: A Comparative View." *Early Science and Medicine* 10 (2005): 341–348.
Davis, Robert C. *Christian Slaves, Muslim Masters: White Slavery in the Mediterranean, the Barbary Coast, and Italy, 1500–1800*. Houndmills, Basingstoke: Palgrave Macmillan, 2003.
———. *Shipbuilders of the Venetian Arsenal: Workers and Workplace in the Preindustrial City*. Baltimore: Johns Hopkins University Press, 1991.
Decker, Michael. "Agriculture and Agricultural Technology." In Jeffreys with Haldon and Cormack, *Oxford Handbook of Byzantine Studies*, 397–405.
———. "Plants and Progress: Rethinking the Islamic Agricultural Revolution." *Journal of World History* 20 (June 2009): 187–206.
———. *Tilling the Hateful Earth: Agricultural Production and Trade in the Late Antique East*. Oxford: Oxford University Press, 2009.

Déléry, Claire. "Using *Cuerda Seca* Ceramics as a Historical Source to Evaluate Trade and Cultural Relations between Christian Ruled Lands and Al-Andalus, from the Tenth to Thirteenth Centuries." *Al-Masāq* 21, no. 1 (2009): 31–58.

Delmaire, Bernard and Fulgence Delleaux. "Northern France, 1000–1700." In Thoen and Soens, *Struggling with the Environment*, 149–184.

Delogu, Paolo. "Reading Pirenne Again." In Hodges and Bowden, *Sixth Century*, 15–40.

De Meulemeester, Johnny and Kieran O'Conor. "Fortifications." In Graham-Campbell with Valor, *Archaeology of Medieval Europe*, vol. 1, *Eighth to Twelfth Centuries, AD*, 316–341.

Dennis, George T. "Byzantine Heavy Artillery: The Helepolis." *Greek, Roman, and Byzantine Studies* 39 (1998): 99–115.

De Vorsey, Louis. "Americas: Exploration Voyages, 1500–1620 (Eastern Coast)." In Hattendorf, *Oxford Encyclopedia of Maritime History*, 1: 28–36.

Devos, Isabelle, Thijs Lambrecht, and Richard Paping, "The Low Countries, 1000–1750." In Vanhaute, Devos, and Lambrecht, *Making a Living*, 157–183.

DeVries, Kelly. *Joan of Arc: A Military Leader*. Stroud, Gloucestershire: History Press, 2011.

DeVries, Kelly and Robert Douglas Smith. *Medieval Military Technology*. 2nd ed. Toronto: University of Toronto Press, 2012.

Devroey, Jean-Pierre, Alexis Wilkin, and Alban Gautier. "Agricultural Production, Distribution and Consumption around the North Sea, 500–1000." In Molle and Segers, *Agro-Food Market: Production*, 12–65.

Devroey, Jean-Pierre and Anne Nissen[-Jaubert], "Early Middle Ages, 500–1000." In Thoen and Soens, *Struggling with the Environment*, 10–68.

Devroey, Jean-Pierre and Anne Nissen Jaubert, "Family Income and Labour around the North Sea, 500–1000." In Vanhaute, Devos, and Lambrecht, *Making a Living*, 4–44.

Dizer, Muammar. "Observatories and Astronomical Instruments." In al-Hassan, Ahmed, and Iskandar, *Science and Technology in Islam*, pt. 1, 235–266.

Dodwell, C. R. "Introduction," ix–lxxvi. In *Theophilus: The Various Arts, De Diversis Artibus*. Oxford: Clarendon Press, 1986.

Dohrn-Van Rossum, Gerhard. *History of the Hour: Clocks and Modern Temporal Orders*. Translated by Thomas Dunlap. Chicago: University of Chicago Press, 1996.

Dolza, Luisa. "*Utilitas et delectatio*: Libri di tecniche e teatri di macchine." In Braunstein and Molà, *Produzione e techniche*, 115–143.

Dolza, Luisa and Hélène Vérin. "Illustrating Mechanics: The Enigma of Theater of Machines during the Renaissance." *Revue d'Histoire Modern et Contemporaine*, 51/52, no. 2 (2004): 7–37.

Dondi, Cristina. "The European Printing Revolution." In Suarez and Woudhuysen, *Oxford Companion to the Book*, 1: 53–67.

Dover, Paul M. *The Information Revolution in Early Modern Europe*. Cambridge: Cambridge University Press, 2021.

Dubourg Glatigny, Pascal and Hélène Vérin. *Réduire en art: La technologie de la Renaissance aux Lumières*. [Paris]: Éditions de la Maison sciences de l'homme, 2008.

Dudbridge, Glen. "Reworking the World System Paradigm." In Holmes and Standen, *Global Middle Ages*, 297–316.

Dunn, Ross E. *The Adventures of Ibn Battuta: A Muslim Traveler of the 14th Century*. Rev. ed. Berkeley: University of California Press, 2012.

Durand, Aline and Philippe Leveau. "Farming in Mediterranean France and Rural Settlement in the Late Roman and Early Medieval Periods: The Contribution from Archaeology and Environmental Sciences in the Last Twenty Years (1980–2000)." In Barceló and Sigaut, *Making of Feudal Agricultures?*, 177–253.

Dyer, Christopher. *Making a Living in the Middle Ages: The People of Britain, 850–1520*. New Haven: Yale University Press, 2002.

Eamon, William. "Corn, Cochineal, and Quina: The 'Zilsel Thesis' in a Colonial Iberian Setting." *Centaurus* 60 (2018): 141–158.

———. *Science and the Secrets of Nature: Books of Secrets in Medieval and Early Modern Culture*. Princeton: Princeton University Press, 1994.

Edgerton, David. *The Shock of the Old: Technology and Global History since 1900*. Oxford: Oxford University Press, 2007.

———. "What is the Historiography of Technology About?" *Technology and Culture* 51 (July 2010): 680–697.

Eichberger, Dagmar. "The Winged Altarpiece in Early Netherlandish Art." In Lindley, *Making Medieval Art*, 152–172.

Eisenstein, Elizabeth L. *The Printing Press as an Agent of Change: Communications and Cultural Transformations in Early Modern Europe*. 2 vols. Cambridge: Cambridge University Press, 1979.

Ellenblum, Ronnie. *The Collapse of the Eastern Mediterranean: Climate Change and the Decline of the East, 950–1072*. Cambridge: Cambridge University Press, 2012.

Ellmers, Detlev. "The Cog as Cargo Carrier." In Gardiner and Unger, *Cogs, Caravels and Galleons*, 29–46.

Eltis, David. *The Military Revolution in Sixteenth-Century Europe*. London: Tauris Academic Studies, 1995.

Epstein, Steven A. *The Economic and Social History of Later Medieval Europe, 1000–1500*. Cambridge: Cambridge University Press, 2009.

———. "Urban Society." In *The New Cambridge Medieval History*, vol. 5, *c. 1198–c. 1300*, 26–37, edited by David Abulafia. Cambridge: Cambridge University Press, 1999.

Epstein, S. R. "Craft Guilds, Apprenticeship, and Technological Change in Preindustrial Europe." *Journal of Economic History* 58 (September 1998): 684–713.

Epstein, S. R. and Maarten Prak, eds. *Guilds, Innovation and the European Economy, 1400–1800*. Cambridge: Cambridge University Press, 2008.

Esposito, Anna. *Donne del Rinascimento a Roma e dintorni*. Rome: Roma nel Rinascimento, 2013.

Estácio dos Reis, António. "Nunes, Pedro (1502–1578)." In Hattendorf, *Oxford Encyclopedia of Maritime History*, 3: 116–117.

Ettinghausen, Richard, Oleg Grabar, and Marilyn Jenkins-Madina. *Islamic Art and Architecture, 650–1250*. 2nd ed. New Haven: Yale University Press, 2001.

Fabijanec, Sabine Florence. "Fishing and the Fish Trade on the Dalmatian Coast in the Late Middle Ages." In Bekker-Nielsen and Gertwagen, *Inland Seas*, 369–386.

Fábregas García, Ádela. "Other Markets: Complementary Commercial Zones in the Nasrid World of the Western Mediterranean (Seventh/Thirteenth to Ninth/Fifteenth Centuries)." *Al-Masāq* 25, no. 1 (2013): 135–155.

Fagan, Brian. *Fishing: How the Sea Fed Civilization*. New Haven: Yale University Press, 2017.

———. *The Little Ice Age: How Climate Made History, 1300–1850*. New York: Basic Books, 2020.

Falkowski, Mateusz. "Fear and Abundance: Reshaping of Royal Forests in Sixteenth-Century Poland and Lithuania." *Environmental History* 22 (October 2017): 618–642.

Farmer, Sharon. *The Silk Industries of Medieval Paris: Artisanal Migration, Technological Innovation, and Gendered Experience*. Philadelphia: University of Pennsylvania Press, 2017.

Febvre, Lucien. "Réflexions sur l'histoire des techniques." *Annales d'histoire économique et histoire* 7 (1935): 531–535.

Febvre, Lucien and Henri-Jean Martin. *The Coming of the Book: The Impact of Printing, 1480–1800.* Translated by David Gerard. London: Verso, 1976.

Fenwick, Corisande. *Early Islamic North Africa: A New Perspective.* London: Bloomsbury, 2020.

———. "From Africa to Ifrīqiya: Settlement and Society in Early Medieval North Africa (650–800)." *Al-Masāq* 25, no. 1 (2013): 9–33.

Fernández-Puertas, Antonio. *The Alhambra.* 2 vols. London: Saqi Books, 1997.

———. *Clepsidras y relojes musulmanes / Muslim Water Clocks and Mechanical Time Pieces.* Granada: Fundación Pública Andaluza El legado Andalusi, 2010.

Filson, Lily. "Magical and Mechanical Evidence: The Late-Renaissance Automata of Francesco I de' Medici." In *Evidence in the Age of the New Sciences,* edited by James A. T. Lancaster and Richard Raiswell, 177–201. Cham: Springer, 2018.

Findlen, Paula, ed. *Early Modern Things: Objects and their Histories, 1500–1800.* London: Routledge, 2013.

Finley, M. I. *The Ancient Economy.* 2nd ed. Updated by Ian Morris. Berkeley: University of California Press, 1999.

Flood, Barry Finbarr. *The Great Mosque of Damascus: Studies on the Makings of an Umayyad Visual Culture.* Leiden: Brill, 2001.

Franci, Raffaella. "Mathematics in the Manuscript of Michael of Rhodes." In Long, McGee, and Stahl, *Book of Michael of Rhodes,* 3: 115–146.

François, Véronique and Jean-Michel Spieser. "Pottery and Glass in Byzantium." In Laiou, *Economic History of Byzantium,* 2: 593–609.

Frankel, Rafael. "Introduction." In Ayalon, Frankel, and Kloner, *Oil and Wind Presses in Israel,* 1–18.

Frantz-Murphy, Gladys. "Land-Tenure in Egypt in the First Five Centuries of Islamic Rule (Seventh–Twelfth Centuries AD)." In Bowman and Rogan, *Agriculture in Egypt,* 237–266.

Frenkel, Miriam. "Medieval Alexandria—Life in a Port City." *Al-Masāq* 26, no. 1 (2014): 5–35.

Freshfield, Edwin H. *Ordinances of Leo VI, c. 897: From the Book of the Eparch.* 1938. Reprint, London: Variorum Reprints, 1970.

Frick, Carole Collier. *Dressing Renaissance Florence: Families, Fortunes, and Fine Clothing.* Baltimore: Johns Hopkins University Press, 2002.

Friedman, John Block and Kristen Mossler Figg, eds. *Trade, Travel, and Exploration in the Middle Ages.* New York: Garland, 2000.

Friel, Ian. "The Carrack: The Advent of the Full Rigged Ship." In Gardiner and Unger, *Cogs, Caravels and Galleons,* 77–90.

———. *The Good Ship: Ships, Shipbuilding and Technology in England, 1200–1520.* Baltimore: Johns Hopkins University Press, 1995.

Frison, Guido and Giulia Brun. "*Compositiones Lucenses* and *Mappae Clavicula*: Two Traditions or One?: New Evidence from Empirical Analysis and Assessment of the Literature." *Heritage Science* 6 (2018), article 24.

Fuga, Antonella. *Artists' Techniques and Material.* Translated by Rosanna M. Giammanco Frongia. Los Angeles: J. Paul Getty Museum, 2006.

Gadeyne, Jan. "Short Cuts: Observations on the Formation of the Medieval Street System in Rome. In Smith and Gadeyne, *Perspectives on Public Space in Rome,* 67–83.

Galilei, Galileo. *Operations of the Geometric and Military Compass.* Translated and introduced by Stillman Drake. Washington, DC: Smithsonian Institution Press, 1978.

Galloway, James A. "Coastal Flooding and Socioeconomic Change in Eastern England in the Later Middle Ages." *Environment and History* 19 (May 2013): 173–207.

Galloway, James A., Derek Keene, and Margaret Murphy. "Fueling the City: Production and

Distribution of Firewood and Fuel in London's Region, 1290–1400." *Economic History Review* 49 (August 1996): 447–472.

Galluzzi, Paolo. *The Italian Renaissance of Machines*. Cambridge, MA: Harvard University Press, 2020.

Gans, Paul J. "The Medieval Horse Harness: Revolution or Evolution? A Case Study in Technological Change." In *Villard's Legacy: Studies in Medieval Technology, Science and Art in Memory of Jean Gimpel*, edited by Marie-Thérèse Zenner, 175–187. Aldershot: Ashgate, 2004.

Gardiner, Mark. "Hythes, Small Ports, and Other Landing Places in Later Medieval England." In Blair, *Waterways and Canal-Building*, 85–109.

———. "Timber Churches in Medieval England: A Preliminary Study." In Khodakovsky and Lexau, *Historic Wooden Architecture in Europe and Russia*, 28–41.

Gardiner, Robert and Richard W. Unger, eds. *Cogs, Caravels and Galleons: The Sailing Ship, 1000–1650*. London: Conway Maritime Press, 1994.

Gentilcore, David. "The Cistern-System of Early Modern Venice: Technology, Politics, and Culture in a Hydraulic Society." *Water History* 13 (October 2021): 1–32.

Gerbino, Anthony and Stephen Johnston. *Compass and Rule: Architecture as Mathematical Practice in England*. Oxford: Museum of the History of Science, 2009.

Gerrard, Christopher. "Contest and Cooperation: Strategies for Medieval and Later Irrigation along the Upper Huecha Valley, Aragón, North-East Spain." *Water History* 3 (2011): 2–28.

Gertwagen, Ruthy. "Nautical Technology." In Horden and Kinoshita, *Companion to Mediterranean History*, 154–169.

———. "Towards a Maritime Eco-History of the Byzantine and Medieval Eastern Mediterranean." In Bekker-Nielsen and Gertwagen, *Inland Seas*, 341–368.

Gilbertson, David D. "Explanations: Environment as Agency." In Barker, *Farming the Desert*, 291–317.

Gilbertson, David D. and Chris Hunt. "Romano-Libyan Agriculture: Walls and Floodwater Farming." In Barker, *Farming the Desert*, 191–225.

Gimpel, Jean. *The Medieval Machine: The Industrial Revolution of the Middle Ages*. Hammondsworth, Middlesex: Penguin Books, 1976.

Ginalis, Alkiviadis, et al. *Harbours and Landing Places on the Balkan Coasts of the Byzantine Empire (4th to 12th Centuries)*. European Harbour Data Repository, vol. 4, edited by L. Werther, H. Müller, and M. Foucher. Jena: Friedrich-Schiller-Universität Jena, 2019.

Girouard, Mark. *Elizabethan Architecture: Its Rise and Fall, 1540–1640*. New Haven: Yale University Press, 2009.

Glick, Thomas F. *Islamic and Christian Spain in the Early Middle Ages*. 2nd rev. ed. Leiden: Brill, 2005.

Glick, Thomas F. and Helena Kirchner. "Hydraulic Systems and Technologies of Islamic Spain: History and Archaeology." In Squatriti, *Working with Water in Medieval Europe*, 266–329.

Glick, Thomas F. and Luis Pablo Martinez. "Mills and Millers in Medieval Valencia." In Walton, *Wind and Water*, 189–211.

Gluzman, Renard. *Venetian Shipping from the Days of Glory to Decline, 1453–1571*. Leiden: Brill, 2021.

Goitein, Shelomo Dov. *A Mediterranean Society: The Jewish Communities of the Arab World as Portrayed in the Documents of the Cairo Geniza*. 6 vols. Berkeley: University of California Press, 1967–1993.

Goldberg, Jessica L. *Trade and Institutions in the Medieval Mediterranean: The Geniza Merchants and Their Business World*. Cambridge: Cambridge University Press, 2012.

Goldman, Paul. "The History of Illustration and Its Technologies." In Suarez and Woudhuysen, *Oxford Companion to the Book*, 1: 137–145.

Gómez, Luis. *The Floods of the Tiber with Additional Documents on the Tiber Flood of 1530.* Translated by Chiara Bariviera, Pamela O. Long, and William L. North. New York: Italica Press, 2023.

Goodman, David. *Power and Penury: Government, Technology, and Science in Philip II's Spain.* Cambridge: Cambridge University Press, 1988.

Goodson, Caroline. *Cultivating the City in Early Medieval Italy.* Cambridge: Cambridge University Press, 2021.

Gorelik, Michael. "Arms and Armour in South-Eastern Europe in the Second Half of the First Millennium AD." In Nicolle, *Companion to Medieval Arms and Armour*, 127–147.

Goskar, Tehmina. "Material Culture." In Horden and Kinoshita, *Companion to Mediterranean History*, 281–295.

———. "Material Worlds: The Shared Cultures of Southern Italy and Its Mediterranean Neighbours in the Tenth to Twelfth Centuries." *Al-Masāq* 23, no. 3 (2011): 189–204.

Goubitz, Olaf, Carol van Driel-Murray, and Willy Groenman-van Waateringe. *Stepping through Time: Archaeological Footwear from Prehistoric Times until 1800.* Zwolle: Stichting Promotie Archeologie, 2001.

Grafton, Anthony. *Commerce with the Classics.* Ann Arbor: University of Michigan Press, 1997.

———. *Inky Fingers: The Making of Books in Early Modern Europe.* Cambridge, MA: Harvard University Press, 2020.

———. "Premodern Regimes and Practices." In Blair et al., *Information*, 3–20.

Graham-Campbell, James with Magdalena Valor, eds. *The Archaeology of Medieval Europe.* Vol. 1, *Eighth to Twelfth Centuries AD.* Aarhus: Aarhus University Press, 2006.

Graulau, Jeannette. *The Underground Wealth of Nations: On the Capitalist Origins of Silver Mining, A.D. 1150–1450.* New Haven: Yale University Press, 2019.

Graves, Margaret S. *Arts of Allusion: Object, Ornament, and Architecture in Medieval Islam.* New York: Oxford University Press, 2018.

———. "Ceramics: The Art of Being Human." In Graves et al., *Ceramic Art*, 19–46.

Graves, Margaret S., et al., eds. *Ceramic Art.* Princeton: Princeton University Press, 2023.

Green, Monica H. "The Four Black Deaths." *American Historical Review* 125 (December 2020): 1601–1631.

———, ed. *Pandemic Disease in the Medieval World: Rethinking the Black Death.* Kalamazoo, MI: Arc Medieval Press, 2015.

Greene, Kevin. "Technological Innovation and Economic Progress in the Ancient World: M. I. Finley Reconsidered." *Economic History Review* 53 (2000): 29–59.

Grenville, Jane. "Urban and Rural Houses and Households in the Late Middle Ages: A Case Study from Yorkshire." In Kowaleski and Goldberg, *Medieval Domesticity*, 92–123.

Griffith, Paddy. *The Viking Art of War.* London: Greenhill Books, 1998.

Grove, A. T. and Oliver Rackham. *The Nature of Mediterranean Europe: An Ecological History.* New Haven: Yale University Press, 2001.

Gruber, Elisabeth. "Meeting Water Needs as a Major Challenge in an Urban Context: Examples from the Danube Region (1300–1600)." In Chiarenza, Haug, and Müller, *Power of Urban Water*, 179–195.

Gruendler, Beatrice. "Aspects of Craft in the Arabic Book Revolution." In *Globalization of Knowledge in the Post-Antique Mediterranean, 700–1500*, edited by Sonja Brentjes and Jürgen Renn, 31–65. London: Routledge, 2016.

Guérin, Sahrah M. "Forgotten Routes? Italy, Ifrīqiya and the Trans-Saharan Ivory Trade," *Al-Masāq* 29, no. 1 (2013): 70–91.

Gupta, Vivek. "Routes of Translation: Connected Book Histories and al-Jazari's Robotic Wonders from the Mamluks to Mandu," *South Asian Studies* 39, no. 2 (2023): 1–22.

Gutas, Dimitri. *Greek Thought, Arabic Culture: The Graeco-Arabic Translation Movement in Baghdad and Early Abbasid Society (2nd–4th / 8th–10th Centuries).* London: Routledge, 1998.

Guyon, Marc. *Les fondations des ponts en France: Sabot métalliques des piux de foundation, de l'antiquité à l'époque modern.* Montagnac: Éditions Monique Mergoil, 2000.

Halasz-Csiba, Eva. "Peaux et cuirs: Méthode d'investigation de la dimension historique du tannage en France (XIVᵉ–XVIIIᵉ siècles)." In Audoin-Rouzeau and Beyries, *Le travail du cuir de la Préhistoire à nos jours,* 387–398.

Haldon, John. "Some Aspects of Early Byzantine Arms and Armour." In Nicolle, *Companion to Medieval Arms and Armour,* 65–79.

——. *Warfare, State and Society in the Byzantine World.* London: University College London Press, 1999.

Hall, Bert S. "Lynn White's *Medieval Technology and Social Change* after Thirty Years." In *Technological Change: Methods and Themes in the History of Technology,* edited by Robert Fox, 85–101. Amsterdam: Overseas Publishers Association, 1996.

——. *Weapons and Warfare in Renaissance Europe.* Baltimore: Johns Hopkins University Press, 1997.

Hall, Robert. "Mechanics." In al-Hassan, Ahmed, and Iskandar, *Science and Technology in Islam,* pt. 1, 297–336.

Halleux, Robert. *Le savoir de la main: Savants et artisans dans l'europe pré-industrielle.* Paris: Armand Colin, 2009.

Hamadeh, Sirine and Çiğdem Kafescioğlu, eds. *A Companion to Early Modern Istanbul.* Leiden: Brill, 2022.

Hamel, Christopher de. *The British Library Guide to Manuscript Illumination: History and Techniques.* Toronto: University of Toronto Press, 2001.

——. "The European Medieval Book." In Suarez and Woudhuysen, *Oxford Companion to the Book,* 1: 38–51.

Hamerow, Helena. *Early Medieval Settlements: The Archaeology of Rural Communities in Northwest Europe, 400–900.* Oxford: Oxford University Press, 2002.

Hamil, Les. *Bridge Hydraulics.* Boca Raton, FL: CRC Press, 1999.

Hanawalt, Barbara A. and Anna Dronzek. "Women in Medieval Urban Society." In Mitchell, *Women in Medieval Western European Culture,* 30–45.

Harjula, Janne. *Before the Heels: Footwear and Shoemaking in Turku in the Middle Ages and the Beginning of the Early Modern Period.* Turku: Archaeologia Medii Aevi Finlandiae, 2008.

——. *Sheaths, Scabbards and Grip Coverings: The Use of Leather for Portable Personal Objects in the 14th–16th Century Turku.* Turku: Archaeologia Medii Aevi Finlandiae, 2005.

Harland, Jennifer F., et al. "Fishing and Fish Trade in Medieval York: The Zooarchaeological Evidence." In Barrett and Orton, *Cod and Herring,* 172–204.

Harris, W. V., ed. *Rethinking the Mediterranean.* Oxford: Oxford University Press, 2005.

Harrison, David. *The Bridges of Medieval England: Transport and Society 400–1800.* Oxford: Clarendon Press, 2004.

Hassall, Catherine. "Alchemy, Chemistry and Chemical Technology." In al-Hassan, Ahmed, and Iskandar, *Science and Technology in Islam,* pt. 1, 41–83.

al-Hassan, Ahmad Y. "Military Fires, Gunpowder and Firearms." In al-Hassan, Ahmed, and Iskandar, *Science and Technology in Islam,* pt. 2, 107–134.

——. "Mining and Metallurgy." In al-Hassan, Ahmed, and Iskandar, *Science and Technology in Islam,* pt. 2, 85–106.

——. "Textiles and Other Manufacturing Industries." In al-Hassan, Ahmed, and Iskandar, *Science and Technology in Islam,* pt. 2, 135–163.

al-Hassan, Ahmad Y., Maqbul Ahmed, and A. Z. Iskandar, eds. *Science and Technology in Islam.*

Pt. 1, *The Exact and Natural Sciences*, pt. 2, *Technology and Applied Science*. Paris: UNESCO, 2001.

Hattendorf, John B., ed. *The Oxford Encyclopedia of Maritime History*. 4 vols. Oxford: Oxford University Press, 2007.

Haug, Brendan James. "Watering the Desert: Environment, Irrigation, and Society in the Premodern Fayyūm, Egypt." PhD diss., University of California, Berkeley, 2012.

Hawthorne, John G. and Cyril Stanley Smith. "Introduction." In Theophilus, *On Diverse Arts*, xv–xxxv.

Heck, Gene W. "Gold Mining in Arabia and the Rise of the Islamic State." *Journal of the Economic and Social History of the Orient* 42, no. 3 (1999): 364–395.

Heebøll-Holm, Thomas, Philipp Höhn, and Gregor Rohmann, eds. *Merchants, Pirates, and Smugglers: Criminalization, Economics and the Transformation of the Maritime World (1200–1600)*. Frankfurt: Campus Verlag, 2019.

Hemming, John. "Americas: Exploration Voyages, 1500–1616 (South America)." In Hattendorf, *Oxford Encyclopedia of Maritime History*, 1: 36–42.

Heng, Geraldine. *The Invention of Race in the European Middle Ages*. New York: Cambridge University Press, 2018.

Henning, Joachim. "Did the 'Agricultural Revolution' Go East with Carolingian Conquest? Some Reflections on Early Medieval Rural Economics of the Baiuvarii and Thuringi." In *The Baiuvarii and Thuringi: An Ethnographic Perspective*, edited by Janine Fries-Knoblach and Heiko Steuer with John Hines, 331–359. San Marino: Boydell Press, 2014.

Herreld, Donald J. *A Companion to the Hanseatic League*. Leiden: Brill, 2015.

Hess, Catherine, ed. *The Arts of Fire: Islamic Influences on Glass and Ceramics of the Italian Renaissance*. Los Angeles: J. Paul Getty Museum, 2004.

———. "Brilliant Achievements: The Journey of Islamic Glass and Ceramics to Renaissance Italy." In Hess, *Arts of Fire*, 1–32.

Hettinger, Madonna J. "So Strategize: The Demands of the Day of the Peasant Woman in Medieval Europe." In Mitchell, *Women in Medieval Western European Culture*, 47–63.

Hicks, Robert D. "Nautical Astronomy and Celestial Navigation." In Hattendorf, *Oxford Encyclopedia of Maritime History*, 2: 619–638.

Hilaire-Pérez, Liliane and Catherine Verna. "Dissemination of Technical Knowledge in the Middle Ages and the Early Modern Era: New Approaches and Methodological Issues." *Technology and Culture* 47 (July 2006): 536–565.

Hill, Donald R. *A History of Engineering in Classical and Medieval Times*. 1984. Reprint, London: Routledge, 1996.

———. *Islamic Science and Engineering*. Edinburgh: Edinburgh University Press, 1993.

———. "Mechanical Technology." In al-Hassan, Ahmed, and Iskandar, *Science and Technology in Islam*, pt. 2, 165–192.

———. "Military Technology." In al-Hassan, Ahmed, and Iskandar. *Science and Technology in Islam*, pt. 2, 247–270.

Hillenbrand, Robert. *Islamic Art and Architecture*. London: Thames & Hudson, 1999.

Hills, Richard L. *Power from the Wind: A History of Windmill Technology*. Cambridge: Cambridge University Press, 1994.

Hindle, Paul. *Medieval Roads and Tracks*. 3rd ed. Buckinghamshire: Shire Publications, 1998.

Hocker, Frederick M. "Postmedieval Ships and Seafaring in the West." In Catsambis, Ford, and Hamilton, *Oxford Handbook of Maritime Archaeology*, 445–472.

Hocker, Frederick M. and John M. McManamon. "Mediaeval Shipbuilding in the Mediterranean and Written Culture at Venice." *Mediterranean Historical Review* 21, no. 1 (2006): 1–37.

Hodges, Richard. "Henri Pirenne and the Question of Demand in the Sixth Century." In Hodges and Bowden, *Sixth Century*, 2–14.

Hodges, Richard and William Bowden. *The Sixth Century: Production, Distribution and Demand.* Leiden: Brill, 1998.

Hoffmann, Richard C. *The Catch: An Environmental History of Medieval European Fisheries.* Cambridge: Cambridge University Press, 2023.

———. *An Environmental History of Medieval Europe.* Cambridge: Cambridge University Press, 2014.

———. *Fishers' Craft and Lettered Art: Tracts on Fishing from the End of the Middle Ages.* Toronto: University of Toronto Press, 1997.

———. "Medieval Fishing." In Squatriti, *Working with Water in Medieval Europe*, 331–393.

Hohti Erichsen, Paula. *Artisans, Objects, and Everyday Life in Renaissance Italy: The Material Culture of the Middling Class.* Amsterdam: Amsterdam University Press, 2020.

Holm, Poul. "Commercial Sea Fisheries in the Baltic Region, c. AD 1000–1600." In Barrett and Orton, *Cod and Herring*, 12–22.

Holmes, Catherine and Naomi Standen, eds. *The Global Middle Ages.* Special issue, *Past & Present*, supplement 13. Oxford: Oxford University Press, 2018.

———. "Introduction: Towards a Global Middle Ages." In Holmes and Standen, *Global Middle Ages*, 2–44.

Holt, Richard. "Mechanization and the Medieval English Economy." In Smith and Wolfe, *Technology and Resource Use*, 139–157.

Homsi, Hikmat. "Navigation and Ship-building." In al-Hassan, Ahmed, and Iskandar, *Science and Technology in Islam*, pt. 2, 217–246.

Hooke, Delle. "Use of Waterways in Anglo-Saxon England." In Blair, *Waterways and Canal Building in Medieval England*, 37–54.

Hoppenbrouwers, Peter. "Agricultural Production and Technology in the Netherlands, c. 1000–1500." In Astill and Langdon, *Medieval Farming and Technology*, 89–113.

Horden, Peregrine and Sharon Kinoshita, eds. *A Companion to Mediterranean History.* Chichester, West Sussex: John Wiley & Sons, 2014.

Horden, Peregrine and Nicholas Purcell. *The Boundless Sea: Writing Mediterranean History.* London: Routledge, 2020.

———. *The Corrupting Sea: A Study of Mediterranean History.* Oxford: Blackwell, 2000.

Howell, Martha C. *Commerce before Capitalism in Europe, 1300–1600.* Cambridge: Cambridge University Press, 2010.

Huppert, Ann C. *Becoming an Architect in Renaissance Italy: Art, Science, and the Career of Baldassarre Peruzzi.* New Haven: Yale University Press, 2015.

Husain, Faisal. "Sediment of the Tigris and Euphrates Rivers: An Early Modern Perspective." *Water History* 13 (2021): 13–32.

Hutchins, Edwin. *Cognition in the Wild.* Cambridge, MA: MIT Press, 1995.

Hutchinson, Gillian. *Medieval Ships and Shipping.* Rutherford, PA: Fairleigh Dickinson University Press, 1994.

Ibn Baṭṭūṭa. *The Travels of Ibn Baṭṭūṭa, A.D. 1325–1354.* Translated by H. A. R. Gibb with revisions and notes from the Arabic text edited by C. Defrémery and B. R. Sanguinetti. Vol. 1. 1958. Reprint, Milkwood, NY: Kraus Reprint, 1986.

Ikhwān al-Ṣafā. "Epistle 8 on the Practical Crafts." Translated by Nader El-Bizri, in *Epistles of the Brethren of Purity: On Composition and the Arts: An Arabic Critical Edition and English Translation of Epistles 6–8*, edited and translated by Nader El-Bizri and Godefroid de Callataÿ, 137–165. Oxford: Oxford University Press, 2018.

İlter, Fügen. "The Main Features of the Seljuk, the Beylik and the Ottoman Bridges of the

Turkish Anatolian Architecture from the XIIth to the XVIth Centuries," *Belleten* 57, no. 219 (1994): 481–494.

Jacoby, David. "Silk Production." In Jeffreys with Haldon and Cormack, *Oxford Handbook of Byzantine Studies*, 421–428.

James-Raoul, Danièle and Claude Thomasset, eds. *Le Ponts au Moyen Âge*. Paris: Presses de l'Université-Sorbonne, 2006.

Jansen, Philippe. "La mobilité des maîtres-maçons en Italie au Moyen Âge: Une Mobilité technique ou culturelle?" In *Les systèmes de mobilité de la préhistoire au Moyen Âge*, edited by Nicholas Naudinot et al., 305–318. Antibes: Éditions APDCA, 2015.

al-Jazarī, Ibn al-Razzaz. *The Book of Knowledge of Ingenious Mechanical Devices*. Translated by Donald R. Hill. Dordrecht: Reidel, 1974.

Jeffreys, Elizabeth with John Haldon and Robin Cormack, eds. *The Oxford Handbook of Byzantine Studies*. Oxford: Oxford University Press, 2008.

Jenkins, David, ed., *The Cambridge History of Western Textiles*. 2 vols. Cambridge: Cambridge University Press, 2003.

Jensenius, Jergen H. "Wooden Churches in Viking and Medieval Norway: Two Geometric and Static Strategies." In Khodakovsky and Lexau, *Historic Wooden Architecture in Europe and Russia*, 20–27.

Johns, Adrian. *The Nature of the Book: Print and Knowledge in the Making*. Chicago: University of Chicago Press, 1998.

Jongepier, Iason, et al. "The Brown Gold: A Reappraisal of Medieval Peat Marshes in Northern Flanders (Belgium)." *Water History* 3 (2011): 73–93.

Jordan, William Chester. *The Great Famine: Northern Europe in the Early Fourteenth Century*. Princeton: Princeton University Press, 1996.

Jørgensen, Lise Bender, et al. "Textile Industries of the Early Medieval World." In Jenkins, *Cambridge History of Western Textiles*, 1: 118–175.

Kaijser, Arne. "System Building from Below: Institutional Change in Dutch Water Control Systems." *Technology and Culture* 43 (July 2002): 521–548.

Kale, Gül. "From Measuring to Estimation: Definitions of Geometry and Architect-Engineer in Early Modern Ottoman Architecture." *Journal of the Society of Architectural Historians* 79 (June 2020): 132–151.

Kalinowski, Lech. "Medieval Stained Glass in Poland." In *Ars Vitrea: Collected Writings on Mediaeval Stained Glass*, by Lech Kalinowski and Helena Małkiewiczówna, edited by Dobrosława Horzela and Joanna Utzig, translated by Piotr Mizia and Zbigniew Suszcyński, 69–126. Kraków: Polish Academy of Arts and Sciences, 2016.

Kang, Hyeok Hweon. "Crafting Knowledge: Artisan, Officer, and the Culture of Making in Chosŏn Korea, 1392–1910." PhD diss., Harvard University, 2020.

———. "Difference in an Age of Parity: Technology and Global Military History." In *The Military Revolution and Revolutions in Military Affairs*, edited by Mark C. Fissel, 29–64. Berlin: De Gruyter, 2022.

Kaplan, Michel. *Les hommes et la terre à Byzance du VIe au XIe siècle: Propriété et exploitation du sol*. Paris: Publication de la Sorbonne, 1992.

Karagianni, Flora. "Networks of Medieval City-Ports on the Black Sea (7th–15th Century): The Archaeological Evidence." In Preiser-Kapeller and Daim, *Harbours and Maritime Networks as Complex Adaptive Systems*, 83–104.

Karakaş, Deniz. "Water for the City: Builders, Technology, and Private Initiative." In Hamadeh and Kafescioğlu, *Companion to Early Modern Istanbul*, 308–338.

Kavey, Allison. *Books of Secrets: Natural Philosophy in England, 1550–1600*. Urbana: University of Illinois Press, 2007.

Kazhdan, Alexander P. "The Peasantry." In Cavallo, *Byzantines*, 43–73.

Keen, Maurice, ed. *Medieval Warfare: A History*. Oxford: Oxford University Press, 1999.

Keenan, James G. "Fayyum Agriculture at the End of the Ayyubid Era: Nabulsi's *Survey*." In Bowman and Rogan, *Agriculture in Egypt*, 287–299.

Keene, Derek. "Issues of Water in Medieval London to c. 1300." *Urban History* 28 (August 2001): 161–179.

Kemp, Barry. "Soil (including Mudbrick Architecture)." In *Ancient Egyptian Materials and Technology*, edited by Paul T. Nicholson and Ian Shaw, 78–103. Cambridge: Cambridge University Press, 2000.

Khan, Geoffrey. "Arabic Documents from the Early Islamic Period." In Wissa, *Scribal Practices*, 69–90.

Khodakovsky, Evgeny. "Introduction: Wood in the Architecture of Europe and Russia: National Specifics and International Research." In Khodakovsky and Lexau, *Historic Wooden Architecture in Europe and Russia*, 6–17.

Khodakovsky, Evgeny and Siri Skjold Lexau, eds. *Historic Wooden Architecture in Europe and Russia: Evidence, Study and Restoration*. Basel: Birkhäuser, 2016.

King, David A. *In Synchrony with the Heavens: Studies in Astronomical Timekeeping and Instrumentation in Medieval Islamic Civilization*. Vol. 2, *Instruments of Mass Calculation*. Leiden: Brill, 2005–2014.

Kinney, Dale. "Mosaics at Ravenna." In Lindley, *Making Medieval Art*, 81–90.

Kirchweger, Franz. "Goldsmiths in Medieval Vienna (1150–1527): Their Trade and Their Art." In Zapke and Gruber, *Companion to Medieval Vienna*, 468–496.

Klápště, Jan and Nissen Jaubert. "Rural Settlement." In Graham-Campbell with Valor, *Archaeology of Medieval Europe*, vol. 1, *Eighth to Twelfth Centuries AD*, 76–110.

Knoll, Martin and Reinhold Reith, eds. *An Environmental History of the Early Modern Period: Experiments and Perspectives*. Zurich: LIT Verlag, 2014.

Ko, Dorothy. *The Social Life of Inkstones: Artisans and Scholars in Early Qing China*. Seattle: University of Washington Press, 2017.

Koder, Johannes. "The Authority of the *Eparchos* in the Markets of Constantinople (according to the *Book of the Eparch*)." In *Authority in Byzantium*, edited by Pamela Armstrong, 83–110. Farnham, Surrey: Ashgate, 2013.

Komaroff, Linda. "Color, Precious Metal, and Fire: Islamic Ceramics and Glass." In Hess, *Arts of Fire*, 34–53.

Kostof, Spiro. *The City Shaped: Urban Patterns and Meanings Through History*. Boston: Little, Brown, 1991.

Kostof, Spiro with Greg Castillo. *The City Assembled: The Elements of Urban Form through History*. New York: Thames & Hudson, 2005.

Kowaleski, Maryanne. "The Early Documentary Evidence for the Commercialisation of the Sea Fisheries in Medieval Britain." In Barrett and Orton, *Cod and Herring*, 23–35.

———. "The Maritime Trade Networks of Late Medieval London." In Blockmans, Krom, and Wubs-Mrozewicz, *Routledge Handbook of Maritime Trade*, 383–410.

Kowaleski, Maryanne and P.J.P. Goldberg, eds. *Medieval Domesticity: Home, Housing, and Household in Medieval England*. Cambridge: Cambridge University Press, 2008.

Krause, Heike, Paul Mitchell, and Christoph Sonnlechner. "The Urban Waterscape in Medieval Vienna." In Zapke and Gruber, *Companion to Medieval Vienna*, 222–264.

Kray, Jill, ed. *The Cambridge Companion to Renaissance Humanism*. Cambridge: Cambridge University Press, 2004.

Kucher, Michael P. *The Water Supply System of Siena, Italy: The Medieval Roots of the Modern Networked City*. New York: Routledge, 2005.

Kusukawa, Sachiko. *Picturing the Book of Nature: Image, Text, and Argument in Sixteenth-Century Human Anatomy and Medical Botany.* Chicago: University of Chicago Press, 2012.

Kwakkel, Erik. *Books before Print.* Leeds: Arc Humanities Press, 2018.

Laiou, Angeliki E. "The Agrarian Economy, Thirteenth–Fifteenth Centuries." In Laiou, *Economic History of Byzantium,* 1: 311–375.

———, ed. *The Economic History of Byzantium from the Seventh through the Fifteenth Century.* 3 vols. Washington, DC: Dumbarton Oaks, 2002.

———. "Women in Byzantine Society." In Mitchell, *Women in Medieval Western European Culture,* 81–94.

Laiou, Angeliki E. and Cécile Morrisson. *The Byzantine Economy.* Cambridge: Cambridge University Press, 2007.

Lanaro, Paola, ed. *At the Centre of the Old World: Trade and Manufacturing in Venice and the Venetian Mainland, 1400–1800.* Toronto: Centre for Reformation and Renaissance Studies, 2006.

———. "Le donne velere nell'arsenale di Venezia: Donne e lavoro operaio in una società pre-industriale." In *L'Arsenale di Venezia: Da grande complesso industrial a risorsa patrimoniale,* edited by Paola Lanaro and Christophe Austruy, 57–82. Venice: Marsilio, 2020.

Landau, David and Peter Parshall. *The Renaissance Print, 1470–1550.* New Haven: Yale University Press, 1994.

Langdon, John. "The Efficiency of Inland Water Transport in Medieval England." In Blair, *Waterways and Canal-Building,* 110–130.

———. *Horses, Oxen, and Technological Innovation: The Use of Draught Animals in English Farming from 1066 to 1550.* Cambridge: Cambridge University Press, 1986.

———. *Mills in the Medieval Economy: England 1300–1540.* Oxford: Oxford University Press, 2004.

La Salvia, Vasco and Ľubonír Mihok. "The Smithy Workshop of the Medieval Site of Leopoli-Cencelle (Viterbo, Italy)." In Nørbach, *Prehistoric and Medieval Direct Iron Smelting,* 125–128.

Lavan, Luke, et al., eds. *Technology in Transition, A.D. 300–650.* Leiden: Brill, 2007.

LeCain, Timothy J. *The Matter of History: How Things Create the Past.* Cambridge: Cambridge University Press, 2017.

Lee, John S. *The Medieval Clothier.* Woodbridge: Boydell Press, 2018.

Lefèvre, Wolfgang, ed. *Picturing Machines, 1400–1700.* Cambridge, MA: MIT Press, 2004.

Lefort, Jacques. "The Rural Economy, Seventh–Twelfth Centuries." In Laiou, *Economic History of Byzantium,* 1: 231–310.

Leitão, Henrique. "Instruments and Artisanal Practices in Long Distance Oceanic Voyages." *Centaurus* 60 (2018): 189–202.

Leitão, Henrique and Antonio Sánchez. "Zilsel's Thesis, Maritime Culture, and Iberian Science in Early Modern Europe. *Journal of the History of Ideas* 78 (April 2017): 191–210.

Leng, Rainer. Ars belli: *Deutsche taktische und kriegstechnische Bilderhandschriften und Traktate im 15. Und 16. Jahrhundert.* 2 vols. Wiesbaden: Reichert, 2002.

L'Engle, Susan and Gerald B. Guest, eds. *Tributes to Jonathan J. G. Alexander: The Making and Meaning of Illuminated Medieval and Renaissance Manuscripts, Art and Architecture.* London: Harvey Miller, 2006.

Leong, Elaine Yen Tien. *Recipes and Everyday Knowledge: Medicine, Science, and the Household in Early Modern England.* Chicago: University of Chicago Press, 2018.

Leong, Elaine Yen Tien and Alisha Rankin, eds. *Secrets and Knowledge in Medicine and Science, 1500–1800.* Farnham, Surrey: Ashgate, 2011.

Levanoni, Amalia. "Water Supply in Medieval Middle Eastern Cities: The Case of Cairo." *Al-Masāq* 20, no. 2 (2008): 179–205.

Lewcock, Ronald. "Materials and Techniques." In *Architecture of the Islamic World: Its History and Social Meaning*, edited by George Michell, 129–143. London: Thames & Hudson, 1978.

Liang, Yuen-Gen, et al. "Unity and Disunity across the Strait of Gibraltar." *Medieval Encounters* 19 (2013): 1–40.

Lincoln, Evelyn. *The Invention of the Italian Renaissance Printmaker*. New Haven: Yale University Press, 2000.

Lindley, Phillip. "Gothic Sculpture: Studio and Workshop Practice." In Lindley, *Making Medieval Art*, 54–80.

———, ed. *Making Medieval Art*. Donington, Lincolnshire: Shaun Tyas, 2003.

Ling, Roger and Lesley Ling. "Fresco and Other Forms of Wall Painting." In Lindley, *Making Medieval Art*, 4–29.

Little, Lester K., ed. *Plague and the End of Antiquity: The Pandemic of 541–750*. Cambridge: Cambridge University Press, 2007.

———. "Plague Historians in Lab Coats." *Past & Present*, no. 213 (November 2011): 267–290.

Loengard, Janet. "Lords' Rights and Neighbors' Nuisances: Mills and Medieval English Law." In Walton, *Wind and Water*, 129–152.

Long, Pamela O. "The *Annales* and the History of Technology." *Technology and Culture* 46 (January 2005): 177–186.

———. *Artisan/Practitioners and the Rise of the New Sciences, 1400–1600*. Corvallis: Oregon State University Press, 2011.

———. "The Craft of Premodern European History of Technology." *Technology and Culture* 51 (July 2010): 698–714.

———. *Engineering the Eternal City: Infrastructure, Topography, and the Culture of Knowledge in Late Sixteenth-Century Rome*. Chicago: University of Chicago Press, 2018.

———. "Introduction: The World of Michael of Rhodes, Venetian Mariner." In Long, McGee and Stahl, *The Book of Michael of Rhodes*, 3: 1–33.

———. "Manuals." In Blair et al., *Information*, 589–593.

———. *Openness, Secrecy, Authorship: Technical Arts and the Culture of Knowledge from Antiquity to the Renaissance*. Baltimore: Johns Hopkins University Press, 2001.

———. "Picturing the Machine: Francesco di Giorgio and Leonardo da Vinci in the 1490s." In Lefévre, *Picturing Machines*, 117–141.

———. "Power, Patronage, and the Authorship of *Ars*: From Mechanical Know-How to Mechanical Knowledge in the Last Scribal Age." *Isis* 88 (March 1997): 1–41.

———. "Responses to a Recurrent Disaster: Flood Writings in Rome, 1476–1606." In *Disaster in the Early Modern World: Examinations, Representations, Interventions*, edited by Ovanes Akopyan and David Rosenthal, 225–245. Abingdon: Routledge, 2024.

———. "Trading Zones in Early Modern Europe." *Isis* 106 (December 2015): 840–847.

———. "Work, Power, Energy." In Soens, *Cultural History of the Environment*. Forthcoming.

Long, Pamela O., David McGee, and Alan M. Stahl, eds. *The Book of Michael of Rhodes: A Fifteenth Century Maritime Manuscript*. 3 vols. Cambridge, MA: MIT Press, 2009.

Lopez, Robert S. *The Commercial Revolution of the Middle Ages, 950–1350*. Cambridge: Cambridge University Press, 1976.

Lorenzen-Schmidt, Klaus-Joachim. "Northwest Germany 1000–1750." In Thoen and Soens, *Struggling with the Environment*, 309–338.

Lowden, John. "Book Production." In Jeffreys with Haldon and Cormack, *Oxford Handbook of Byzantine Studies*, 462–471.

Lucas, Adam Robert. "Industrial Milling in the Ancient and Medieval Worlds: A Survey of the Evidence for an Industrial Revolution in Medieval Europe." *Technology and Culture* 46 (January 2005): 1–30.

———. "Narratives of Technological Revolution in the Middle Ages." In *Handbook of Medieval Studies: Terms, Methods, Trends*, edited by Albrecht Classen, 967–990. Berlin: De Gruyter, 2010).

———. "The Role of the Monasteries in the Development of Medieval Milling." In Walton, *Wind and Water*, 89–127.

———. *Wind, Water, Work: Ancient and Medieval Milling Technology*. Leiden: Brill, 2006.

Lucassen, Jan, Tine De Moor, and Jan Luiten van Zanden. "The Return of the Guilds: Towards a Global History of the Guilds in Pre-industrial Times." *International Review of Social History*, supplement 16, *The Return of the Guilds*, 53 (2008): 5–18.

Lyngstrøm, Henriette. "Farmers, Smelters, Smiths: Relations between Production, Consumption, and Distribution of Iron in Denmark, 500 BC–AD 1500." In Nørbach, *Prehistoric and Medieval Direct Iron*, 21–25.

Lynn, John A., II. *Women, Armies, and Warfare in Early Modern Europe*. Cambridge: Cambridge University Press, 2008.

Macleod, Catriona, Alexandra Shepard, and Maria Ågren, *The Whole Economy: Work and Gender in Early Modern Europe*. Cambridge: Cambridge University Press, 2023.

MacMurdie, Meekyung. "The Manuscript Machine: Assemblages and Divisions in Jazari's Compendium." In *Destroyed—Disappeared—Lost—Never Were*, edited by Beate Fricke and Aden Kumler, 113–138. University Park: Penn State University Press, 2022.

Maffioli, Cesare S. "A Fruitful Exchange/Conflict: Engineers and Mathematicians in Early Modern Italy." *Annals of Science* 70, no. 2 (2013): 197–228.

Magdalino, Paul. "Medieval Constantinople: Built Environment and Urban Development." In Laiou, *Economic History of Byzantium*, 2: 529–537.

Magnusson, Roberta J. "Public and Private Urban Hydrology: Water Management in Medieval London." In Walton, *Wind and Water*, 171–187.

———. *Water Technology in the Middle Ages: Cities, Monasteries, and Water Works after the Roman Empire*. Baltimore: Johns Hopkins University Press, 2001.

Maier, Jessica. *Rome Measured and Imagined: Early Modern Maps of the Eternal City*. Chicago: University of Chicago Press, 2015.

Maitte, Corine. "L'arte del vetro: Innovazione e trasmissione delle techniche." In Braunstein and Molá, *Produzione e techniche*, 235–259.

Makrypoulias, Christos G. "'Our Engines Are Better Than Yours': Perception and Reality of Late Byzantine Military Technology." In *Byzantium and the West: Perception and Reality (11th–15th c.)*, edited by Nikolaos G Chrissis, Athina Kolia-Dermitzaki, and Angeliki Papageorgiou, 306–315. Abingdon: Routledge, 2019.

Maniatis, George C. *Guilds, Price Formation, and Market Structures in Byzantium*. Farnham, Surrey: Ashgate/Variorum, 2009.

Mango, Marlia Mundell. "Metalwork." In Jeffreys with Haldon and Cormack, *Oxford Handbook of Byzantine Studies*, 444–451.

Marcus, Joyce and Jeremy A. Sabloff. "Introduction." In *The Ancient City: New Perspectives on Urbanism in the Old and New World*, edited by Joyce Marcus and Jeremy A. Sabloff, 3–26. Santa Fe, NM: School for Advanced Research Press, 2008.

Margolis, Nadia. *An Introduction to Christine de Pizan*. Gainesville: University Press of Florida, 2011.

Mark, Robert. "Technological Innovation in High Gothic Architecture." In Smith and Wolfe, *Technology and Resource Use in Medieval Europe*, 11–25.

Marks, Richard. "Window Glass." In Blair and Ramsay, *English Medieval Industries*, 265–294.

Marr, Alexander. *Between Raphael and Galileo: Mutio Oddi and the Mathematical Culture of Late Renaissance Italy*. Chicago: University of Chicago Press, 2011.

Mårtelius, Johan. "Sinan's Ablution Fountains." In Shilling and Stephenson, *Fountains and Water Culture in Byzantium*, 324–340.

Martinez-Medina, Ramón, Encarnación Gil-Meseguer, and José Gómez-Espin. "Research on Qanats in Spain." *Water History* 10 (2018): 339–355.

Masonen, Pekka. "The Sahara as Highway for Trade and Knowledge." In *Highways, Byways, and Road Systems in the Pre-Modern World*, edited by Susan E. Alcock, John Bodel, and Richard J. A. Talber, 168–184. Chichester, West Sussex: Wiley-Blackwell, 2012.

Massing, Ann. "A Short History of Tempera Painting." In Lindley, *Making Medieval Art*, 30–41.

Matschke, Klaus-Peter. "The Late Byzantine Urban Economy, Thirteenth–Fifteenth Centuries." In Laiou, *Economic History of Byzantium*, 2: 464–496.

———. "Mining." In Laiou, *Economic History of Byzantium*, 1: 115–120.

Mattingly, David. "Explanations: People as Agency." In Barker, *Farming the Desert*, 319–342.

Mattingly, David, with a contribution from Crispin Flower. "Romano-Libyan Settlement: Site Distribution and Trends." In Barker, *Farming the Desert*, 159–190.

Mazzaoui, Maureen Fennell. "The First European Cotton Industry: Italy and Germany, 1100–1800." In *The Spinning World: A Global History of Cotton Textiles, 1200–1850*, edited by Giorgio Riello and Prasannan Parthasarathi, 63–88. Oxford: Oxford University Press, 2009.

McCormick, Michael. *Origins of the European Economy: Communications and Commerce, AD 300–900*. Cambridge: Cambridge University Press, 2001.

McCray, Patrick. *Glassmaking in Renaissance Venice: The Fragile Craft*. Aldershot: Ashgate, 1999.

McKee, Sally. "Slavery." In Bennett and Karras, *Oxford Handbook of Women and Gender in Medieval Europe*, 281–294.

McKitterick, Rosamond., ed. *The New Cambridge Medieval History*, Vol. 2, *c. 700–c. 900*. Cambridge: Cambridge University Press, 1995.

McRee, Benjamin R. and Trisha K. Dent. "Working Women in the Medieval City." In Mitchell, *Women in Medieval Western European Culture*, 241–256.

McSweeney, Anna. "The Tin Trade and Medieval Ceramics: Tracing the Sources of Tin and Its Influence on Mediterranean Ceramics Production." *Al-Masāq* 23, no. 3 (2011): 155–169.

Méheut, Constant. "Facing a Future of Drought, Spain Turns to Medieval Solutions and 'Ancient Wisdom.'" *New York Times*, July 20, 2023, https://www.nytimes.com/2023/07/19/world/europe/spain-drought-acequias.html.

Meyer, Werner, et al. "Archaeologies of Coercion." In Carver and Klápště, *Archaeology of Medieval Europe*, vol. 2, *Twelfth to Sixteenth Centuries*, 230–276.

Mikhail, Alan. *Nature and Empire in Ottoman History: An Environmental History*. Cambridge: Cambridge University Press. 2011.

———. *Under Osman's Tree: The Ottoman Empire, Egypt and Environmental History*. Chicago: University of Chicago Press, 2017.

Miller, Maureen C. *The Bishop's Palace: Architecture and Authority in Medieval Italy*. Ithaca: Cornell University Press, 2000.

Miller, Sequoia. "Ceramic Art: An Introduction." In Graves et al., *Ceramic Art*, 1–18.

Milne, Gustaf. *The Port of Medieval London*. Stroud: Tempus, 2003.

Mitchell, Linda E., ed. *Women in Medieval Western European Culture*. New York: Garland Publishing, 2016.

Mocarelli, Luca. "Guilds Reappraised: Italy in the Early Modern Period." *International Review of Social History*, supplement 16, *The Return of the Guilds*, 53 (2008): 159–178.

Molà, Luca. *The Silk Industry of Renaissance Venice*. Baltimore: Johns Hopkins University Press, 2000.

———. "Stato e impresa: Privilegi per l'introduzione di nuove arti e brevetti." In Braunstein and Molá, *Produzione e techniche*, 533–572.

Molà, Luca, Reinhold C. Mueller, and Claudio Zainer, eds. *La seta in Italia dal Medioevo al Seicento: Dal baco al drappo*. Venice: Marsilio, 2000.

Molle, Leen Van and Yves Segers, eds. *The Agro-Food Market: Production, Distribution and Consumption*. Vol. 2 in *Rural Economy and Society in North-Western Europe, 500–2000*, ed. Erik Thoen. Turnhout: Brepols, 2013.

Montemezzo, Stefania. "Ships and Trade: The Role of Public Navigation in Renaissance Venice." In Nigro, *Reti maritime / Maritime Networks*, 473–484.

Montgomery, David R. *Dirt: The Erosion of Civilizations*. Berkeley: University of California Press, 2007.

Moore-Scott, Terry. "Medieval Fish Weirs on the Mid-Tidal Reaches of the Severn River (Ashleworth-Arlingham)." *Glevensis* 42 (2009): 31–44.

Morel, Thomas. *Underground Mathematics: Craft Culture and Knowledge Production in Early Modern Europe*. Cambridge: Cambridge University Press, 2023.

Morgan, Martha E. "Reconstructing Early Islamic Maghribi Metallurgy." PhD diss., University of Arizona, 2009.

Mosely, James. "The Technologies of Print." In Suarez and Woudhuysen, *Oxford Companion to the Book*, 1: 89–104.

Mosley, Stephen. *The Environment in World History*. London: Routledge, 2010.

Mott, Lawrence V. *Sea Power in the Medieval Mediterranean: The Catalan-Aragonese Fleet in the War of the Sicilian Vespers*. Gainesville: University Press of Florida, 2003.

Mould, Quita, Ian Carlisle, and Esther Cameron. *Craft, Industry and Everyday Life: Leather and Leatherworking in Anglo-Scandinavian and Medieval York*. York: Council for British Archaeology, 2003.

Mozzato, Andrea. "The Production of Woollens in Fifteenth- and Sixteenth-Century Venice." In Lanaro, *At the Centre of the Old World*, 73–107.

Muhanna, Elias. "Realms of Information in the Medieval Islamic World." In Blair et al., *Information*, 21–37.

Mukerji, Chandra. *Impossible Engineering: Technology and Territoriality on the Canal du Midi*. Princeton: Princeton University Press, 2009.

Munro, John H. "Medieval Woollens: Textiles, Textile Technology and Industrial Organisation, c. 800–1500." In Jenkins, *Cambridge History of Western Textiles*, 1: 181–227.

Muthesius, Anna. *Byzantine Silk Weaving, A.D. 400 to A.D. 1200*. Vienna: Verlag Fassbaender, 1997.

———. "Silk in the Medieval World." In Jenkins, *Cambridge History of Western Textiles*, 1: 344–345.

———. *Studies in Medieval and Islamic Silk Weaving*. London: Pindar Press, 1995.

Nagy, Balázs, et al., eds., *Medieval Buda in Context*. Leiden: Brill, [2016].

Navarro Palazón, Julio and Pedro Jiménez Castillo. "Southerners: House and Garden in Al-Andalus." Pt. 3 of "Housing," in Carver and Klápště, *Archaeology of Medieval Europe*, vol. 2, *Twelfth to Sixteenth Centuries*, 176–188.

Necipoğlu, Gülru. *The Age of Sinan: Architectural Culture in the Ottoman Empire*. London: Reaktion Books, 2005.

Nedkvitne, Arnved. "The Development of the Norwegian Long-Distance Stockfish Trade." In Barrett and Orton, *Cod and Herring*, 50–59.

Nevola, Fabrizio. *Street Life in Renaissance Italy*. New Haven: Yale University Press, 2020.

Newman, Paul B. *Travel and Trade in the Middle Ages*. Jefferson, NC: McFarland, 2011.

Ney, Andreas. *Wasser- und Windmühlen in Europa in der Spätantike und dem Mittelalter nach archäologischen, bildlichen und schriftlichen Quellen*. Detmold: Verlag Moritz Schafer, [2019].

Niazi, Kaveh. "Karajī's Discourse on Hydrology." *Oriens* 44 (2016): 44–68.

Nicholas, David. *The Growth of the Medieval City: From Late Antiquity to the Early Fourteenth Century*. London: Longman, 1990.

——. *The Later Medieval City, 1300–1500*. London: Longman, 1997.

Nicolle, David, ed. *A Companion to Medieval Arms and Armour*. Woodbridge: Boydell Press, 2002.

Nielssen, Alf Ragnar. "Early Commercial Fisheries and the Interplay among Farm, Fishing Station and Fishing Village in North Norway." In Barrett and Orton, *Cod and Herring*, 42–49.

Nigro, Giampiero ed. *Reti maritime come fattori dell'integrazione Europea / Maritime Networks as a Factor in European Integration*. Florence: Firenze University Press, 2019.

Nilsen, Andrine. *Vernacular Buildings and Urban Social Practice: Wood and People in Early Modern Swedish Society*. Oxford: Archaeopress, 2020.

Nørbach, Lars Christian, ed. *Prehistoric and Medieval Direct Iron Smelting in Scandinavia and Europe: Aspects of Technology and Society*. Acta Jutlandica 76, special issue (2003).

Nowacki, Horst and Matteo Valleriani, eds. *Shipbuilding Practice and Ship Design Methods from the Renaissance to the 18th Century: A Workshop Report*. Berlin: Max Planck Institute for the History of Science, 2003. Preprint 245.

Nye, David E. *Technology Matters: Questions to Live With*. Cambridge, MA: MIT Press, 2006.

O'Connell, Monique. "Venice: City of Merchants or City for Merchandise?" In Blockmans, Krom, and Wubs-Mrozewicz, *Routledge Handbook of Maritime Trade*, 103–120.

Oestmann, Günther. *The Astronomical Clock of Strasbourg Cathedral: Function and Significance*. Translated by Bruce W. Irwin. Leiden: Brill, 2020.

Ogilvie, Sheilagh. *The European Guilds: An Economic Analysis*. Princeton: Princeton University Press, 2019.

Oikonomidès, Nicholas. "Entrepreneurs." In Cavallo, *Byzantines*, 144–171.

O'Malley, Michelle. "A Pair of Little Gilded Shoes: Commission, Cost, and Meaning in Renaissance Footwear." *Renaissance Quarterly* 63 (Spring 2010): 45–83.

Oosterhoff, Richard J., José Ramón Marcaida, and Alexander Marr, eds. *Ingenuity in the Making: Matter and Technique in Early Modern Europe*. Pittsburgh: University of Pittsburgh Press, 2021.

Opll, Ferdinand. "The Heritage of Maps and City Views." In Zapke and Gruber, *Companion to Medieval Vienna*, 135–159.

Orlandi, Angela. "Between the Mediterranean and the North Sea: Networks of Men and Ports (14th–15th Centuries)." In Nigro, *Reti maritime / Maritime Networks*, 49–67.

Oschinsky, Dorothy. *Walter of Henley and Other Treatises on Estate Management and Accounting*. Oxford: Clarendon Press, 1971.

Ousterhout, Robert. *Master Builders of Byzantium*. Princeton: Princeton University Press, 1999.

Overton, Mark. *Agricultural Revolution in England: The Transformation of the Agrarian Economy, 1500–1800*. Cambridge: Cambridge University Press, 1996.

Palliser, D. M. *The Cambridge Urban History of Britain*. Vol. 1, *600–1540*. Cambridge: Cambridge University Press, 2000.

Papathanassiou, Maria K. "Metallurgy and Metalworking Techniques." In Laiou, *Economic History of Byzantium*, 1: 121–127.

Park, Katharine and Lorraine Daston, eds. *The Cambridge History of Science*. Vol. 3, *Early Modern Science*. Cambridge: Cambridge University Press, 2006.

Parker, Geoffrey. *The Military Revolution: Military Innovation and the Rise of the West, 1500–1800*. 2nd ed. Cambridge: Cambridge University Press, 1996.

Paselk, Richard A. "Navigational Instruments: Measurement of Altitude." In Hattendorf, *Oxford Encyclopedia of Maritime History*, 3: 29–42.

Pedersen, Anne. "Scandinavian Weaponry in the Tenth Century: The Example of Denmark." In Nicolle, *Companion to Medieval Arms and Armour*, 25–35.

Pentz, Peter. "Ships and the Vikings." In Williams, Pentz, and Wemhoff, *Vikings: Life and Legend*, 202–233.

Perry, Craig, et al. *World History of Slavery*. Vol. 2, *AD 500–AD 1420*. Cambridge: Cambridge University Press, 2021.

Petersen, Andrew. "Medieval Bridges of Palestine." In *Egypt and Syria in the Fatimid, Ayyubid and Mamluk Eras*, vol. 6, edited by U. Vermeulen and K. D'Hulster, 291–306. Leuven: Peeters, 2010.

———. "Roman, Medieval or Ottoman: Historic Bridges of the Lebanon Coast." In *Bridge of Civilizations: The Near East and Europe c. 1100–1300*, edited by Peter Edbury, Denys Pringle, and Balázs Major, 175–203. Oxford: Archaeopress Publishing, 2019.

Pettegree, Andrew. *The Book in the Renaissance*. New Haven: Yale University Press, 2010.

Philippe, Michel. *Naissance de la verrerie moderne, xiie–xvie siècles: Apects économiques, techniques et humains*. Turnhout: Brepols, 1998.

Phillips, Carla Rahn. "Americas: Exploration Voyages, 1492–1509." In Hattendorf, *Oxford Encyclopedia of Maritime History*, 1: 23–28.

———. "The Caravel and the Galleon." In Gardiner and Unger, *Cogs, Caravels and Galleons*, 91–114.

Phillips, Jonathan P. *Holy Warriors: A Modern History of the Crusades*. New York: Random House, 2010.

Picard, Christophe. *Sea of the Caliphs: The Mediterranean in the Medieval Islamic World*. Translated by Nicholas Elliot. Cambridge, MA: Harvard University Press, 2018.

Pirenne, Henri. *Mohammed and Charlemagne*. Translated by Bernard Miall. London: Allen & Unwin, 1968.

Pleiner, Radomír. *Iron in Archaeology: Early European Blacksmiths*. Prague: Archeologický Ústav AV ČR, 2006.

———. *Iron in Archaeology: The European Bloomery Smelters*. Prague: Archeologický Ústav AV ČR, 2000.

Popplow, Marcus. "Formalization and Interaction." *Isis* 106 (December 2015): 848–856.

———. "Why Draw Pictures of Machines? The Social Contexts of Early Modern Machine Drawings." In Lefèvre, *Picturing Machines*, 17–48.

Portuondo, María M. *Secret Science: Spanish Cosmography and the New World*. Chicago: University of Chicago Press, 2009.

Prak, Maarten. "Preface: S. R. Epstein (1960–2007) and the Guilds." *International Review of Social History*, supplement 16, *The Return of the Guilds*, 53 (2008): 1–3.

Preiser-Kapeller, Johannes. "A Collapse of the Eastern Mediterranean?" *Jahrbuch Österreichischen Byzantinistik* 65 (2015): 195–242.

———. "Harbours and Maritime Mobility: Networks and Entanglements." In Preiser-Kapeller and Daim, *Harbours and Maritime Networks as Complex Adaptive Systems*, 119–139.

Preiser-Kapeller, Johannes and Falko Daim, eds. *Harbours and Maritime Networks as Complex Adaptive Systems*. Mainz: Verlag des Römisch-Germanisches Zentralmuseum, 2015.

Preiser-Kapeller, Johannes, Taxiarchis G. Kolias, and Falko Dam, eds. *Seasides of Byzantium: Harbours and Anchorages of a Mediterranean Empire*. Mainz: Verlag des Römisch-Germanisches Zentralmuseum, 2021.

Preiser-Kapeller, Johannes and Ekaterini Mitsiou. "The Little Ice Age and Byzantium within the

Eastern Mediterranean, ca. 1200–1350: An Essay on Old Debates and New Scenarios." In Bauch and Schenk, *Crisis of the 14th Century*, 190–220.

Preiser-Kapeller, Johannes, Lucian Reinfandt, and Yannis Stouratis, eds. *Migration Histories of the Medieval Afroeurasian Transition Zone: Aspects of Mobility between Africa, Asia, and Europe, 300–1500 C.E.* Leiden: Brill, 2020.

———. "Migration History of the Afro-Eurasian Transition Zone, c. 300–1500: An Introduction (with a Chronological Table of Selected Events of Political and Migration History)." In Preiser-Kapeller, Reinfandt, and Stouratis, *Migration Histories of the Medieval Afroeurasian Transition Zone*, 1–48.

Pressouyre, Léon, et al., eds., *L'hydraulique monastique: Milieux, reseaux, usage.* Grâne: Créaphis, 1996.

Pritchard, Frances. "The Uses of Textiles, c. 1000–1500." In Jenkins, *Cambridge History of Western Textiles*, 1: 355–377.

Prouteau, Nicolas. "L'artilleur et l'artillerie avant le temps des canons." In Prouteau, Crouy-Chanel, and Faucherre, *Artillerie et fortification, 1200–1600*, 23–32.

Prouteau, Nicolas, Emmanuel de Crouy-Chanel, and Nicolas Faucherre. *Artillerie et fortification, 1200–1600.* Rennes: Presses Universitaires de Rennes, 2011.

Pryor, John H. "The Mediterranean Round Ship." In Gardiner and Unger, *Cogs, Caravels and Galleons*, 59–77.

———. "Shipping and Seafaring." In Jeffreys with Haldon and Cormack, *Oxford Handbook of Byzantine Studies*, 482–491.

Pryor, John H. and Elizabeth M. Jeffreys. *The Age of the Dromōn: The Byzantine Navy ca. 500–1204.* Leiden: Brill, 2006.

Quilici, Lorenzo. "Land Transport, Part I: Roads and Bridges." In *The Oxford Handbook of Engineering and Technology in the Classical World*, edited by John Peter Oleson, 551–579. New York: Oxford University Press, 2008.

Rackham, Oliver. *Trees and Woodland in the British Landscape: The Complete History of Britain's Trees, Woods, and Hedgerows.* Rev. ed. London: J. M. Dent, 1990.

Raepsaet, Georges. *Attelages et technique de transport dans le monde gréco-romain.* Brussels: Laboratoire d'Archéologie Classique de l'Université Libre de Bruxelles, 2002.

———. "The Development of Farming Implements between the Seine and the Rhine from the Second to the Twelfth Centuries." In Astill and Langdon, *Medieval Farming and Technology*, 40–68.

———. "Les prémices de la mécanisation agricole entre Seine et Rhin de antiquité au 13ᵉ siècle." *Annales. Histoire, Sciences Sociales* 50 (July–August 1995): 911–942.

Raj, Kapil. "Thinking without the Scientific Revolution: Global Interactions and the Construction of Knowledge." *Journal of Early Modern History* 21 (2017): 445–458.

Ramey, Lynn T. *Black Legacies: Race and the European Middle Ages.* Gainesville: University Press of Florida, 2014.

Raphael, Renée. "Toward a Critical Transatlantic History of Early Modern Mining: Depiction, Reality, and Readers' Expectations in Álvaro Alonso Barba's 1640 *El arte de los metals.*" *Isis* 14 (June 2023): 341–358.

Rapoport, Yossef. *Rural Economy and Tribal Society in Islamic Egypt. A Study of al-Nābulusī's Villages of the Fayyum.* Turnhout: Brepols, 2018.

Rapoport, Yossef and Ido Shahar. "Irrigation in the Medieval Islamic Fayyum: Local Control in a Large-Scale Hydraulic System." *Journal of Economic and Social History of the Orient* 55 (2012): 1–31.

———, eds. *The Villages of the Fayyum: A Thirteenth-Century Register of Rural, Islamic Egypt.* Turnhout: Brepols, 2018.

Raven, James. "Book." In Blair et al., *Information*, 333–338.

———, ed. *The Oxford Illustrated History of the Book*. Oxford: Oxford University Press, 2020.

———. *What Is the History of the Book?* Cambridge: Polity Press, 2018.

Raven, James and Joran Proot. "Renaissance and Reformation." In Raven, *Oxford Illustrated History of the Book*, 137–168.

Raymond, André. "The Ottoman Conquest and the Development of the Great Arab Towns." In Raymond, *Arab Cities in the Ottoman Period: Cairo, Syria and the Maghreb*, 17–34. Aldershot: Ashgate, 2002.

Rees Jones, Sarah. "Building Domesticity in the City: English Urban Housing before the Black Death." In Kowaleski and Goldberg, *Medieval Domesticity*, 66–91.

Reigniez, Pascal. "Histoire et techniques: L'outil agricole dans la periode du Haut Moyen-Âge (V^e–Xe S.)." In Barceló and Sigaut, *Making of Feudal Agricultures?*, 33–113.

Rėklaitytė, Ieva. "The Rumor of Water: A Key Element of Morrish Granada." In Boloix-Gallardo, *Companion to Islamic Granada*, 441–462.

Renes, Hans, et al. "Water Meadows as European Agricultural Heritage." In *Adaptive Strategies for Water Heritage: Past, Present and Future*, edited by Carola Hein, 107–130. Cham: Springer, 2020.

Reyerson, Kathryn. "Urban Economies." In Bennett and Karras, *Oxford Handbook of Women and Gender in Medieval Europe*, 295–307.

Reynolds, Susan. *Fiefs and Vassals: The Medieval Evidence Reinterpreted*. Oxford: Oxford University Press, 1994.

Riello, Giorgio. *Cotton: The Fabric That Made the Modern World*. Cambridge: Cambridge University Press, 2013.

Rieth, Eric. "Mediterranean Ship Design in the Middle Ages." In Catsambis, Ford, and Hamilton, *Oxford Handbook of Maritime Archaeology*, 406–425.

Rinne, Katherine Wentworth. *The Waters of Rome: Aqueducts, Fountains, and the Birth of the Baroque City*. New Haven: Yale University Press, 2010.

Rippon, Stephen. "Waterways and Water Transport on Reclaimed Coastal Marshlands: The Somerset Levels and Beyond." In Blair, *Waterways and Canal-Building in Medieval England*, 207–227.

Roberts, Lissa, Simon Schaffer, and Peter Dear, eds. *The Mindful Hand: Inquiry and Invention in the Late Renaissance to Early Industrialisation*. Amsterdam: Koninklijke Nederlandse Akademie van Wetenschappen, 2007.

Roberts, Owain T. P. "Descendants of Viking Boats." In Gardiner and Unger, *Cogs, Caravels and Galleons*, 11–28.

Rodrigues, Ana Duarte and Magdalena Merlos Romero. "Noras, Norias and Technology-of-Use." In Rodrigues and Toribio Marín, *History of Water Management*, 331–350.

Rodrigues, Ana Duarte and Carmen Toribio Marín, eds. *The History of Water Management in the Iberian Peninsula between the 16th and 18th Centuries*. Cham: Springer, 2020.

Rodríguez-Arribas, Josefina, et al., eds. *Astrolabes in Medieval Cultures*. Leiden: Brill, 2019.

Roesdahl, Else and Frans Verhaeghe. "Material Culture—Artifacts and Daily Life." In Carver and Klápště, *Archaeology of Medieval Europe*, vol. 2, *Twelfth to Sixteenth Centuries*, 189–225.

Rogers, Clifford J. "The Artillery and Artillery Fortress Revolutions Revisited." In Prouteau, Crouy-Chanel, and Faucherre, *Artillerie et fortification, 1200–1600*, 75–80.

———. *The Military Revolution Debate: Readings on the Military Transformation of Early Modern Europe*. Boulder, CO: Westview Press, 1995.

Rohr, Christian. "Floods of the Upper Danube River and Its Tributaries and Their Impact on Urban Economies (c. 1350–1600): The Examples of the Towns of Krems/Stein and Wels (Austria)." *Environment and History* 19 (May 2013): 133–148.

———. "Ice Jams and Their Impact on Urban Communities from a Long-Term Perspective (Middle Ages to the 19th Century)." In Chiarenza, Haug, and Müller, *Power of Urban Water,* 197–212.

Rohr, René R. J. *Sundials: History, Theory, and Practice.* Translated by G. Godin. 1970. Reprint, New York: Dover, 1996.

Roland, Alex. "Once More in the Stirrups: Lynn White jr. [*sic*], Medieval Technology, and Social Change." *Technology and Culture* 44 (July 2003): 574–585.

———. "Secrecy, Technology, and War: Greek Fire and the Defense of Byzantium, 678–1204." *Technology and Culture* 33 (October 1992): 655–679.

Romizi, Donata, Monika Wulz, and Elisabeth Nemeth, eds. *Edgar Zilsel: Philosopher, Historian, Sociologist.* Cham: Springer, 2022.

Roper, Geoffrey. "The History of the Book in the Muslim World." In Suarez and Woudhuysen, *Oxford Companion to the Book,* 1: 321–339.

Rose, Susan. "Medieval Ships and Seafaring." In Catsambis, Ford, and Hamilton, *Oxford Handbook of Maritime Archaeology,* 426–444.

Rosser, Gervase. *The Art of Solidarity in the Middle Ages: Guilds in England, 1250–1550.* Oxford: Oxford University Press, 2015.

Rossi, Franco. "L'Arsenale: I quadri direttivi." In Tenenti and Tucci, *Storia di Venezia,* vol. 5, *Il Rinascimento,* 593–639.

Rotman, Youval. *Byzantine Slavery and the Mediterranean World.* Translated by Jane Marie Todd. Cambridge, MA: Harvard University Press, 2009.

———. "Migration and Enslavement: A Medieval Model." In Preiser-Kapeller, Reinfandt, and Stouraitis, *Migration Histories of the Medieval Afroeurasian Transition Zone,* 387–412.

Rotolo, Antonio. "Drainage Galleries in the Iberian Peninsula during the Islamic Period." *Water History* 6 (2014): 191–210.

Rubin, Miri. *Cities of Strangers: Making Lives in Medieval Europe.* Cambridge: Cambridge University Press, 2020.

Rublack, Ulinka. *Dressing Up: Cultural Identity in Renaissance Europe.* Oxford: Oxford University Press, 2010.

Ruggles, D. Fairchild. *Islamic Gardens and Landscapes.* Philadelphia: University of Pennsylvania Press, 2008.

———. "Waterwheels and Garden Gizmos: Technology and Illusion in Islamic Gardens." In Walton, *Wind and Water in the Middle Ages,* 69–88.

Rundle, David. "Medieval Western Europe." In Raven, *Oxford Illustrated History of the Book,* 113–136.

Rusk, Bruce. "Xylography." In Blair et al., *Information,* 828–831.

Russell, Andrew L. and Lee Vinsel. "After Innovation, Turn to Maintenance." *Technology and Culture* 29 (January 2018): 1–25.

Saliba, George. *Islamic Science and the Making of the European Renaissance.* Cambridge, MA: MIT Press, 2011.

Sanaan Bensi, Negar. "The Qanat System: A Reflection on the Heritage of the Extraction of Hidden Waters." In *Adaptive Strategies for Water Heritage: Past, Present and Future,* edited by Carola Hein, 41–56. Cham: Springer, 2020.

Sánchez Martínez, Antonio. "Los artífices del Plus Ultra: Pilotos, cartógrafos y cosmógrafos en la Casa de la Contratación de Sevilla durante el siglo XVI." *Hispania* 70 (September–December 2010): 607–632

———. "Charts for an Empire: A Global Trading Zone in Early Modern Portuguese Cartography." *Centaurus* 60 (2018): 173–188.

———. "Cosmography, Maritime Culture, and Practical Knowledge in the Early Modern

Spanish Empire." In *The Routledge Hispanic Studies Companion to Early Modern Spanish Literature and Culture*, edited by Rodrigo Cacho Casal and Caroline Egan, 79–92. Abingdon: Routledge, 2022.

———. "Practical Knowledge and Empire in the Early Modern Iberian World. Towards an Artisanal Turn." *Centaurus* 61 (2019): 268–281.

Sandman, Alison. "Mirroring the World: Sea Charts, Navigation, and Territorial Claims in Sixteenth-Century Spain." In Smith and Findlen, *Merchants and Marvels*, 83–108.

Sapoznik, Alexandra. "Britain, 1000–1750." In Thoen and Soens, *Struggling with the Environment*, 71–107.

al-Sarraf, Shihab. "Close Combat Weapons in the Early ʿAbbāsid Period: Maces, Axes and Swords." In Nicolle, *Companion to Medieval Arms and Armour*, 149–178.

Sawday, Jonathan. *Engines of the Imagination: Renaissance Culture and the Rise of the Machine*. London: Routledge, 2007.

Schäfer, Dagmar, Annapurna Mamidipudi, and Marius Buning, eds. *Ownership of Knowledge: Beyond Intellectual Property*. Cambridge, MA: MIT Press, 2023.

Schafer, Daniel E. "Caravans." In *Travel, Trade, and Exploration in the Middle Ages: An Encyclopedia*, edited by John Block Friedman and Kristen Mossler Figg, 94–96. New York: Garland, 2000.

Schatzberg, Eric. *Technology: Critical History of a Concept*. Chicago: University of Chicago Press, 2018.

Schofield, John and Heiko Steuer, "Urban Settlement." In Graham-Campbell with Valor, *Archaeology of Medieval Europe*, vol. 1, *Eighth to Twelfth Centuries AD*, 111–153.

Schofield, Phillipp and Jane Whittle. "Britain, 1000–1750." In Vanhaute, Devos, and Lambrecht, *Making a Living*, 47–69.

Schönfield, Martin. "Was There a Western Inventor of Porcelain?" *Technology and Culture* 39 (October 1998): 716–727.

Schotte, Margaret E. *Sailing School: Navigating Science and Skill, 1550–1800*. Baltimore: Johns Hopkins University Press, 2019.

Schwinges, Rainer Christoph, ed. *Strassen- und Verkehrswesen in hohen und späten Mittelalt*. Ostfildern: Jan Thorbecke Verlag, 2007.

Segarra Lagunes, Maria Margarita. *Il tevere e Roma: Storia di una simbiosi*. Rome: Gangemi Editore, 2004.

Sennett, Richard. *The Craftsman*. New Haven: Yale University Press, 2018.

Shapin, Steven. *The Scientific Revolution*. 2nd ed. Chicago: University of Chicago Press, 2018.

Shapiro, Lawrence and Shannon Spaulding. "Embodied Cognition." In *The Stanford Encyclopedia of Philosophy*, edited by Edward N. Zalta, June 25, 2021, https://plato.stanford.edu/archives/win2021/entries/embodied-cognition/.

Shatzmiller, Maya. *Labour in the Medieval Islamic World*. Leiden: Brill, 1994.

Shelby, Lon R. and Jacques Heyman. "Mason (i)." In *Grove Art on Line*, 2023, https://doi.org/10.1093/gao/9781884446054.article.T054911.

Shepard, Jonathan. "Networks." In Holmes and Standen, *Global Middle Ages*, 116–157.

Shilling, Brooke and Paul Stephenson, eds. *Fountains and Water Culture in Byzantium*. Cambridge: Cambridge University Press, 2016.

Shulman, Jaime-Chaim. "The Groundbreaking Water Supply Systems of Central and Eastern European Cities, 1300–1580." *Technology and Culture* 60 (July 2019): 726–769.

———. *A Tale of Three Thirsty Cities: The Innovative Water Supply Systems of Toledo, London and Paris in the Second Half of the Sixteenth Century*. Leiden: Brill, 2018.

Sigaut, François. "Crops and Agricultural Developments in Western Europe." In *Early Agricultural Remnants and Technical Heritage (EARTH): 8,000 Years of Resilience and Innovations*, edited by Patricia C. Anderson and Leonor Peña-Chocarro, 107–112. Oxford: Oxbow Books, 2014.

———. "L'Evolution des techniques." In Barceló and Sigaut, *Making of Feudal Agricultures?*, 1–31.

Sindbaek, Søren M. "Northern Emporia and Maritime Networks: Modelling Past Communication Using Archaeological Network Analysis." In Preiser-Kapeller and Daim, *Harbours and Maritime Networks as Complex Adaptive Systems*, 105–117.

Sistrunk, Tim. "The Right to Wind in the Later Middle Ages." In Walton, *Wind and Water in the Middle Ages*, 153–167.

Slavin, Philip, *Experiencing Famine in Fourteenth-Century Britain*. Turnhout: Brepols, 2019.

Smil, Vaclav. *Energy and Civilization: A History*. Cambridge, MA: MIT Press, 2017.

Smith, Elizabeth Bradford and Michael Wolfe, eds. *Technology and Resource Use in Medieval Europe: Cathedrals, Mills, and Mines*. Aldershot: Ashgate, 1997.

Smith, Gregory and Jan Gadeyne, eds. *Perspectives on Public Space in Rome, from Antiquity to the Present Day*. Farnham, Surrey: Ashgate, 2013.

Smith, Julia M. H. "Review of McCormick, *Origins of the European Economy*." *Speculum* 78, no. 3 (July 2003): 956–959.

Smith, Pamela H. *The Body of the Artisan: Art and Experience in the Scientific Revolution*. Chicago: University of Chicago Press, 2004.

———. *From Lived Experience to the Written Word: Reconstructing Practical Knowledge in the Early Modern World*. Chicago: University of Chicago Press, 2022.

———, ed. "The Making and Knowing Project." https://www.makingandknowing.org/.

Smith, Pamela H. and Paula Findlen, eds. *Merchants and Marvels: Commerce, Science, and Art in Early Modern Europe*. New York: Routledge, 2002.

Smith, Pamela H., Amy R. W. Meyers, and Harold J. Cook, eds. *Ways of Making and Knowing: The Material Culture of Empirical Knowledge*. New York: Bard Graduate Center, 2014.

Soens, Tim, ed. *A Cultural History of the Environment*. Vol. 3, *The Oceanic Age, 1200–1550*. London: Bloomsbury, [forthcoming].

———. "Flood Security in the Medieval and Early Modern North Sea Area: A Question of Entitlement?" *Environment and History* 19 (May 2013): 209–232.

Sørheim, Helge. "The Birth of Commercial Fisheries and the Trade of Stockfish in the Borgundfjord, Norway." In Barrett and Orton, *Cod and Herring*, 60–70.

Sonnlechner, Christoph, Severin Hohensinner, and Gertrud Haidvogle. "Floods, Fights and a Fluid River: The Viennese Danube in the Sixteenth Century." *Water History* 5 (2013): 173–194.

Sourdel-Thomine, J. "Khat (A) Writing." In Bearman et al., *Encyclopaedia of Islam*, 4: 1113–1122.

Squatriti, Paolo. "'Advent and Conquests' of the Water Mill in Italy." In Smith and Wolf, *Technology and Resource Use in Medieval Europe*, 125–138.

———. "Of Seeds, Seasons, and Seas: Andrew Watson's Medieval Agrarian Revolution Forty Years Later." *Journal of Economic History* 74 (December 2014): 1205–1220.

———. *Water and Society in Early Medieval Italy, AD 400–1000*. Cambridge: Cambridge University Press, 1998.

———. *Weeds and the Carolingians: Empire, Culture, and Nature in Frankish Europe, AD 750–900*. Cambridge: Cambridge University Press, 2022.

———, ed. *Working with Water in Medieval Europe: Technology and Resource Use*. Leiden: Brill, 2000.

Stahl, Alan M. "Michael of Rhodes: Mariner in Service to Venice." In Long, McGee and Stahl, *Book of Michael of Rhodes*, 3: 35–98.

———. *Zecca: The Mint of Venice in the Middle Ages*. Baltimore: Johns Hopkins University Press, 2000.

Stanton, Charles D. *Roger of Lauria (c. 1250–1305): Admiral of Admirals*. Woodbridge, Suffolk: Boydell Press, 2019.

Stone, David. *Decision-Making in Medieval Agriculture*. Oxford: Oxford University Press, 2005.

Strickland, Matthew and Robert Hardy. *The Great Warbow: From Hastings to the Mary Rose*. Thrupp, Stroud: Sutton Publishers, 2005.

Strobl, Sebastian. *Glastechnik des Mittelalters*. Stuttgart: Gentner, 1990.

Suarez, Michael F. and H. R. Woudhuysen, eds. *The Oxford Companion to the Book*. 2 vols. Oxford: Oxford University Press, 2010.

Szabó, Péter. *Woodland and Forests in Medieval Hungary*. BAR S1348. Oxford: Archaeopress, 2005.

Szabó, Thomas, ed. *Die Welt der europäischen Strassen: Von der Antike bis in die Frühe Neuzeit*. Cologne: Böhlau Verlag, 2009.

Szilágyi, Magdolna. *On the Road: The History and Archaeology of Medieval Communication Networks in East-Central Europe*. Budapest: Archaeolingua, 2014.

Talbot, Alice-Mary. "Women." In Cavallo, *Byzantines*, 117–143.

Taragan, Hana. "Constructing a Visual Rhetoric: Images of Craftsmen and Builders in the Umayyad Palace at Qusayr 'Amra." *Al-Masāq* 20, no. 2 (2008): 141–160.

Tarantino, Giovanni and Paola von Wyss-Giacosa, eds. *Twelve Cities—One Sea: Early Modern Mediterranean Port Cities and Their Inhabitants*. Naples: Edizione Scientifiche Italiana, 2023.

Tarver, W.T.S. "The Traction Trebuchet: A Reconstruction of an Early Medieval Siege Engine." *Technology and Culture* 36 (January 1995): 136–167.

Taylor, Jerome, ed. and trans. *The* Didascalicon *of Hugh of St. Victor: A Medieval Guide to the Arts*. New York: Columbia University Press, 1961.

Tchikine, Anatole. " '*L'anima del giardino*': Water, Gardens, and Hydraulics in Sixteenth-Century Florence and Naples." In *Technology and the Garden*, edited by Michael G. Lee and Kenneth I. Helphand, 129–153. Washington, DC: Dumbarton Oaks, 2014.

———. "Technology of Grandeur: Early Modern Aqueducts in Portugal." In Rodrigues and Toribio Marín, *History of Water Management,* 139–158.

TeBrake, William H. "Taming the Waterwolf: Hydraulic Engineering and Water Management in the Netherlands during the Middle Ages." *Technology and Culture* 43 (July 2002): 475–499.

Teixeira, André and Rodrigo Banha da Silva. "The Water Supply and Sewage Networks in Sixteenth Century Lisbon: Drawing the Renaissance City." In Rodrigues and Toribio Marín, *History of Water Management*, 3–24.

Tenenti, Alberto and Ugo Tucci, eds. *Storia di Venezia dalle origini alla caduta della Serenissima*. Vol. 5, *Il Rinascimento società ed economia*. Rome: Enciclopedia Italiana Fondata da Giovanni Treccani, 1996.

Themudo Barata, Filipe. "Portugal and the Mediterranean Trade: A Prelude to the Discovery of the 'New World.'" *Al-Masāq* 17, no. 2 (2005): 205–219.

Theophilus. *On Divers Arts: The Foremost Medieval Treatise on Painting, Glassmaking and Metalwork*. Translated by John G. Hawthorne and Cyril Stanley Smith. New York: Dover, 1979.

Thoen, Erik. "The Birth of 'The Flemish Husbandry': Agricultural Technology in Medieval Flanders." In Astill and Langdon, *Medieval Farming and Technology*, 69–88.

Thoen, Erik, et al. "Series Introduction: A New Rural History of North-West Europe." In Vanhaute, Devos, and Lambrecht, *Making a Living: Family, Income and Labour*, xi–xv.

Thoen, Erik and Tim Soens. "The Low Countries, 1000–1750." In Thoen and Soens, *Struggling with the Environment*, 221–258.

———, eds. *Struggling with the Environment: Land Use and Productivity*. Vol. 1 in *Rural Economy and Society in North-Western Europe, 500–2000*, edited by Erik Thoen. Turnhout: Brepols, 2015.

Thomas, Anabel. *The Painter's Practice in Renaissance Tuscany*. Cambridge: Cambridge University Press, 1995.

Thompson, Daniel V., Jr., trans. *The Craftsman's Handbook: "Il Libro dell' Arte" Cennino d'Andrea Cennini*. 1933. Reprint, New York: Dover Publications, 1954.

Thomson, Roy. "Leather Working Process." In *Leather and Fur: Aspects of Early Medieval Trade and Technology*, edited by Esther Cameron, 1–9. London: Archetype Publications, 1998.

Thrower, Norman J. W. "Mercator, Gerardus (1512–1594)." In Hattendorf, *Oxford Encyclopedia of Maritime History*, 2: 555–556.

Thys-Şenocak, Lucienne. "Women in the City." In Hamadeh and Kafescioğlu, *Companion to Early Modern Istanbul*, 86–113.

Timmermann, Achim. "The Workshop Practice of Medieval Painters." In Lindley, *Making Medieval Art*, 42–53.

Toch, Michael. "Agricultural Progress and Agricultural Technology in Medieval Germany: An Alternative Model." In Smith and Wolfe, *Technology and Resource Use*, 158–169.

Toda, Oana. "Economic and Material Aspects of the Late Medieval Bridges from Transylvania: The Written Sources." *Banatica* 27 (2017): 361–397.

———. "Evidence on the Engineering and Upkeep of Roads in Late Medieval Transylvania." *Annales Universitatis Apulensis, Series Historica* 17, no. 2 (2013): 173–200.

Tomory, Leslie. *The History of the London Water Industry, 1580–1820*. Baltimore: Johns Hopkins University Press, 2017.

Trivellato, Francesca. *Fondamenta dei vetrai: Lavoro, tecnologia e mercato a Venezia tra Sei e Settecento*. Rome: Donzelli Editore, 2000.

———. "Murano Glass, Continuity and Transformation (1400–1800)." In Lanaro, *At the Centre of the Old World*, 143–184.

Truitt, Elly R. *Medieval Robots: Mechanism, Magic, Nature, and Art*. Philadelphia: University of Pennsylvania Press, 2015.

Uekoetter, Frank, ed. "Mining in Central Europe: Perspectives from Environmental History," *Rachel Carson Center Perspectives* 2012, no 10, doi.org/10.5282/rcc/5600.

Unger, Richard W. "Ships and Sailing Routes in Maritime Trade around Europe, 1300–1600." In Blockmans, Krom, and Wubs-Mrozewicz, *Routledge Handbook of Maritime Trade*, 17–35.

Valenti Marco. "Architecture and Infrastructure in the Early Medieval Village: The Case of Tuscany." In Lavan et al., *Technology in Transition*, 451–489.

Valleriani, Matteo. "The Garden of Pratolino: Ancient Technology Breaks through the Barriers of Modern Iconology." In Ludi Naturae: *Spiele der Natur in Kunst und Wissenschaft*, edited by Natascha Adamowsky, Hartmut Böhme, and Robert Felfe, 121–141. Munich: Weilhelm Fink, 2011.

———. "Structure and Epistemology of the Practical Knowledge of the Venetian Arsenal." In *Les ingénieurs, des intermédiaires? Transmission et cooperation à l'épreuve du terrain (Europe, XV–XVIII siècle)*, edited by Stéphane Blond et al., 31–52. Toulouse: Presses Universitaires du Midi, 2021.

Vanhaelen, Angela. "Strange Things for Strangers: Transcultural Automata in Early Modern Amsterdam." *Art Bulletin* 103, no. 3 (2021): 42–68.

Vanhaute, Eric, Isabelle Devos, and Thijs Lambrecht, eds. *Making a Living: Family, Income, and Labour*. Vol. 3 in *Rural Economy and Society in North-Western Europe, 500–2000*, edited by Erik Thoen. Turnhout: Brepols, 2011.

Veen, Marijke van der, Annie Grant, and Graeme Barker. "Romano-Libyan Agriculture: Crops and Animals." In Barker, *Farming the Desert*, 228–263.

Veikou, Myrto. "Mediterranean Byzantine Ports and Harbours in the Complex Interplay between Environment and Society: Spatial, Socioeconomic and Cultural Considerations based on Archaeological Evidence from Greece, Cyprus and Asia Minor." In Preiser-Kapeller and Daim, *Harbours and Maritime Networks as Complex Adaptive Systems*, 39–60.

Vergani, Raffaello. "L'attivitá mineraria e metallurgica: Argento e rame." In Braunstein and Molá, *Produzione e techniche*, 217–233.

Verhulst, Adriaan. *The Carolingian Economy*. Cambridge: Cambridge University Press, 2002.

———. "Economic Organization." In McKitterick, *New Cambridge Medieval History*, vol. 2, *c. 700–c. 900*, 481–509.

Vermeir, Koen. "Openness versus Secrecy? Historical and Historiographical Remarks." *British Journal for the History of Science* 45 (June 2012): 165–188.

Vigil-Escalera Guirado, Alfonso, Giovanna Bianchi, and Juan Antonio Quirós, eds. *Horrea, Barns and Silos: Storage and Incomes in Early Medieval Europe*. [Bilbao]: Universidad del País Vasco, 2013.

Vitruvius. *De l'architecture, Livre X*. Edited and translated by Louis Callebat, commentary by Philippe Fleury. Paris: Les Belles Lettres. 1986.

———. *Ten Books on Architecture*. Translated by Ingrid D. Rowland, commentary and illustrations by Thomas Noble Howe. Cambridge: Cambridge University Press, 1999.

Wagner, Esther-Miriam. "Scribal Practice in the Jewish Community of Medieval Egypt." In Wissa, *Scribal Practices*, 91–110.

Walton, Penelope. "Textiles." In Blair and Ramsay, *English Medieval Industries*, 332–337.

Walton, Steven A. "Proto-Scientific Revolution or Cookbook Science? Early Gunnery Manuals in the Craft Treatise Tradition." In Córdoba, *Craft Treatises and Handbooks*, 221–236.

———. "Technologies of Pow(d)er: Military Mathematical Practitioners' Strategies and Self-Presentation." In Cormack, Walton, and Schuster, *Mathematical Practitioners*, 87–113.

———, ed. *Wind and Water in the Middle Ages: Fluid Technologies from Antiquity to the Renaissance*. Tempe: Arizona Center for Medieval and Renaissance Studies, 2006.

Ward, Gerald W. R., ed. *The Grove Encyclopedia of Materials and Techniques in Art*. Oxford: Oxford University Press, 2008.

Ward, Rachel. *Islamic Metalwork*. New York: Thames & Hudson, 1993.

Waring, Judith. "Byzantine Book Culture." In *A Companion to Byzantium*, edited by Liz James, 275–288. Chichester, West Sussex: Wiley-Blackwell, 2010.

Watson, Andrew M. *Agricultural Innovation in the Early Islamic World: The Diffusion of Crops and Farming Techniques, 700–1100*. Cambridge: Cambridge University Press, 1983.

Weeks, Robert. "Transport and Trade in South Wales c. 1100–c1400: A Study in Historical Geography." PhD diss., University of Wales College, 2003.

West, Charles. *Reframing the Feudal Revolution: Political and Social Transformation between the Marne and Moselle, c. 800–c. 1100*. Cambridge: Cambridge University Press, 2013.

White, Lynn, Jr. "The Historical Roots of Our Ecologic Crisis." *Science*, March 10, 1967.

———. *Machina ex Deo: Essays in the Dynamism of Western Culture*. Cambridge, MA: MIT Press, 1968.

———. *Medieval Technology and Social Change*. Oxford: Oxford University Press, 1962.

Whitewright, Julian. "Technological Continuity and Change: The Lateen Sail of the Medieval Mediterranean." *Al-Masāq* 24, no. 1 (2012): 1–19.

Whitney, Elspeth. *Paradise Restored: The Mechanical Arts from Antiquity through the Thirteenth Century*. Transactions of the American Philosophical Society, vol. 80, pt.1. Philadelphia: American Philosophical Society, 1990.

Whittle, Jane. "Rural Economies." In Bennett and Mazo Karras, *Oxford Handbook of Women and Gender in Medieval Europe*, 311–326.

Whittow, Mark. "Geographic Survey." In Jeffreys with Haldon and Cormack, *Oxford Handbook of Byzantine Studies*, 219–231.

Wickham, Chris. *Framing the Early Middle Ages: Europe and the Mediterranean, 400–800*. Oxford: Oxford University Press, 2005.

———. "Rural Society in Carolingian Europe." In McKitterick, *New Cambridge Medieval History*, vol. 2, *c. 700–c. 900*, 510–537.

Wiesner-Hanks, Mary E. *Women and Gender in Early Modern Europe*. 4th ed. Cambridge: Cambridge University Press, 2019.

Wikander, Örjan. "Sources of Energy and Exploitation of Power." In *Engineering and Technology in the Classical World*, edited by John Peter Oleson, 136–157. Oxford: Oxford University Press, 2008.

———. "The Watermill." In *Handbook of Ancient Water Technology*, edited by Örjan Wikander, 371–410. Leiden: Brill, 2000.

Wilfong, Terry G. *Women of Jeme: Lives in a Coptic Town in Late Antique Egypt*. Ann Arbor: University of Michigan Press, 2002.

Williams, Alan. *The Knight and the Blast Furnace: A History of the Metallurgy of Armour in the Middle Ages and the Early Modern Period*. Leiden: Brill, 2003.

———. "The Metallurgy of Medieval Arms and Armour." In Nicolle, *Companion to Medieval Arms and Armour*, 45–54.

Williams, Gareth, Peter Pentz, and Matthias Wemhoff, eds. *Vikings: Life and Legend*. Ithaca: Cornell University Press, 2014.

Wilson, N. G. "The History of the Book in Byzantium." In Suarez and Woudhuysen, *Oxford Companion to the Book*, 1: 35–37.

Wing, John T. *Roots of Empire: Forests and State Power in Early Modern Spain, c. 1500–1750*. Leiden: Brill, 2015.

Wissa, Myriam, ed. *Scribal Practices and the Social Construction of Knowledge in Antiquity, Late Antiquity and Medieval Islam*. Leuven: Peeters, 2017.

Wittfogel, Karl. *Oriental Despotism: A Comparative Study of Total Power*. New Haven: Yale University Press, 1956.

Wolfe, Michael. *Walled Towns and the Shaping of France: From the Medieval to the Early Modern Era*. New York: Palgrave Macmillan, 2009.

Worthen, Shana. "The Influence of Lynn White, jr.'s [*sic*] *Medieval Technology and Social Change*." *History Compass* 7, no. 4 (2009): 1201–1217.

Wu, Nancy Y., ed. *Ad Quadratum: The Practical Application of Geometry in Medieval Architecture*. Aldershot: Ashgate, 2002.

Yawn, Lila. "Public Access, Action, and Display in Rome of the Later *anni mille*," In Smith and Gadeyne, *Perspectives on Public Space in Rome*, 85–105.

Zanetti, Cristiano. *Janello Torriani and the Spanish Empire: A Vitruvian Artisan at the Dawn of the Scientific Revolution*. Leiden: Brill, 2017.

Zanini, Enrico. "Constantinople: Building and Maintenance." In Bassett, *Cambridge Companion to Constantinople*, 102–116.

———. "Technology and Ideas: Architects and Master-Builders in the Early Byzantine World." In Lavan et al., *Technology in Transition*, 381–405.

Zanoboni, Maria Paola. *Donne al lavoro nell'Italia e nell'Europa medievali (secoli XIII–XV)*. Milan: Editorale Jouvence, 2016.

Zapke, Susana and Elisabeth Gruber, eds. *A Companion to Medieval Vienna*. Leiden: Brill, 2021.

Zilsel, Edgar. *The Social Origins of Modern Science*, ed. Diederick Raven, Wolfgang Krohn, and Robert S. Cohen. Dordrecht: Kluwer Academic, 2000.

Index

Page numbers in **bold** indicate illustrations.